AGRICULTURE IN MEDITERRANEAN EUROPE: BETWEEN OLD AND NEW PARADIGMS

RESEARCH IN RURAL SOCIOLOGY AND DEVELOPMENT

Series Editor: Terry Marsden

Recent Volumes:

Volume 1: Focus on Agriculture

Volume 2: Focus on Communities

Volume 3: Third World Contexts

Volume 4: Rural Labor Markets

Volume 5: Household Strategies

Volume 6: Sustaining Agriculture & Rural Community

Volume 7: Focus on Migration

Volume 8: Dairy Industry Restructuring

Volume 9: Walking Towards Justice: Democratization in Rural Life

Volume 10: Nature, Raw Materials and Political Economy

Volume 11: New Directions in the Sociology of Global Development

Volume 12: Between the Local and the Global

Volume 13: Gender Regimes, Citizen Participation and Rural Restructuring

Volume 14: Beyond the Rural-Urban Divide: Cross-Continental Perspectives on the Differentiated Countryside and Its Regulation

Volume 15: Welfare Reform in Rural Places: Comparative Perspectives

Volume 16: From Community to Consumption: New and Classical Themes in Rural Sociological Research

Volume 17: Globalization and the Time-Space Reorganisation: Capital Mobility in Agriculture and Food in the Americas

Volume 18: Rethinking Agricultural Policy Regimes: Food Security, Climate Change and the Future Resilience of Global Agriculture

RESEARCH IN RURAL SOCIOLOGY AND DEVELOPMENT
VOLUME 19

AGRICULTURE IN MEDITERRANEAN EUROPE: BETWEEN OLD AND NEW PARADIGMS

EDITED BY

DIONISIO ORTIZ-MIRANDA
Universitat Politècnica de València, València, Spain

ANA MORAGUES-FAUS
School of Planning and Geography, Cardiff University, UK

ELADIO ARNALTE-ALEGRE
Universitat Politècnica de València, València, Spain

United Kingdom – North America – Japan
India – Malaysia – China

Emerald Group Publishing Limited
Howard House, Wagon Lane, Bingley BD16 1WA, UK

First edition 2013

British Library Cataloguing in Publication Data
A catalogue record for this book is available from the British Library

ISBN: 978-1-78190-597-5
ISSN: 1057-1922 (Series)

Printed and bound by CPI Group (UK) Ltd, Croydon, CR0 4YY

ISOQAR certified Management System, awarded to Emerald for adherence to Environmental standard ISO 14001:2004.

ISOQAR
REGISTERED
Certificate Number 1985
ISO 14001

INVESTOR IN PEOPLE

CONTENTS

LIST OF CONTRIBUTORS

Eladio Arnalte-Alegre	Department of Economics and Social Sciences, Universitat Politècnica de València, València, Spain
Elisa Ascione	Italian National Institute of Agricultural Economics (INEA), Rome, Italy
Biagia De Devitiis	Università degli Studi di Foggia, Foggia, Italy
Ramona Franić	Faculty of Agriculture, University of Zagreb, Zagreb, Croatia
Sérgio Godinho	Instituto de Ciências Agrárias e Ambientais Mediterrânicas, University of Évora, Évora, Portugal
Roberto Henke	Italian National Institute of Agricultural Economics (INEA), Rome, Italy
Charalambos Kasimis	Department of Agricultural Economics and Rural Development, Agricultural University of Athens, Athens, Greece
Catherine Laurent	French National Institute for Agronomic Research (INRA), Paris, France
Edelmiro López-Iglesias	Department of Applied Economics, University of Santiago de Compostela, Santiago de Compostela, Spain
Roberto Lorenzana-Fernández	Department of Applied Economics, University of Santiago de Compostela, Lugo, Spain
Ornella Wanda Maietta	Università degli Studi di Napoli "Federico II", Naples, Italy

Terry Marsden	School of Planning and Geography, Cardiff University, Cardiff, UK
Ornella Mikuš	Faculty of Agriculture, University of Zagreb, Zagreb, Croatia
Ana Moragues-Faus	School of Planning and Geography, Cardiff University, Cardiff, UK
Olga M. Moreno-Pérez	Department of Economics and Social Sciences, Universitat Politècnica de València, València, Spain
Dionisio Ortiz-Miranda	Department of Economics and Social Sciences, Universitat Politècnica de València, València, Spain
Apostolos G. Papadopoulos	Department of Geography, Harokopio University of Athens, Athens, Greece
Teresa Pinto-Correia	Instituto de Ciências Agrárias e Ambientais Mediterrânicas, University of Évora, Évora
Cristina Salvioni	Department of Economics, University of Chieti-Pescara, Pescara, Italy
Francisco Sineiro-García	Department of Applied Economics, University of Santiago de Compostela, Lugo, Spain

CHAPTER 1

INTRODUCTION: REFRAMING AGRICULTURE IN MEDITERRANEAN EUROPE

Dionisio Ortiz-Miranda, Ana Moragues-Faus and Eladio Arnalte-Alegre

ABSTRACT

This introductory chapter evidences the need to push again to the fore research undertaken in Southern European countries, highlighting its Mediterranean features and how they relate to old and new theoretical and political debates. Consequently, in this first chapter we describe the main aim of the book as well as how the subsequent chapters contribute to fulfill this quest outlining the structure of the book.

Keywords: Mediterranean Europe; agricultural pathways; theories of agrarian change

PURPOSE OF THE BOOK

By and large, Southern European agriculture has been envisaged as a landscape of small holdings with a low level of technological equipment and

---

Agriculture in Mediterranean Europe: Between Old and New Paradigms
Research in Rural Sociology and Development, Volume 19, 1–8

ISSN: 1057-1922/doi:10.1108/S1057-1922(2013)0000019003

managed by part-time and aging farmers. These characteristics became a common picture in many analyses – particularly throughout the process of integration into the European single market – reinforcing the stereotype of 'delay' from the modernization developments spreading in Northern countries, and therefore constituting a common concern for policy makers. However, under this apparent common Mediterranean portrait there are manifold expressions of rural and agrarian processes, which challenge classical and contemporary conceptualizations. This diversity not only relates with the structural dynamics of holdings but, perhaps more importantly, it relies on disparate forms of farm management, different modes of articulation among actors, and diverse representations and discourses, which altogether shape distinctive agricultural pathways. It is not to say that these processes are only affecting Mediterranean countries, however, the Mediterranean feature introduces singular territorial, ecological and cultural attributes in these pathways that call for a comprehensive contemporary 'picture' of Southern European agriculture.

In the last decades, the academic literature has identified these pathways developed by agrarian holdings as the expressions of different and competing paradigms. These theoretical contributions have evolved from conceptualizing the 'desired' process of modernization and competiveness enhancement of European agriculture to embrace a wider range of trajectories. However, the modernization paradigm is still present, spreading and dominating relevant farming systems in Southern Europe. Indeed, the traditional characteristics of this 'old' model are being reproduced, taking new forms that are currently conceptualized as neo-productivist developments, weak ecological modernization processes or as part of the bioeconomy model. In contrast, other types of initiatives are also proliferating in many European regions, based on new forms of organization of agricultural production linked to the emergence of what has been called 'sustainable food systems'. These later initiatives have been considered as expressions of a new rural development paradigm based on the multifunctionality of agriculture and more recently as the building blocks of an eco-economy. Nevertheless, between these two competing models of development there are not only hybrid expressions and contradictions but also gaps materialized through processes of farm and land abandonment, reflecting the failures and the limitations of markets and policies, and affecting negatively to the social and environmental rural fabric.

In this context, the main aim of the book is to illustrate and deepen into the understanding of these current agrarian dynamics developing in Mediterranean countries in the light of recent theoretical contributions.

Throughout the analysis of a set of Mediterranean case studies, this book attempts to compile the range of transformations that are shaping contemporary agriculture in Southern Europe. In this regard, this analysis pursues to show not only the diversity of pathways and situations, but also the interdependences among them; that is, how these models interact and relate, influencing each other evolvement in specific situations. In addition, special attention is paid to how these pathways take place territorially, that is, in a territorial (social, cultural, ecological) context affecting and being affected by such changes. Furthermore, the analysis of such processes leads to identifying the driving forces behind them: social demands, market drivers, policies, ecological constraints, etc. The breakdown of these new and old trends of agricultural change aims to contribute to the actual theoretical conceptualizations. Precisely, the richness and multiplicity of case studies contained in the book allows considering the usefulness of recent theoretical frameworks in explaining the array of dynamics developing in Southern European agriculture; and, subsequently, contribute to the refinement of these conceptualizations.

STRUCTURE OF THE BOOK

The first chapter of the book starts reviewing the sequence of theoretical approaches formulated around the study of agrarian change in the last decades. In this regard, *Moragues-Faus*, *Ortiz-Miranda* and *Marsden* highlight the evolution of the different paradigms, pointing out essential features as well as criticism received. Special attention is paid to the reactions and contribution to mainly Anglo-Saxon formulations from Mediterranean-based research, but also political and practical aspects are contrasted with the academic evolution on the field of agrarian change. This overview allows identifying innovations introduced throughout time but also the reproduction of conceptual frameworks, on the one hand from modernization and productivism to neo-productivism and bio-economy approaches; and on the other to multifunctionality and sustainable rural development to eco-economy formulations. The second part of the chapter describes a set of transversal issues that arise when analyzing Mediterranean agricultures through these theoretical approaches; such as the blurring of dualist strategies, the relationship between agrarian and non-agrarian dynamics, the importance of part-time non-pluriactive holdings and the spreading of a 'Mediterranean delay' stereotype through theoretical and political contexts. The current crisis, as a paramount issue in Southern Europe, is also discussed from an agrarian perspective. These elements constitute an essential guide to deepen into the subsequent chapters of the volume and therefore provide the basis for a more integrative understanding of old and new dynamics of Mediterranean agriculture.

In Chapter 3, *Arnalte-Alegre* and *Ortiz-Miranda* present an overview of the 'big' data of agriculture in the four countries (Portugal, Spain, Italy and Greece) that have been traditionally included into the 'Southern' model of European agriculture. This provides a historical backdrop for the reading and the interpretation of rest of national, regional and farming system cases analysed in the subsequent chapters of the volume. In this global picture, these authors discuss two preconceived ideas about Mediterranean agriculture. On the one hand, the assumption that farming systems in the South have not followed the process of 'productivist modernization' that has been the main pattern of transformation of Northern Europe agriculture along the second half of the past century. On the other hand, the thesis that, precisely because they have been by-passed or marginalized by the 'productivist' rationale, most holdings and regions from the South would have more possibilities to adapt to new approaches of multifunctional rural development. The authors end up showing the heterodox character of some transformation pathways undergone by these agricultures. In addition, these authors claim for the need of a longer historical perspective to fully understand the current situation and trends of agriculture in Mediterranean countries.

The first case study revolves around the sustainability of a high-value natural agro-ecosystem, the *montado*, which expands through vast areas of Portugal. Throughout this chapter, *Pinto-Correia* and *Godinho* reveal how this essentially multifunctional farming system is exposed to distinct threats that are finally reducing dramatically the *montado* landscape. In their analysis, they focus on the different land management options, highlighting the importance among farmers of productivist ideals and practices despite the traditional multifunctional character of the *montado* (which is based in a careful balance between farming practices and forestry). In this regard, they show the dichotomies and co-existence of paradigms and how they are expressed at different levels, in this case that of farmers and territories. Furthermore, they draw attention into the influence of the Common Agricultural Policy in shaping the current trajectories of the *montado* farms and the potential new directions that can be envisaged to change these deteriorating trends.

López-Iglesias, *Sineiro-García* and *Lorenzana-Fernández* focus on one of the hidden faces of the process of agricultural modernization: farmland abandonment. And they deal with this subject on the North-Western Spanish region of Galicia, where agro-ecological conditions, productive distinctiveness and rural dynamics converge to shape a paradigmatic example of a region prone to farmland abandonment. This is a process that, as these authors discuss, is not easy to characterize and, particularly, to quantify. The chapter provides a historical perspective to the evolution of abandonment, based on the exploitation of official statistical sources. This has led authors to work with extensive bases of farm-level microdata in order to build cross-sectional statistic universes and to isolate explanatory farmland abandonment variables. Their results show how abandonment is expressed in two ways, both invariably related to the progress of the modernization of Galician agriculture: regionally, with the re-localization and concentration of agricultural activity in the areas more favourable for intensive livestock production; and locally, due to the abandonment of the traditional agricultural-related land uses (extensive grazing, firewood and compost provision), which dissociated *stricto sensu* farmland and the management of other rural spaces. López-Iglesias, Sineiro-García and Lorenzana-Fernández also discuss the role played by three key drivers of abandonment: the typical structure of Mediterranean holdings (small-sized and fragmented), the deficiencies of land markets (due to a combination of economic, social and political factors) and productive changes (namely intensive livestock specialization).

The other, and more visible, facet of productive modernization is analysed also for Spain by *Moreno-Pérez*. Indeed, this chapter shows how the classical process of structural adjustment of agriculture – mainly based upon the physical growth of farms – has been giving way to a process more focused on the economic growth of holdings by means of intensification. This transition would be contributing to swell and reshape the profile of holdings included in the 'hard-core' of Spanish agriculture. The chapter illustrates this process by analysing in depth three relevant farming systems, where the integration of productivist and post-productivist dynamics acquires different nature. First, intensive horticulture in Southern Spain, able to adopt new production methods (integrated and organic farming) in a context of advance of productivism. Second, the author tackles vineyard production, where the development of territorial quality labels has been parallel to processes of strong intensification which have resulted in deep modifications of vineyard traditional landscapes. Finally, the chapter focuses on a traditionally extensive production (often a key component of less favoured and mountain areas), small ruminant system, where strong intensification practices, gradual abandonment of grazing and marginalization of weaker holdings have combined in complex forms.

The availability of foreign migrant labour has been crucial in explaining the recent evolution and transformation of Southern European agricultures. This is illustrated in Chapter 7, where *Catherine Laurent* tackles agrarian dynamics in the Mediterranean regions of France, which could be considered, at first sight, an idyllic example of multifunctional agriculture. Nevertheless, this chapter presents and analyses the internal contradictions that can underpin in some cases this seemingly win-win model of agricultural development. And it does after a suggestive question: What exactly is French Mediterranean agriculture? Is the region emblematic of a successful development strategy based on recognition of the multifunctionality of agriculture, or is it a special experimental device to test new recipes for social dumping? The chapter deals with how the consolidation of this model has been parallel to (and supported by) the deterioration of working conditions of foreign migrant employees. In addition, the author argues that the interests of both territorial and sectoral agricultural developments have converged in hiding the working conditions of these workers. Therefore, a 'virtual farmer' emerges, a farmer who meets simultaneously the functions expected from a multifunctional agricultural holding and the profit maximization demands of intensive productive orientations.

In Chapter 8, *De Devitiis* and *Maietta* conduct a regional analysis of structural change in Italian agriculture including not only classical variables on the evolution of agricultural area and number of farms, but also relating these regional trends with human capital assets as well as different types of innovations of Italian farms. In this sense, they combine different national and regional databases to tackle the spread of initiatives such as place-based certification of food-stuffs, organic farming, agri-tourism and the different food supply chains, where Italian farms are articulated. Through this study they show the marked duality North-South of Italian agriculture, mostly in terms of physical and economic dimensions. But, more importantly, they highlight the intertwinement between different structural trajectories of farms and the type of innovations they develop, challenging theoretical dualisms. In fact, they show how certain innovations such as origin certifications (i.e. PDO, PGI) are more important in number and economic dimension in Northern regions and associated to modernization developments; and how other initiatives such as short food supply chains and organic farming prevail in Southern farms, mostly by-passed by productivist premises.

The dynamics of Italian agriculture are further explored by *Salvioni*, *Henke* and *Ascione*. Their chapter constitutes a contribution to the analysis of two major phenomena in Italian agriculture: non-agricultural diversification and participation in food quality schemes. They conduct this analysis upon the basis of two very relevant and not often considered dimensions. On the one hand, they tackle the farm-level economic dimension of these practices for the whole country, using the Farm Accountancy Data Network (FADN). On the other hand, they adopt a dynamic cross-sectional perspective, which allows analysing the evolution of farms developing different combinations of these practices. Their results show how conventional (i.e. neither diversifying nor differentiating production) commercial farms maintain higher levels of unitary economic outcomes (income per unit of land or work), than those adopting one of these strategies. However, diversification and/or differentiation practices have resulted in more sustained and stable income growth paths than that of conventional ones. In their analysis, these authors also discuss the role played by CAP subsidies (both Pillar I payments and rural development schemes), which leads them to claim for more targeted and flexible policy instruments to strengthen the resilience potential of these non-conventional strategies.

The case of Croatia shows a history that still is familiar for other EU Southern countries, that of the accession to the EU (to take place in 2013) and the need to adapt to the huge EU *acquis* (related to farms' performance, health and food quality standards, rural development, etc.), while at the same time certain contradictions between EU agricultural objectives and national ones arise. This is the framework where *Franić* and *Mikuš* place their analysis of agriculture and agricultural policy in Croatia, a country where the evolution of agriculture is still strongly conditioned by the transition to a market economy, affecting mainly land tenure issues. Furthermore, the effects of the independence war of early 90s – including among others the destruction of capital assets, rural population displacement and loss of agricultural potential – are an active force shaping current agrarian dynamics in the country. These authors show how the necessity of pursuing standards and reforms of EU policies has sometimes resulted in lack of coordination among public bodies, low adaptability of policies and programs to diverse territorial needs and the elaboration of different strategies and programmes however never implemented.

The chapter of *Kasimis* and *Papadopoulos* shows how the transformations of rural society has been a process associated to the way rural household carried out complex family strategies, aiming to take advantage of the new economic opportunities arising from the past development of Greek economy. These patterns of transformation led to a sort of reflexive de-agriculturalization of family economic rationalities and, aggregately, of rural Greece. In addition, this process intertwines with the growing and overlapped models of rural mobilities (what these authors refer to made up as a sort of perplexity), most of which flow through the still maintained family rural–urban linkages. This has particularly affected the evolution of agriculture, since out-rural migration did not resulted in the release of land for farms' restructuring. This is precisely one of the scenarios where the deep economic crisis is developing in Greece. In this sense, Kasimis and Papadopoulos discuss the seemingly more resilience of rural areas and agricultural activity within the generalized deterioration of national economy and climbing unemployment; and to what extent the very characteristics of the evolution Greek family farming are playing in favour of such resilience as well as facilitating the 'return to the countryside' phenomenon.

Finally, in the concluding chapter *Ortiz-Miranda*, *Moragues-Faus* and *Arnalte-Alegre* extract from the chapters of the volume a set of transversal elements that challenge aspects of current theoretical approaches and policy formulations from a Mediterranean perspective. The authors argue that studies illustrate the 'impure' nature (where both productivist and non-productivist practices intertwine) of agricultural transformations at both farm and territorial levels. In this complex reality, the chapter tackles three interconnected elements of the dynamic of Southern European agriculture: the dynamics of family farming and the role played by family networks, the position of migrants in the evolution of agriculture and some hypothesis and clues about the implications of the current economic crisis for Southern farming systems. Finally, the authors stress the by-pass of small-scale farming systems in theoretical frameworks and political strategies, and highlight the challenge and the need of embracing these Mediterranean realities in order to build a more sustainable and inclusive agri-food system orchestrating practice, theory and policy domains.

CHAPTER 2

BRINGING MEDITERRANEAN AGRICULTURE INTO THE THEORETICAL DEBATES

Ana Moragues-Faus, Dionisio Ortiz-Miranda and Terry Marsden

ABSTRACT

This chapter aims to analyse the evolution of competing paradigms and theoretical frameworks that have pervaded the debates on the present and future of agricultural and food systems and their associated rural areas. From this global overview, we will extract common features of paradigms that are being reproduced over time as well as highlight the innovations introduced. Particular attention will be paid to discuss the responses and contributions inspired by European Mediterranean-based research, setting up the framework that underlines the subsequent chapters of the volume.

Keywords: Mediterranean agriculture; productivism; post-productivism; rural development; bio-economy; eco-economy

---

Agriculture in Mediterranean Europe: Between Old and New Paradigms
Research in Rural Sociology and Development, Volume 19, 9–35

ISSN: 1057-1922/doi:10.1108/S1057-1922(2013)0000019004

INTRODUCTION

Mediterranean agriculture is by and large envisaged as a landscape of small farms of high nature value producing worldwide recognisable quality food products that make up the basis of the famous Mediterranean diet and shape Southern European cultures. However, the dynamics developing in the Mediterranean countryside are further complex and diverse, comprising differentiated agricultural systems – from highly intensive vegetable production to extensive cereal farms – which have been scarcely analysed in an integrated fashion.

Additionally, from a theoretical perspective, Mediterranean agriculture has had a difficult assembly into the paradigms that pervade the European countryside. On the one hand, it has long been conceived as structurally 'delayed' through the lens of the modernisation paradigm. On the other hand, there is a contested debate around its adaptation to the new paradigms formulated in the last decades, from new forms of neo-productivist modernisation to the emergence of 'sustainable food systems'.

In this line, this chapter firstly introduces the sequence of theoretical debates on agrarian change and associated criticisms, ranging from more classical conceptualisations – such as the modernisation paradigm – to more recent theoretical formulations such as the bio- and eco-economy models. This theoretical framework is followed by the identification of useful elements to assess and discuss the claimed Mediterranean differentiation in the domain of rural and agrarian studies. These elements constitute a guide to better understand the contribution of the subsequent chapters to the main aim of the book.

OVERVIEW ON THE EVOLUTION OF AGRARIAN CHANGE PARADIGMS

Modernisation, Productivism, Intensification … or Marginalisation: The Mediterranean Corrective

The study of agricultural change (especially in Europe) has been characterised by a long series of exchanges between 'competing paradigms' that have led to conceptual transitions. From the 1950s until the late 1980s agricultural restructuring and science were pervaded by a productivist ethos. This productivist regime, expanding through most developed countries, was *conceptualised* as process of farm modernization aiming to increase food production through an 'intensive, industrially driven and expansionist agriculture with state support based primarily on output and increased productivity' (Lowe, Murdoch, Marsden, Munton, & Flynn, 1993, p. 221). By and large, the modernization of farming has been characterised by three main processes (Ilbery & Maye, 2010): intensification (through mechanisation, use of chemicals and variety selection), specialisation (farmers concentrate in few products with higher returns) and concentration (production comes from fewer farms and specific regions). Modernisation gradually led to the industrialization of farming, including the adoption of manufacturing industry practices (further specialisation of labour and product and assembly-line type production) and integration of agriculture into the food supply chain which progressively gave rise to the emergence of large agribusiness (Ilbery & Maye, 2010; Wilson, 2001).

The prevalence of the productivist agricultural regime or intensive food regime (Friedmann & McMichael, 1989) was fuelled by a devastated Europe after the Second World War when different policies and institutions were set up to foster domestic food production in order to assure food security, including not only public institutions but also financial institutions, inputs suppliers, R&D centres, etc. (Lowe et al., 1993; Marsden, Murdoch, Lowe, Munton, & Flynn, 1993). These circumstances coincide with what has been conceptualised as a Fordist mode of regulation, where state intervention in the agricultural sector was conceived as part of a larger social contract between capital and labour, and therefore the state played a necessary role for economic development (Potter & Tilzey, 2005). Consequently, in areas where agriculture was unable to intensify or use economies of scale in order to respond to the agrarian cost-price squeeze (enlarge holdings, increase inputs and technology, etc.) farms became increasingly dependent on state support, finding it increasingly difficult to compete under this rationale and being pushed finally to abandon commercial agriculture (Marsden et al., 1993; Van der Ploeg & Renting, 2000), what Marsden (2003) called a 'race to the bottom'. As Lang (1999) puts it, the productivist paradigm message from policies to farmers mainly through subsidies was: intensify, get larger or get out.

Furthermore, this period of productivist enhancement was further reinforced by the role of the agricultural economist shaping *scientific spheres* and mainly analysing rural areas and agrarian change through a modernisation lens (Newby, 1982). As Newby puts it, rural sociologists were conceived as 'researchers charged with the tasks of overcoming the "social problems" that interfere with cost-efficiency in agriculture' (1983, p. 69). An *academic* turning point was posited in the late 1980s by Bowler and Ilbery (1987), calling for theoretical innovations in the study of agricultural change given the redundancy of research topics that at that time mainly focused on 'regional changes around farm inputs, farm-size structures, farm incomes and agricultural marketing' (Morris & Evans, 1999, p. 349).

This call coincides with a political crisis of productivism in the mid-1980s related to the increase in production and its associated policy budgetary problems in the European Community, a rise of public health and environmental concerns associated to intensive agriculture (e.g. BSE) and international market tensions due to state intervention (Lowe, Marsden, & Whatmore, 1990; Patterson, 1997). Furthermore, Potter and Tilzey (2005) also identify a transition in this period from a Fordist mode of regulation to the progressive incorporation of corporatist interests (including farmers and environmental interests groups) in the design and enact of policies, which they call corporatist political productivism.

Significant mismatches also occurred between actual agricultural dynamics and productivist interpretations both in academic and in political spheres. The dualist 'modernise or disappear' message inherent to the productivism paradigm was obscuring a much more complex and rich farming reality. For instance, Fabiani and Scarano (1995) rejected, on the basis of the Italian agricultural census of 1990, the validity of a dualist conceptualisation which confronted 'laggard' vs. 'productive' holdings. They argued that the data were reflecting a more complex structure, which would be better explained by a continuum rather than a polarisation of differential situations. According to these authors, this differentiation was responding to the plurality of functions performed by diverse types of farms (i.e. residential function, disengagement for retiring farms, income complement under several forms of pluriactivity). Furthermore, this differentiation process resulted from the adaptation of agriculture to the broader diversified model of territorial development in Italy.

Another example comes from Greece, a country also affected by 'serious' structural problems under the lens of modernisation and characterised by a marked territorial duality between littoral and plain agriculture and mountains and islands farming. In this context, Greek authors avoided a simplistic view in terms of delay and that of confronting modern vs. traditional agriculture. Rather, they point out the hybridity of small farmers' production activities as well as the coexistence in the same holding of heterogeneous traits of relations of production. In addition, these authors highlight the increase in pluriactivity of farm families and the myriad of differences between rural and urban spaces. Furthermore, these studies highlight the importance of social networks (i.e. family, relatives) in the context where these processes unfold. Succinctly, the multiple transformation processes of Greek agriculture included the preservation of some specific features intimately conditioned by its social structure and inevitably complicating the conversion of farms into simple 'benefit generating business' (Beopoulos & Damianakos, 1997; Damianakos, 1997).

These pitfalls prompted the rise of political economy in the study of agrarian change. However, despite early precedents of the use of political economy framework in the field (see for instance Bertolini & Meloni, 1978; Marsden, Munton, Whatmore, & Little, 1986; Servolin, 1972), this approach consolidated in the Anglo-Saxon literature through the 1990s (Morris & Evans, 1999), emphasising the importance of state and policies, macro-economic factors of decision-making processes and focused on food production and global market regimes (Marsden et al., 1993; Wilson, 2001). The analytical framework provided by critical political economy prompted the development of food regime scholarship and also impregnated conceptualisations on productivism and post-productivism (Buttel, 2001; McMichael, 1997).

The Post-Productivism Allegory

Post-productivism arose as a concept from outside agriculture that was spreading through political spheres and different academic fields (Tovey, 2001). The term 'post-productivism' (or post-productivist transition) emerged in the agrarian domain early 1990s as an attempt to capture the crisis on agrarian policies and the incorporation of health, environmental and rural development concerns (Walford, 2003; Ward, 1993) as well as the cost-price squeeze faced by farmers (Evans, Morris, & Winter, 2002). This concern was channelled through new rules and regulations for farming – constituting a bureaucratic hygienic mode (Marsden, Banks, Renting, & Van Der Ploeg, 2001) – that sometimes constrain traditional and ecological practices, constituting a barrier to access markets by some producers. Furthermore, these new norms (that are eventually formalised as certifications) entail different forms of private–public modes of regulation where mostly large retailers hold a privileged position. Nevertheless, this re-regulation process not only affects the agro-food chain, but also implies new norms in order to modify the use of rural space usually directed to consumption practices, such as landscape, agri-tourism or rural housing. An important part of environmental regulation in particular conceives nature as a consumable good to be enjoyed by urban dwellers, while agriculture represents a 'dirty-business' (Marsden, 2003).

The concept and subsequent debates benefited from a large attention in British academic spheres. The emergence of post-productivism as a term to explain actual agrarian dynamics led to refine productivism to a point that these concepts were completely opposite; if productivism was represented by intensification, specialisation and concentration, post-productivism was characterised by extensification, dispersion and diversification (Ilbery & Bowler, 1998). Furthermore, the post-productivist transition was conceived as a progressive reversal of trends that dominated the preceding productivist era, embodying a change of agricultural regime. Contributions under the umbrella of post-productivism evolved leading to an agitated debate not only on the actual reach of the phenomenon but also on the usefulness of the concept itself.

Throughout this debate, empirical evidence of the vitality of productivism practices in the European countryside jointly with a limited spread of post-productivist activities started to accumulate. Processes of intensification and concentration were in force, as Walford (2003) demonstrates for the UK recording a continued replacement of labour by capital inputs accompanied by the acquisition of land, an increasing proportion of production generated by large farms and also continued adoption of mechanisation and automation processes in the agricultural sector. Also different studies showed that changes derived from the participation in agri-environmental schemes, quality production or organic agriculture did not imply a major change in farmers' practices and behaviours; and in fact, in many cases they were operated under productivist practices (Evans et al., 2002).

The artificial dualism of productivist and post-productivist conceptualisations leads to the proposal of a new concept, non-productivism as the true opposite of productivism (Wilson, 2007). However, the dualist and simplistic nature of the concept was not the only theoretical criticism, the linearity that was implicit in the term regarding the substitution of regimes and the spatial homogeneity also received wide disapproval (Ward, Jackson, Russell, & Wilkinson, 2008; Wilson, 2007). These elements were progressively tinged with recognition of the co-existence of different farm pathways within the European countryside (Walford, 2003). Nevertheless, they did not succeed in rescuing the concept beyond its usefulness as a descriptive category (Evans et al., 2002). In fact, critiques also pointed out the lack of agency, actor-oriented and behavioural analysis in post-productivist studies in order to tackle grass-roots and intermediate levels (Wilson, 2001), falling into an excessive structuralist approach (Marsden et al., 1993). However, some of the works under actor-oriented approaches faced later criticism for taking in their analysis exogenous policy drivers and other structural influences for granted obscuring wider trends of market productivism and commodification, what Potter and Tilzey (2005) called 'conceptual evasion'.

One of the strongest critiques or fallacies of the productivism and post-productivism conceptualisation has been related to its UK-centrism, failing to discuss whether the concept has wider applicability within Europe and beyond (Wilson, 2001). Accordingly, this debate has timidly spread through academic and political debates particularly in Southern Europe (Ortiz & Moreno, 2007). Mismatches between theory and practice and also between Northern and Southern European dynamics have also been pointed out in the wider rural restructuring conceptualisations coming from Northern latitudes. In fact, Hoggart and Paniagua challenged the actual extent and depth of changes occurring in the British countryside under the post-productivist label, calling for a more critical and strong theoretical evaluation of rural change (2001b). In their analysis they also tackled the Spanish case (Hoggart & Paniagua, 2001a), not finding signs of diversification, professionalisation or environmentalism in the agrarian sector, which led them to deny any sign of post-productivism and to emphasise the quantitative rather than qualitative nature of the changes occurring.

Breaking the Deadlock: Multifunctionality and the New Rural Development Paradigm

A way forward from this dualistic theoretical and empirical debate was the formulation of a third paradigm strongly linked to the emergence of multifunctionality as a conceptual, practical and political device (see below) (Marsden et al., 2001; Van der Ploeg et al., 2000). In this regard, Marsden (2003) identified three models shaping changes in rural areas and agriculture, not only as expressions of specific socio-economic dynamics but also involving political and scientific conceptualisations where the relationships between society and nature constitute a defining element. Firstly, the *agro-industrial mode*, strongly linked to the productivist agricultural regime described above and therefore underpinned by modernisation theory, in which intensification and economies of scale in agriculture constitute the dominant development pathway. Secondly, the *post-productivist model* – which under this classification included aspects related to the *countryside consumption model* or the *bureaucratic hygienic mode* (Marsden et al., 2001). According to Marsden (2003), this model does not imply a break with the agro-industrial dynamic, but rather a continuity and attempt to correct some of its perceived deficiencies, not least through the proliferation of agri-environmental schemes and protectionist forms of planning. As Marsden recently put it, this model implied a compromise '*whereby environmental protection, amenity pressures, as well as food production demands on agricultural land could be assuaged by increasingly cheap imports of both temperate and exotic foodstuffs*' (2011, p. 1).

Finally, the *new rural development model* was identified as a real shift from previous dynamics. The new model emerged from the constrictions that former models were imposing over farmers, that is mainly as a response to the cost-price squeeze in agriculture and to the 'regulatory treadmill' fuelled by the bureaucratic hygienic model (Ward, 1993). According to several authors, these processes result in lower revenues for farmers which in turn mobilise new sources of income to set up different rural development practices in order to maintain the farming activity, and thus constitute livelihood strategies (Kinsella, Wilson, de Jong, & Renting, 2000; Van der Ploeg et al., 2000). Therefore, under this model farms were considered to be pushed to transform into *multifunctional enterprises* in order to survive (Knickel & Renting, 2000). Some authors state that this model may owe its origins particularly to those regions which have been largely 'passed by' other development logics (Marsden, 2003), areas that traditionally were termed 'peripheral rural regions' and conceived as lagged behind under the productivist modernisation model.

The different pathways that farms could adopt under this model were grouped into three processes: deepening, broadening and re-grounding (Van der Ploeg & Roep, 2003). This framework led to a reconceptualisation of diversification and pluriactivity phenomena that were devised as a marginalisation trait and part of a transition towards farm disappearance under the modernisation lens. The rural development model conceives these processes as an opportunity to build up new consumption–production relationships and spill over benefits on rural communities (Kinsella et al., 2000; Ventura & Milone, 2000). This new approach fuelled an important body of work on how these three processes take hold, conducting comparative case studies from different European countries. However, the study of deepening processes, mostly in the form of establishing short food supply chains and constructing quality food products, constituted a major keystone and finally a distinctive body of work in itself, which has constituted a new food geography (Goodman, DuPuis, & Goodman, 2012). In this sense, the broad literature on 'short food supply chains' focusing on farmers evolved to the term of 'alternative food networks' and more recently 'civic food networks', in order to include consumer and producers and the possibility of becoming mainstream (Seyfang, 2006). These 'alternative' food networks are being characterised through the notions of 'quality' (Ilbery & Kneafsey, 2000), 'trust' (Murdoch, Marsden, & Banks, 2000) and 'place' (Bowen, 2010; Moragues-Faus & Sonnino, 2012), and recently in close relationship to the construction of 'sustainable food systems' (Feenstra, 2002; Lang, Barling, & Caraher, 2009).

As these pathways reveal, this new rural development model was mainly based on a new agricultural model (which constituted a 'bull's eye' target for critics on its usefulness as a rural development theory) rooted in the multifunctionality of agriculture, which bridged with the broader implications and impacts of agrarian activities. However, the concept of multifunctionality emerged earlier, in 1993 in the European Council for Agricultural with the aim to harmonise European legislation and ground the notion of sustainable agriculture (Marsden & Sonnino, 2008) or, as other authors point out, as a social welfare justification for state assistance since the early years of the CAP (Potter & Tilzey, 2005). The multifunctionality of agriculture has received much attention by academics and policy makers, mostly in the European context and related to the political construct of the European model of agriculture (Buller, 2001). Despite these conversations, multifunctionality is not a uniform or agreed concept, in fact assumptions such as 'all agriculture its multifunctional' have led to try to clarify debates through more complex conceptualisations such as the distinction between weak and strong multifunctionality (Wilson, 2007, 2008). Furthermore, the concept has also been continuous victim of discursive appropriation (Tilzey, 2006). For instance, Marsden and Sonnino (2008) identify three different conceptualisations of the multifunctionality of agriculture under the three models described above:

- Under a productivist regime, multifunctionality is a transition process of farms that cannot adapt to the cost-price squeeze via modernisation, mostly focused on pluriactivity strategies that will finally lead to abandon farm business.
- Under the post-productivist lens multifunctionality is not related to farming but to rural space. This multifunctionality of the countryside is to be exploited by industrial capital and also by urban dwellers.
- Under the sustainable rural development model *multifunctional agriculture acquires its most comprehensive meaning and displays its highest integrative development potential (…) This emerging paradigm considers multifunctional agriculture no longer simply as a 'survival strategy' for*

> *farmers. Rather, multifunctionality is a proactive development tool to promote more sustainable economies of scope and synergy* (Marsden & Sonnino, 2008, p. 423). This conceptualisation of agriculture and multifunctionality establishes clear linkages to strong multifunctionality formulations and particularly agroecology theory, regarding the new reconnections between nature and society and their implications for development processes (Alonso Mielgo, Sevilla Guzmán, Jiménez Romera, & Guzmán Casado, 2001; Marsden et al., 2001).

This new sustainable rural development model – and its associated agrarian model – received much attention by European academic spheres, resulting in a very large number of case studies and publications including research on pluriactivity, agri-tourism and mainly developing an extensive body of work on quality food and alternative food networks.[1] However, many of these contributions also received criticism from different fronts. Conceptually, several authors point out the benign view and normative assumptions underlying some of the formulations of this new paradigm. As Goodman (2004) puts it, the model is not only aligned with a sectoral agrarian agenda; furthermore, it is 'aligned with an idealised vision of a rural Europe of resourceful yeoman farmers and the era of high farming' (p. 8). He also acknowledges an insufficiently developed method to assess the impact of these new rural development practices (in his view also vaguely defined), in particular the incorporation of elements such as equity and power distribution. In this regard, several scholars state that the enthusiasm of some European academics is leading to uncritical analysis of some realities, for example assuming organic is good when the labour conditions of the farmers are not assessed or local is good when it might be reproducing relations of domination (DuPuis & Goodman, 2005; Goodman, 2004; Sayer, 2001).

What is clear is that during the 2000s especially, the new rural development paradigm sparked a proliferation in a wider and deeper concern among European scholars for understanding alternative and differentiated agri-food networks, and the ways in which these were linked to wider rural development pathways. Despite the numerous case studies developed throughout Europe, which importantly and partially overcame the Northern emphasis of previous conceptualisations, many of the empirical studies underpinning these theoretical contributions have generally focused on individual niche initiatives, such as the development of quality products, agritourism or pluriactivity farms that stand out for their exemplifying nature. This focus might prevent a more integrated understanding of actual dynamics and therefore result in a lack of comprehensive data to support the level of expansion of the new paradigm. It is particularly remarkable that the attempt of measuring the spread and impact of initiatives under the new rural development paradigm throughout Europe found limits in using reliable statistical data (Van der Ploeg & Roep, 2003). This analysis relied on proxies such as number of farmers involved in Protected Designations of Origin (PDO) which potentially occlude dynamics operating under productivist or post-productivist logics. In this regard, there are works that emphasise the hybridity of processes and initiatives taking place (Sonnino & Marsden, 2006), for instance merging industrial and artisanal modes of production under PDO products in Italy (Trabalzi, 2007) or the lack of farming-based activities and products on agri-tourism initiatives as well as low synergies with the rural fabric in Greece (Kizos & Iosifides, 2007). Furthermore, we need to recognise that pluriactivity is an old strategy and more traditional concept that is once more re-labelled under this paradigm, and therefore it might contain elements of productivism – as a transition to abandon agriculture – or an innovative response to maintain agricultural production, expressing hybridity and continuity of some processes. As a consequence, critics have disputed the fact that this new paradigm was actually emerging – as formerly happened with post-productivism conceptualisations – and constituted a consistent break with previous dynamics (Goodman, 2004).

New Concepts and Debates, Old Dualisms

These debates have recently broadened by maintaining the basic traits of these paradigms but shifting the focus from agrarian and rural development issues to the food system, from seed to plate, and in fact increasingly integrating the role of food consumption. The expansion of this conceptual 'battleground' is inevitably related to what has been labelled as the 'new food equation' (Morgan & Sonnino, 2010). The rising significance of food security following the 2008 food price surges, and the multiple and interconnected problems of food insecurity including hunger as well as epidemic of obesity and other diet-related diseases, led to the recognition of food security as a national and international security issue, now combining with the effects of climate change in agro-food systems and vice versa, and the growing incidence of land conflicts around the globe. This has forced more attention upon connecting the rural and the urban realms in understanding agrarian and rural change. Moreover, one of the essential aspects of this period of unresolved contestation is the recognition of the multifunctional character of the food system by political spheres, and therefore its strategic value in resolving a range of problems, from environmental impacts to public health costs (Morgan & Sonnino, 2010). This context has inevitably fuelled more profound debates about what role agriculture and rurality play in contemporary society, considering not only productivity but also how to 'optimise across a far more complex landscape of production, rural development, environmental, social justice and food consumption' (Pretty et al., 2010, p. 220).

Despite the multifaceted processes and the complexity that characterizes the current situation, the solutions and conceptualizations – envisaged from policy, academic spheres and lobby groups – have revolved around two opposite narratives that partially reproduce some of the old conceptual dichotomies exposed above. These narratives have been identified under different flags such as the productivity (or efficiency) narrative and the sufficiency narrative (Freibauer et al., 2011; Huber, 2000), or the bio-economy and eco-economy paradigms (Kitchen & Marsden, 2009), or weak and strong ecological modernisation. Some authors precisely situate ecological modernisation as an overarching theoretical concept underpinning both narratives or models, since the strategy of ecological modernisation aims for the improvement of both ecological and economical efficiency including the use of innovative technology and the steering role of the state (Horlings & Marsden, 2011).

In this regard, ecological modernisation involves a positive framing of the relationship between ecology, society and economic development, involving '*a gradual re-embedding of ecology in the institutions of economy, creating the spaces for an ecological as well as economic rationality*' (Marsden, 2004, p. 3). The innovative aspects of ecological modernisation jointly with the varied criticism that it has received – from theoretical to pragmatic aspects (Buttel, 2000; Fisher & Freudenburg, 2001; Gibbs, 2000) – have been continually expanding the theoretical construction of the term. As happened before with multifunctionality and sustainability, there has been a distinction between weak and strong strands of ecological modernisation (EM). However, in this case the distinction revolves around the stress of EM on the critical role of political processes and practices to move ecological phenomena into the modernisation processes (Buttel, 2000; Mol, 1995). Consequently, scholars relate weak EM to a corporativist interpretation based on the economisation of nature through elitist decision-making structures; meanwhile, strong EM relates to changes towards sustainable production and consumption patterns through further democratisation, redistribution and consideration of social justice issues (Gibbs, 2000; Hajer, 1995). Under this framework, weak strands will prompt technological solutions for environmental challenges based on elite technocratic decision making processes, in front of changes in the institutional and economic structure of society underpinned by open, democratic decision making processes posited by strong EM (Christoff, 1996).

According to Horlings, Kitchen, Marsden, and Bristow (2010) the bio-economy and the eco-economy paradigms can be seen as examples of weak and strong ecological modernisation, respectively. The bio-economy paradigm refers to 'those economic activities that capture latent value in biological processes and renewable bio-resources to produce improved health and sustainable growth and development' (Horlings et al., 2010, p. 7). The dominant narrative under this paradigm is a new strategy for sustainable capital as a politico-economic strategy operating globally to sustain capital accumulation through nature's modification and commoditisation at different levels (Birch, Levidow, & Papaioannou, 2010; Kitchen & Marsden, 2011). In this regard, several institutions are actively involved in promoting a bio-economy. For instance, the European Union is putting a special emphasis on this model to address current environmental, social and economic challenges and defines Knowledge Based Bio-Economy (KBBE) as 'transforming life sciences knowledge into new, sustainable, eco-efficient and competitive products' (EU Presidency, 2007, p. 2). Similarly the OCDE defines the bio-economy 'as a world where biotechnology contributes to a significant share of economic output. The emerging bio-economy is likely to be global and guided by principles of sustainable development and environmental sustainability' (OECD, 2009, p. 29). This model is mostly aligned with what has been identified as the productivity or efficiency narrative, which relies on scientific advances to face food insecurity and environmental problems through technologies that boost productivity while addressing resource constrains and reducing negative environmental impacts (Freibauer et al., 2011).

The eco-economy is presented as an alternative paradigm cutting across production and also consumption spheres. It is defined as a set of complex networks or webs of new viable businesses and economic activities that utilise the varied and differentiated forms of environmental resources in more sustainable ways. These do not result in a net depletion of resources but provide cumulative net benefits that add value to rural and regional spaces in both ecological and economic ways (Horlings & Marsden, 2011; Kitchen & Marsden, 2009). Distinctively from the previous paradigm, the eco-economy stresses the embeddedness of social and ecological dimensions in place-based constructions of economic relations (Marsden, 2010), as well as places emphasis on 'the recalibration of micro-behaviour and practices that, added together, can potentially realign production-consumption chains and capture local and regional value between rural and urban spaces' (Horlings et al., 2010, p. 8). This paradigm shares principles with the sufficiency narrative that envisages a response to the new food equation not only through scientific advances applied to food production but mainly through behavioural and structural changes in food systems that change current production and consumption patterns (Freibauer et al., 2011).

Despite broadening the focus to the entire food system and increasing the type and number of participants on these debates, both paradigms not only maintain the dualism and relevance of former conceptualizations (see table on differences in Marsden, 2012), but also reproduce mainly their basic elements. On the one hand, the defining traits and the type of innovations aligned with the eco-economy paradigm are by and large related to the actions formerly described under the processes of deepening, broadening and re-grounding of the sustainable rural development paradigm (see Kitchen & Marsden, 2009, for an adaptation of the framework) and also undoubtedly linked to the different dimensions identified under the 'rural web'[2] (Horlings & Marsden, 2012; Marsden, 2010). On the other hand, bio-economy principles are reproducing the logics of productivism, placing the increase in productivity as core solution and main aim of economic (and agricultural) development. This neo-productivism is envisaging new strategies labelled by and large under (conventional) sustainable intensification, although this concept has also been subject to discursive appropriation (Pretty, Toulmin, & Williams, 2011). Such abstract polarisations can of course once more lead to simplifications of reality but they also act as heuristic devices (Tscharntke et al., 2012), since most farmers strategies lie in the spectrum between these two categories.

The urgency and importance of the challenges that societies face under the new food equation are leading to involve increasing types and number of participants on this debates that are commonly aligned with confronted interests. In this regard, the eco-economy represents a more fragmented but grounded arena than the dominant bio-economy paradigm. For instance, under a transitions theory lens (see Spaargaren, Oosterveer, & Loeber, 2012), the expression of an eco-economy model will represent constellations of agri-food niches that might potentially scale up while the bio-economy principles will be more aligned with the changes in the socio-technical dominant regime as it has to adjust to resource scarcity and climate change.

The theoretical review of this chapter clearly unveils that what the reformulations of the bio-economy and the eco-economy represent is the evolution and development of earlier productivist and rural development paradigms. They do not exclude these earlier formulations, but rather demonstrate how they now need embedding in wider societal, political and economic settings. They represent in one important respect the re-definition of rural resources and their management into a wider societal agenda associated with food security, resource depletion and the transitions to a post-carbon society. Also they provide opportunities for different types of rural regions in Southern Europe, as elsewhere to experiment with a wider vector of differentiated and place-based development pathways.

CHANGING THE LENSES: MANIFOLD EXPRESSIONS OF MEDITERRANEAN AGRICULTURE

The previous review on dominant paradigms of agrarian change leads us to highlight two main aspects in the light of Mediterranean agrarian dynamics. Firstly, some of the major paradigms on agricultural change – and the subsequent debate generated from their assumptions – have traditionally being inspired in an Anglo-Saxon context. As shown above, this has noticeably compromised their capacity to consider farming systems and their dynamics beyond this geographical scope, despite the effort of some authors to developing further these approaches to adapt Mediterranean agricultures (see e.g. Moragues-Faus, 2011). Secondly, it is remarkable that some of these new Northern-constructed paradigms now seem to give preference to a specific type of agriculture (and even a type of rural area, those that were largely by-passed by productivist and post-productivist development models) which initially fits with the Mediterranean stereotype. In other words, while Southern Europeans (most agricultural stakeholders, policy-makers and an important share of public opinion) were still concerned about the economic viability of most of their farming systems and rural areas, these were being pointed out as constituting a sort of role model or at least having an advantaged position under the lens of these new theoretical approaches.

With these elements as a backdrop, this section introduces a set of transversal issues that, on the one hand, provide some keys for a more integrative conceptual and empirical understanding of the subsequent chapters; and on the other hand, introduce certain elements to assess and discuss the claimed Mediterranean differentiation and distinctiveness in the domain of rural and agrarian studies.

Beyond Duality

As discussed above, one of the main and longstanding criticisms the paradigms have received from Mediterranean researchers is that they tend to be too dualistic, confronting two (or three) models of agrarian change often presented as mutually exclusive. Although more recent studies (see above) acknowledge that patterns of transformation can coexist at several scales – regionally, locally and even at the farm level, by and large they tend to keep being considered as dissociated, that is they are interpreted as a mere juxtaposition of activities. This approach obscures the existing interconnections among processes, strategies and activities that, within the frame of dominant paradigms, would correspond to distinct models. That is to say, the underlying economic rationale of such coexistence has not being properly addressed. Furthermore, these connections also operate at several scales: at the farm level (Trabalzi, 2007) and locally (Moragues-Faus, 2011).

Moreover, the interrelations between models not only do constitute a theoretical and analytical issue to be addressed (e.g. as it makes it difficult to ascribe either a specific holding or a farming system to a unique model undermining the usefulness of the models and the potential to quantify the phenomenon); but also challenge normative assumptions underlying the main paradigms. In this regard, farmers are often accommodating under their farm management strategy practices that are claimed to correspond to confronted paradigms.

On the Relationship between Agrarian and Non-Agrarian Dynamics

The impact of different processes of agrarian change on the economy of rural territories has been a key aspect in the field of European rural studies from early 1990s, not only fuelled in the academic domain, but gaining a growing relevance in the framework of the emerging rural policy (particularly the Common Agricultural Policy).

In this regard, the way different types of agrarian holdings ascribed to distinct paradigms contribute to the rural economy has also been controversial. While the OECD (2006) mobilised arguments like 'agriculture, and particularly productive agriculture, is a major purchaser of local inputs, not only farm related but also business services' to assert that 'farm businesses that will be able to reach a level of productivity and/or quality of production that will make them competitive in a subsidy-free scenario could play a particularly important role in rural economies' (p. 42); others argued that the farms following the path of modernisation would be reducing their territorial integration, and that more multifunctional enterprises, focused on the exploitation of economies of scope, would reinforce their contribution to the rural economy and the maintenance of the rural fabric. As some commentators assert 'rural development can be seen as the search for a new agricultural development model' (van der Ploeg et al., 2000, p. 392). In this sense, a number of analyses have been carried out to quantify the impact of this model of agrarian change on the rural economy (see for instance Van der Ploeg & Renting, 2000).

However, these linkages have been only partially – namely unidirectionally – tackled, that is in terms of the spill-over effects of agriculture upon the rural economy. The importance of these impacts is widely acknowledged; however, it is necessary to also critically tackle the interrelations and dependencies of agriculture on the rest of the economy. In this regard, it is paramount to consider not only the linkages of agrarian dynamics with the rural non-agricultural economy, but also the close relations to regional and national economies, particularly including the consolidation or otherwise of the welfare state.

Despite the fact that these interrelations are a common and straightforward trait of agriculture, this articulation is especially relevant in the case of Mediterranean countries. This means that, in certain instances, agriculture has – paradoxically – benefited from a de-agrarianisation of the rural economy. Incomes from non-agricultural (via pluriactivity) and even non-rural sources (thanks to the increases of the rural–urban mobility), as well as direct transfers from the welfare state (e.g. retirement pensions), have contributed to maintaining low-profitability farming systems that would have been otherwise economically unviable. However, this maintenance also relies upon sets of embedded cultural and social dimensions that underpin farmers' motivations to remain in the business of farming and actively shape their particular strategies. This cultural and social dimension of farming is also strengthened by the traditional characteristics of many Mediterranean agricultural systems.

On Pluriactivity and Part-Time Farming

As explained before, pluriactivity – traditionally a feature of marginalisation and insufficient farming – acquired a renewed interpretation as a major component of the new rural development paradigm. This reading of pluriactivity has led to use the variable as a means to quantify the diffusion of this paradigm of agrarian/rural development in some Mediterranean countries (Van der Ploeg & Roep, 2003). Nevertheless, similarly to earlier modernisation assumptions, the ascription of pluriactivity to a specific paradigm has been made often uncritically, without considering the rationale of several 'pluriactivities' that should not be muddled up. In this sense, more nuanced specifications of which conditions particular practices have to comply with to be an expression of a new paradigm would be particularly welcome. Following Marsden's proposal (2003, p. 186), these practices would be showing new forms and potentially diversified forms of rural resource exploitation, as well as a new sorts of integration of agriculture in society. However, many forms of pluriactivity do not fit these 'novelty' requirements, so they could hardly be read in terms of a new component of such a rural development model.

Despite the fact that pluriactivity is a common feature of many European countries, a specific characteristic of Mediterranean regions is the presence of part-time farming, or more precisely, of part-time *non-pluriactive* holdings. Part-time in these countries is mostly the expression of small-scale farming where holdings are unable to provide employment to a full-time worker. In many of these holdings with an insufficient economic dimension the main source of income is retirement pensions received by an aged farming population. This ties farm households' continuity and occupancy not only to the vagaries of the CAP subsidy system, but in many cases more importantly to national welfare systems.

On the Delay, Southern Perspectives and Policy Idiosyncrasy

One of the main components of the Mediterranean agricultural stereotype was that of 'delay'. These agricultural systems were said to be – due to their structural features (small holdings, old and non-full time farmers and the like) – delayed in the path towards modernisation. Actually, this delay became a central concern for Mediterranean policy makers particularly when these countries initiated their process of integration into the EU single market in the 1980s. Consequently, the design of agricultural and rural policies focused on reducing the structural gap with Northern European countries to the detriment of other policy goals, namely environmental ones (Paniagua, 2001).

This tension between the goals of, on the one hand, reinforcing the competitiveness of the agricultural sector by means of holdings' structural transformation and, on the other hand, the integration of environmental issues into the agricultural and rural policy has been a constant feature of the policy process. The allegedly weaker environmental concerns of Southern societies, combined with the structural pressures exerted by agricultural stakeholders, have worked against the latter. The impact of these imbalances and tensions has become evident in the differences among EU Member States with regard to the allocation of rural development funds between competitiveness and environmental priorities (Tietz & Grajewski, 2009).

This argument of weak social support to the 'environmentalization' of policies in Southern countries in comparison to other European regions has become a sort of 'new delay' discourse. Indeed, in certain contexts, the stereotype of Mediterranean agriculture as delayed under the modernisation lens seems to have been replaced with a new interpretation: which claims/assumes that the low diffusion of certain post-productivist/multifunctional attributes is a sign of the still weak environmental awareness of Southern societies, a weakness that can be also linked to the degree of economic development of the countries. This lower diffusion of these practices (see Chapter 3) could be, in this context, attributed to a weaker demand of non-conventional farm outputs (either food or non-food) and/or the lower willingness of land-holders to undertake this sort of initiatives. In addition it has also been argued that the lower diffusion of multifunctional behaviour in Southern Europe has been due to the ways national and regional administrative levels have mediated the alleged multifunctional policies. As Wilson (2009, p. 274) states '*Mediterranean countries exemplify particularly well the importance of this spatial layer through frequent failure of implementation of strong multifunctionality policies which are often perceived as leading to a reduction in regional revenue*'. The example of agri-environmental measures has been often utilised to show this sort of Mediterranean exception (Buller, Wilson, & Höll, 2000).

One could argue, from these points, that such a new delay – both now in terms of ecological and agricultural 'modernisation' – could indeed be just a time delay or transitionary issue. In other words, as the country develops, society's environmental concerns will increase, and policies will gradually adopt a more 'balanced' environmental focus.

Nevertheless, as other authors have argued, the very concept and the interpretation of multifunctionality may vary among different regions (Clark, 2005). For instance, Ortiz and Moreno (2007) analyse the discourses utilised by Spanish policy makers to underpin, seemingly, productivist-oriented policies. They conclude that the core reason is not to increase agricultural production itself; rather, the main objective seems to be more territorially based. Indeed, when justifying the maintenance of coupled direct payments, the need to invest in new irrigation districts or the bias of rural development measures towards the reinforcement of holdings' competitiveness, the main underlying reasoning is that this is the best way to maintain the vitality of the rural fabric and to break down rural depopulation.

This seems to be coherent with what is shown by public opinion studies about Europeans' priorities. This is the case regarding the results provided by the Eurobarometers carried out by the European Commission concerning the role of farming and farmers, and the perceptions about the agricultural policy. These data show that there is not a clear North–South growing gradient of environmental awareness regarding the priorities given both to agriculture and the objectives of agrarian policies. Where there is actually a North–South difference, it concerns the relevance assigned by Mediterranean societies to the territorial role of agriculture in maintaining rural areas.

AND NOW, THE CRISIS: A CONCLUSION

The current economic and financial crisis in Southern European countries – as very well known – is having a particularly dramatic impact. However, it is still too early to comprehensively assess the global impact of the crisis in the evolution of Southern European farming systems. For instance, official agricultural statistics have not captured the possible effects of the crisis in the evolution of holdings. Indeed, most agricultural censuses were carried out in 2009 (when the crisis was taking off in some countries). In addition they have tended not to gather information from smallest production units (due to changes in the methodology), whose capacity to adapt shocks constitutes a major question.

Nevertheless, we can point out some preliminary hypotheses. Firstly, the crisis is affecting the very dimension and role of the state, and, in doing so, the very rationale and scope of agricultural and rural policies. The withdrawal of the public sector from some domains, budgetary restrictions to carry out activities and to provide services and an expected renewed emphasis on the utilisation of (allegedly more efficient) marked-based instruments will deeply modify the arena of rural and agricultural development. In addition, there has been a reinforcement of arguments in favour of measures that are expected to have short-run impacts on employment, to the detriment of other long-term (e.g. environmental) purposes. In addition to this, we should not forget that most of these policies (particularly regional and rural policies) are co-funded, so that the budgetary constraints of states and regions will constrain and condition the allocation of the funds.

Secondly, in the frame of the generalised fall of demand, those strategies based on the rural supply of value-added products (quality food, cultural rural tourism and rural amenities) will be confronted to a complicated and potentially shrinking market situation. Territorial and farm-based development strategies have been oriented, precisely in the framework of new rural paradigms, to value-added niche markets, with highly differentiated products and services. The generalised economic crisis will check whether these strategies have resulted in more resilient or more fragile structures. On the contrary, other practices – both individual (e.g. direct selling) and collective (e.g. farmers' search for cooperation and synergies) – could be interpreted to reinforce, under certain circumstances, the resilience of farming. The new rural development paradigm, as discussed above, was a conceptualisation built out of the period of relative economic growth in Europe in the 1990s and 2000s. Now, the changing economic and fiscal conditions will test its degree of vulnerability and resilience as a model for sustainable rural development.

Thirdly, it has been said that a growing movement of return would be taking place in Mediterranean countries as a consequence of the crisis (Brun, 2012). Nevertheless, the lack of updated statistics and the diversity of rural–urban linkages in Mediterranean regions (as this volume shows) require further research to understand the extent and specificities of this phenomenon.

Though it is perhaps too soon to assess the real magnitude and the scope of these processes of economic crisis and austerity, several questions arise: To what extent will the distinctiveness of Mediterranean agricultural/non-agricultural linkages play a role in facilitating and shaping these processes? Which attributes of Southern farming systems could make agriculture and rural areas either more resilient or vulnerable to current economic and social shocks? What role is there for agriculture in contributing to moving beyond the economic crisis in the South of Europe?

We have sought in the chapter to provide a conceptual overview of the relevance of existing agrarian change paradigms and models to the case of Mediterranean agriculture. This is very much a two-way learning process, whereby these models, on the one hand, can provide valuable heuristics for critical understanding of the complexities and distinctive features of contemporary Mediterranean agriculture, while, on the other, the study of the agricultural and agrarian 'condition' in the region can also further improve and enlighten the value of these models themselves. In particular we have argued that these models have tended to be excessively focused on Northern European dynamics, and have been unable to incorporate important explanatory aspects of the complexity and diversity of Mediterranean agriculture. The subsequent chapters in the volume will further conceptually and empirically illuminate these critical two-way processes.

Former conceptualisations of agrarian change in Europe have, as we have argued, originated from a rather hegemonic modernisation paradigm which tended to define agriculture around a dominant set of productivist principles which, in turn, then tended to define many rural regions in Europe as 'peripheral', 'passed-by' or 'delayed'. As we have seen in the treatment of the scholarship outlined in the first part of this chapter, gradually, but not completely, we have now entered a new more emancipated (and less retarded) conceptual terrain whereby there is more acceptance of multiple and more socio-ecological pathways towards modernisation, through critically embracing the wider rural development paradigm and more recently notions of the embedded eco-economy. In addition – and in the context of macro economic and welfare state crisis, combined with resource scarcity and climate change phenomena – we are entering a far more contested period when particularly embedded and regional and place-based models of sustainable development and ecological modernisation will have to come to the fore. In some senses this then frames one significantly new agrarian question for Mediterranean regions. How will they adjust and innovate in ways which will create more socio-ecologically resilient agricultural systems which can contribute to the wider needs of a metabolically and ecologically dependent society?

NOTES

1. See special issue in *Sociologia Ruralis, 40*(4) for a compilation of papers.
2. The rural web is a concept that was proposed in order to enlarge the theoretical understanding of rural development in an academic context of much empirical evidence but low theoretical innovations. Theoretically, the rural web is constituted by six interrelated dimensions: endogeneity, novelty production, sustainability, social capital, institutional arrangements and market governance (Van der Ploeg & Marsden, 2008).

REFERENCES

Alonso Mielgo, A. M., Sevilla Guzmán, E., Jiménez Romera, M., & Guzmán Casado, G. (2001). Rural development and ecological management of endogenous resources: The case of mountain olive groves in Los Pedroches comarca (Spain). *Journal of Environmental Policy and Planning*, *3*, 163–175. doi:10.1002/jepp.80

Beopoulos, N., & Damianakos, S. (1997). Grèce: Le cache-cache entre la modernité et la tradition. In M. Jollivet (Ed.), *Vers un rural postindustriel. Rural et environnement dans huit pays européenes* (pp. 176–231). Paris: L'Harmattan.

Bertolini, P., & Meloni, B. (1978). *Azienda contadina, sviluppo economico e stratificazione sociale*. Torino: Rosenberg & Sellier.

Birch, K., Levidow, L., & Papaioannou, T. (2010). Sustainable capital? The neoliberalization of nature and knowledge in the European "knowledge-based bio-economy". *Sustainability*, *2*, 2898–2918.

Bowen, S. (2010). Embedding local places in global spaces: Geographical indications as a territorial development strategy. *Rural Sociology*, *75*, 209–243. doi:10.1111/j.1549-0831.2009.00007.x

Bowler, I. R., & Ilbery, B. W. (1987). Redefining agricultural geography. *Area*, *19*, 327–332.

Brun, M. (2012). Europe du sud: un retour à la terre pour affronter la crise économique? Les notes d'alerte du CIHEAM 82.

Buller, H. (2001). Is this the European model? In H. Buller & K. Hoggart (Eds.), *Agricultural transformation, food and environment: Perspectives on European rural policy and planning* (Vol. 1, pp. 1–8). Aldershot: Ashgate.

Buller, H., Wilson, G. A., & Höll, A. (Eds.). (2000). *Agri-environmental policy in the European Union*. Aldershot: Ashgate.

Buttel, F. H. (2000). Ecological modernization as social theory. *Geoforum*, *31*, 57–65.

Buttel, F. H. (2001). Some reflections on late twentieth century agrarian political economy. *Sociologia Ruralis*, *41*, 165–181.

Christoff, P. (1996). Ecological modernisation, ecological modernities. *Environmental Politics*, *5*, 476–500. doi:10.1080/09644019608414283

Clark, J. (2005). The 'new associationalism' in agriculture: Agro-food diversification and multifunctional production logics. *Journal of Economic Geography*, *5*, 475–498.

Damianakos, S. (1997). The ongoing quest for a model of Greek agriculture. *Sociologia Ruralis*, *37*, 190–208.

DuPuis, E. M., & Goodman, D. (2005). Should we go "home" to eat?: Toward a reflexive politics of localism. *Journal of Rural Studies*, *21*, 359–371. doi:10.1016/j.jrurstud.2005.05.011

EU Presidency. (2007). En route to the knowledge-based bio-economy. *Cologne Summit of the German Presidency*, EU Council.

Evans, N., Morris, C., & Winter, M. (2002). Conceptualizing agriculture: A critique of post-productivism as the new orthodoxy. *Progress in Human Geography*, *26*, 313–332. doi:10.1191/0309132502ph372ra

Fabiani, G., & Scarano, G. (1995). Una stratificazione socioeconomica delle azienda agricole: Pluralismo funzionale e sviluppo territoriale. La Questione Agraria 59.

Feenstra, G. (2002). Creating space for sustainable food systems: Lessons from the field. *Agriculture and Human Values*, *19*, 99–106.

Fisher, D. R., & Freudenburg, W. R. (2001). Ecological modernization and its critics: Assessing the past and looking toward the future. *Society & Natural Resources*, *14*, 701–709.

Freibauer, A., Mathijs, E., Brunori, G., Damianova, Z., Faroult, E., Girona i Gomis, J., OBrien, L., & Treyer, S. (2011). *Sustainable food consumption and production in a resource-constrained world.* The 3rd SCAR (European Commission–Standing Committee on Agricultural Research) Foresight Exercise.

Friedmann, H., & McMichael, P. (1989). Agriculture and the state system. *Sociologia Ruralis, 29*, 93–117.

Gibbs, D. (2000). Ecological modernisation, regional economic development and regional development agencies. *Geoforum, 31*, 9–19. Retrieved from: http://dx.doi.org/10.1016/S0016-7185(99)00040-8

Goodman, D. (2004). Rural europe redux? Reflections on alternative agro-food networks and paradigm change. *Sociologia Ruralis, 44*, 3–16.

Goodman, D., DuPuis, E. M., & Goodman, M. K. (2012). *Alternative food networks: Knowledge, place and politics.* Oxon, NY: Routledge.

Hajer, M. A. (1995). *The politics of environmental discourse: ecological modernization and the policy process.* Oxford: Clarendon Press.

Hoggart, K., & Paniagua, A. (2001a). The restructuring of rural Spain? *Journal of Rural Studies, 17*, 63–80. doi:10.1016/s0743-0167(00)00037-1

Hoggart, K., & Paniagua, A. (2001b). What rural restructuring? *Journal of Rural Studies, 17*, 41–62. doi:10.1016/s0743-0167(00)00036-x

Horlings, L. G., Kitchen, L. C., Marsden, T. K., & Bristow, G. I. (2010). *Exploring the potential contributions of the bio-economy and the eco-economy to agri-food and rural regional development.* Working Paper Series No. 60. The Centre For Business Relationships, Sustainability and Society (BRASS). Cardiff, UK.

Horlings, L. G., & Marsden, T. K. (2011). Towards the real green revolution? Exploring the conceptual dimensions of a new ecological modernisation of agriculture that could 'feed the world'. *Global Environmental Change, 21*, 441–452. Retrieved from: http://dx.doi.org/10.1016/j.gloenvcha.2011.01.004

Horlings, L. G., & Marsden, T. K. (2012). *Exploring the 'new rural paradigm' in Europe: Eco-economic strategies as a counterforce to the global competitiveness agenda.* European Urban and Regional Studies. In press, published on line in May 2012. doi:10.1177/0969776412441934

Huber, J. (2000). Towards industrial ecology: Sustainable development as a concept of ecological modernization. *Journal of Environmental Policy & Planning, 2*, 269–285. doi:10.1080/714038561

Ilbery, B., & Bowler, I. (1998). From agricultural productivism to post-productivism. In B. Ilbery (Ed.), *The geography of rural change* (pp. 57–84). Harlow: Prentice Hall.

Ilbery, B., & Kneafsey, M. (2000). Producer constructions of quality in regional speciality food production: A case study from south west England. *Journal of Rural Studies, 16*, 217–230.

Ilbery, B., & Maye, D. (2010). Agricultural restructuring and changing food networks in the UK. In N. M. Coe & A. Jones (Eds.), *Reading the economy: The UK in the 21st century* (pp. 166–180). London: Sage.

Kinsella, J., Wilson, S., de Jong, F., & Renting, H. (2000). Pluriactivity as a livelihood strategy in Irish farm households and its role in rural development. *Sociologia Ruralis, 40*, 481–496.

Kitchen, L., & Marsden, T. (2009). Creating sustainable rural development through stimulating the eco-economy: Beyond the eco-economic paradox? *Sociologia Ruralis, 49*, 273–294. doi:10.1111/j.1467-9523.2009.00489.x

Kitchen, L., & Marsden, T. (2011). Constructing sustainable communities: A theoretical exploration of the bio-economy and eco-economy paradigms. *Local Environment, 16*, 753–769. doi:10.1080/13549839.2011.579090

Kizos, T., & Iosifides, T. (2007). The contradictions of agrotourism development in greece: Evidence from three case studies. *South European Society and Politics, 12*, 59–77. doi:10.1080/13608740601155443

Knickel, K., & Renting, H. (2000). Methodological and conceptual issues in the study of multifunctionality and rural development. *Sociologia Ruralis, 40*, 512–528.

Lang, T. (1999). The complexities of globalization: The UK as a case study of tensions within the food system and the challenge to food policy. *Agriculture and Human Values, 16*, 169–185. doi:10.1023/A:1007542605470

Lang, T., Barling, D., & Caraher, M. (2009). *Food Policy: Integrating health, environment and society*. Oxford Oxford University Press.

Lowe, P., Marsden, T., & Whatmore, S. (1990). *Rural restructuring: global processes and their responses*. London: David Fulton.

Lowe, P., Murdoch, J., Marsden, T., Munton, R., & Flynn, A. (1993). Regulating the new rural spaces: The uneven development of land. *Journal of Rural Studies, 9*, 205–222.

Marsden, T. (2003). *The condition of rural sustainability*. Assen: Van Gorcum.

Marsden, T. (2004). The quest for ecological modernisation: Re-spacing rural development and agri-food studies. *Sociologia Ruralis, 44*, 129–146. doi:10.1111/j.1467-9523.2004.00267.x

Marsden, T. (2010). Mobilizing the regional eco-economy: Evolving webs of agri-food and rural development in the UK. *Cambridge Journal of Regions, Economy, and Society, 3*, 225–244. doi:10.1093/cjres/rsq010

Marsden, T. (2011). From post-productionism to reflexive governance: Contested transitions in securing more sustainable food futures. *Journal of Rural Studies*. Retrieved from http://dx.doi.org/10.1016/j.jrurstud.2011.10.001

Marsden, T. (2012). Third natures? Reconstituting space through place-making strategies for sustainability. *XXIV Congress of the European Society for Rural Sociology, Chania, Crete, Greece*, August 2011, pp. 257–274.

Marsden, T., Banks, J., Renting, H., & Van Der Ploeg, J. D. (2001). The road towards sustainable rural development: Issues of theory, policy and research practice. *Journal of Environmental Policy and Planning, 3*, 75–83. doi:10.1002/jepp.77

Marsden, T., Munton, R., Whatmore, S., & Little, J. (1986). Towards a political economy of capitalist agriculture: A British perspective. *International Journal of Urban and Regional Research, 10*, 498–521. doi:10.1111/j.1468-2427.1986.tb00026.x

Marsden, T., Murdoch, J., Lowe, P., Munton, R., & Flynn, A. (1993). *Constructing the countryside*. London: UCL Press Limited.

Marsden, T., & Sonnino, R. (2008). Rural development and the regional state: Denying multifunctional agriculture in the UK. *Journal of Rural Studies, 24*, 422–431.

McMichael, P. (1997). Rethinking globalization: The agrarian question revisited. *Review of International Political Economy, 4*, 630–662.

Mol, A. P. J. (1995). *The refinement of production. Ecological modernization theory and the chemical industry*. Utrecht: Van Arkel.

Moragues-Faus, A. (2011). *How is agriculture sustained? Following farmers from the field to the market. A case study of olive oil in a Mediterranean region*. PhD thesis, Department of Economics and Social Science, Universidad Politècnica de València. Valencia, Spain.

Moragues-Faus, A. M., & Sonnino, R. (2012). Embedding quality in the agro-food system: The dynamics and implications of place-making strategies in the olive oil sector of Alto Palancia, Spain. *Sociologia Ruralis*, *52*, 215–234. doi:10.1111/j.1467-9523.2011.00558.x

Morgan, K., & Sonnino, R. (2010). The urban foodscape: World cities and the new food equation. *Cambridge Journal of Regions, Economy and Society*, *3*, 209–224. doi:10.1093/cjres/rsq007

Morris, C., & Evans, N. J. (1999). Research on the geography of agricultural change: Redundant or revitalized? *Area*, *31*, 349–358.

Murdoch, J., Marsden, T., & Banks, J. (2000). Quality, nature, and embeddedness: Some theoretical considerations in the context of the food sector*. *Economic Geography*, *76*, 107–125.

Newby, H. (1982). Rural Sociology and its relevance to the agricultural economist: A review. *Journal of Agricultural Economics*, *33*, 125–165.

Newby, H. (1983). The sociology of agriculture: Toward a new rural sociology. *Annual Review of Sociology*, *9*, 67–81.

OCDE. (2006). *The new rural paradigm: Policy and governance*. Working paper on territorial policy in rural areas OCDE, Paris.

OECD. (2009). The Bioeconomy to 2030: designing a policy agenda, Arundel, A., and Sawaya, D., OECD, Paris, pp. 322. ISBN 978-92-64-03853-0 (2009).

Ortiz, D. & Moreno, O. (2007). La política agraria en España en el contexto del debate europeo productivismo/post-productivismo. *VI Congreso de Economía Agraria: "De la Economía Agraria a la Economía Rural y Agroalimentaria"*. Spain: Albacete, Spain.

Paniagua, A. (2001). Agri-environmental policy in Spain. The agenda of socio-political developments at the national, regional and local levels. *Journal of Rural Studies*, *17*, 81–97.

Patterson, L. A. (1997). Agricultural policy reform in the European Community: A three-level game analysis. *International Organization*, *51*, 135–165. doi:10.1162/002081897550320

Potter, C., & Tilzey, M. (2005). Agricultural policy discourses in the European post-Fordist transition: Neoliberalism, neomercantilism and multifunctionality. *Progress in Human Geography*, *29*, 581–600. doi:10.1191/0309132505ph569oa

Pretty, J., Sutherland, W. J., Ashby, J., Auburn, J., Baulcombe, D., Bell, M., & …, Pilgrim, S. (2010). The top 100 questions of importance to the future of global agriculture. *International Journal of Agricultural Sustainability*, *8*, 219–236.

Pretty, J., Toulmin, C., & Williams, S. (2011). Sustainable intensification in African agriculture. *International Journal of Agricultural Sustainability*, *9*, 5–24. doi:10.3763/ijas.2010.0583

Sayer, A. (2001). For a critical cultural political economy. *Antipode*, *33*, 687–708.

Servolin, C. (1972). L'absorption de l'agriculture dans le mode de production capitaliste. In Y. Tavernier, M. Gervais & C. Servolin (Eds.), *L'univers politique des paysans dans la France contemporaine* (pp. 41–77). Paris: Presses de la Fondation Nationale de Sciences Politiques.

Seyfang, G. (2006). Ecological citizenship and sustainable consumption: Examining local organic food networks. *Journal of Rural Studies*, *22*, 383–395.

Sonnino, R., & Marsden, T. (2006). Beyond the divide: Rethinking relationships between alternative and conventional food networks in Europe. *Journal of Economic Geography*, *6*, 181–199. doi:10.1093/jeg/lbi006

Spaargaren, G., Oosterveer, P., & Loeber, A. (2012). *Food practices in transition: Changing food consumption, retail and production in the age of reflexive modernity*. New York, NY: Routledge.

Tietz, A., & Grajewski, R. (2009). Rural development programmes 2007–2013: Differences in EU member states' financial allocations. *EuroChoices*, *8*(1), 30–31.

Tilzey, M. (2006). Neo-liberalism, the WTO and new modes of agri-environmental governance in the European Union, the USA and Australia. *International Journal of Sociology of Agriculture and Food*, *14*, 1–28.

Tovey, H. (2001). Agricultural development and environmental regulation in Ireland. In H. Buller & K. Hoggart (Eds.), *Agricultural transformation, food and environment: Perspectives on European rural policy and planning* (Vol. 1, pp. 109–130). Ashgate, Aldershot.

Trabalzi, F. (2007). Crossing conventions in localized food networks: Insights from southern Italy. *Environment and Planning A*, *39*, 283–300.

Tscharntke, T., Clough, Y., Wanger, T. C., Jackson, L., Motzke, I., Perfecto, I., & ..., Whitbread, A. (2012). Global food security, biodiversity conservation and the future of agricultural intensification. *Biological Conservation*, *151*, 53–59. Retrieved from: http://dx.doi.org/10.1016/j.biocon.2012.01.068

Van der Ploeg, J. D., & Marsden, T. (2008). *Unfolding webs*. Assen, The Netherlands: Van Gorcum.

Van der Ploeg, J. D., & Renting, H. (2000). Impact and potential: A comparative review of european rural development practices. *Sociologia Ruralis*, *40*, 529–543. doi:10.1111/1467-9523.00165

Van der Ploeg, J. D., Renting, H., Brunori, G., Knickel, K., Mannion, J., Marsden, T., & ..., Ventura, F. (2000). Rural development: From practices and policies towards theory. *Sociologia Ruralis*, *40*, 391–408.

Van der Ploeg, J. D., & Roep, D. (2003). Multifunctionality and rural development: The actual situation in Europe. In G. Durand & G. van Huylenbroeck (Eds.), *Multifunctional agriculture. A new paradigm for european agriculture and rural development* (pp. 37–53). Hampshire: Ashgate.

Ventura, F., & Milone, P. (2000). Theory and practice of multi-product farms: Farm butcheries in umbria. *Sociologia Ruralis*, *40*, 452–465. doi:10.1111/1467-9523.00160

Walford, N. (2003). Productivism is allegedly dead, long live productivism. Evidence of continued productivist attitudes and decision-making in South-East England. *Journal of Rural Studies*, *19*, 491–502. doi:10.1016/S0743-0167(03)00030-5

Ward, N. (1993). The agricultural treadmill and the rural environment in the post-productivist era. *Sociologia Ruralis*, *33*, 348–364. doi:10.1111/j.1467-9523.1993.tb00969.x

Ward, N., Jackson, P., Russell, P., & Wilkinson, K. (2008). Productivism, post-productivism and european agricultural reform: The case of sugar. *Sociologia Ruralis*, *48*, 118–132. doi:10.1111/j.1467-9523.2008.00455.x

Wilson, G. A. (2001). From productivism to post productivism... and back again? Exploring the (un) changed natural and mental landscapes of European agriculture. *Transactions of the Institute of British Geographers*, *26*, 77–102.

Wilson, G. A. (2007). *Multifunctional agriculture: a transition theory perspective*. Wallingford: CABI Publishing.

Wilson, G. A. (2008). From 'weak' to 'strong' multifunctionality: Conceptualising farm-level multifunctional transitional pathways. *Journal of Rural Studies*, *24*, 367–383.

Wilson, G. A. (2009). The spatiality of multifunctional agriculture: A human geography perspective. *Geoforum*, *40*(2), 269–280.

CHAPTER 3

THE 'SOUTHERN MODEL' OF EUROPEAN AGRICULTURE REVISITED: CONTINUITIES AND DYNAMICS

Eladio Arnalte-Alegre and Dionisio Ortiz-Miranda

ABSTRACT

This chapter presents an overview of the 'big' data of Mediterranean agriculture, with a special focus on the four EU countries (Portugal, Spain, Italy and Greece), in order to provide a backdrop for the rest of cases analysed in the volume. In this regard, two thesis are discussed: the assumption that farming systems in the South have not followed the process of 'productivist modernisation' characterising post-war Northern European agricultural change, and that, precisely due to this reason, most holdings and regions from the South would have more possibilities to adapt to new approaches of multifunctional rural development.

Thus, the chapter tackles both the static and dynamic structural traits of Southern agricultures and their differences with the North, as well as several aspects of the organisation of farming in the Mediterranean and other key components of productivist modernisation: farm intensification and specialisation. Later, the diffusion of multifunctional dynamics is

Agriculture in Mediterranean Europe: Between Old and New Paradigms
Research in Rural Sociology and Development, Volume 19, 37–74

ISSN: 1057-1922/doi:10.1108/S1057-1922(2013)0000019005

addressed, in order to introduce some reflections about their meaning and scope in the Mediterranean regions. The chapter ends with a straightforward typology of Southern farming systems and a concluding section, which goes back to discuss the two initial theses.

Keywords: Southern European agriculture; productivist modernisation; multifunctional development; farm structural change

INTRODUCTION

The diversity of realities within the framework of the European agriculture has been widely supported in the scientific literature. This agricultural differentiation stems from the diverse ecological conditions, still being the base of farming, as well as from the way historical trajectories of land appropriation have shaped different patterns of the classical agrarian question in European countries and regions. Likewise, the different paces and chronology of the national processes of industrialisation and economic development in the last two centuries have impacted differently rural areas and farming systems.

These rural and agrarian diversity explains the lack of success of a political construction like the one of the 'European model of agriculture', introduced in late nineties in the context of the CAP reform and the WTO negotiations (Mahé, 2001), but whose 'limited empirical or analytical value' was properly pointed out (Buller, 2001, p. 2).

The analysis on the European agrarian diversity reveals a centre-periphery differentiation between the core of firstly industrialised countries and the periphery (the 'green ring') characterised, until some decades ago, by the importance of agriculture in the national economies and politics (Granberg, Kovách, & Tovey, 2001). In that North-European core around the North Sea, the so-called Danish model of family farming spread, that of a farm early modernised underpinned by an appropriate institutional support. The main exception was the case of United Kingdom, with a different agrarian system based on large holdings from its transformation of the 18th century, and also characterised by the early diffusion of technology and the development of a modern agriculture (Hoggart, Buller, & Black, 1995, pp. 82–85; Sivignon, 1996; Tracy, 1989, pp. 8, 107–110).

In front of that core-model of European agriculture, the several continental peripheries – Mediterranean, Eastern (marked by the socialist regimes from 1945), and also the Northern-Scandinavian and even the Celtic (Ireland) – would have undergone different trajectories, which have left still visible footprints in their agricultures and rural areas.

In this regard, the existence of a distinctive model of Mediterranean agrarian systems in the South is widely acknowledged in the literature. There is even a consensus to include four countries in such model: Portugal (despite its Atlantic character), Spain, Italy and Greece.[1] They share relatively similar ecological conditions, with the presence of a set of characteristic agrarian systems (Beopoulos, 2003), as well as certain parallelisms regarding their paths of industrialisation. The historical conditions of land appropriation were modified in the 20th century in Greece (where the land reform of the 1920s redistributed 40% of arable land and spread the small holding model), and Italy (also through its post-Word War II and less effective land reform, Fonte, 2001, p. 275). On the contrary, in Spain and Portugal the failure of the several attempts of land reform along the century maintained the dominance of large holdings in many areas of the Southern regions of these countries.

Notwithstanding these disparities, most of the agricultural holdings of these four countries showed, at the end of the 20th century, some common structural traits (dominance of small farms, aged holders and rigid structures that change slowly) clearly differentiated from that of Northern Europe. This allowed referring both a 'Southern model' and a 'Northern model' of farm structures in Europe (European Commission, 1997).

French agriculture, in spite of having large regions with fully Mediterranean agronomic conditions, is as a whole much closer to the structural characteristics and dynamics of Northern-Centre farming systems. The Danish model of family farming spread rapidly in Northern France, although, as Sivignon (1996) pointed out, it is only from 1945 when the model generalised in the country. Also from the 1950s, the French agricultural policy began to include well-defined measures aimed to modernise and consolidate that professional family farming (Delorme, 2000). Regarding Balkan countries, which also include Mediterranean farming systems, their situations are still marked by their recent past of planned economies, where land tenancy regimes and agricultural organisation were substantially altered.

Therefore, taking into account the geographical space we have delimited, the objective of this chapter is to discuss, on the basis of the available empirical information, two hypotheses – well established in the literature – about the agricultures of Southern Europe and about their relationship with the dominant paradigms (see Chapter 2, this volume).

The first one is the assumption that farming systems in the South have not followed the process of 'productivist modernisation' that has been the main pattern of transformation of Northern Europe agriculture along the second half of the past century. This delay in terms of modernisation would explain the structural weaknesses and lack of competitiveness of farms of Mediterranean countries, particularly when – once accessed to the European Union along the 1980s – they were more directly confronted to Northern productive systems. This thesis, widely assumed in Southern countries' agricultural policies, explains the adoption of measures directly aimed to accelerate the structural adjustment of their farms (as illustrated by the Spanish Law of Farm Modernisation in 1995). Paradoxically, the emphasis on these policy stimuli took place precisely when this paradigm and the results of its implementation began to be questioned in Northern Europe.

The second thesis to consider is that most holdings and regions from the South – precisely because in many cases they would correspond to what Marsden (2003) refers as areas 'passed over' by other development models (either productivist of post-productivist), would have more possibilities to adapt and take advantage of the opportunities offered by new approaches of multifunctional rural development. In this line, Laurent (1998) called in the French context for a view of agriculture and agricultural policy not exclusively focused on 'professional' farms; rather, she claimed the importance to consider also the rest of holdings, articulated through other ways to the French rural territories and society.

Our analysis is mostly based on the statistical information available nationally, also including several references from the literature on the development and dynamic of agriculture in regions and systems from Southern Europe. In order to tackle the Northern-Southern differences of European agriculture we focus on EU-12 countries (those making up the EU in late 1980s). This delimitation allows contrasting the agriculture of the four Mediterranean countries and that of Northern-Central European countries, although other peripheries are excluded – both Scandinavian (which accessed in 1995) and Eastern (accessed in 2004) ones, since EUROSTAT does not provide time series long enough to allow identifying some aspect of the transformations at play.

The rest of the chapter is structured as follows. The two first sections focus on both the static and dynamic structural traits of Southern agricultures and their differences with the North. Regarding the later one (the evolution of farm structures) we discuss some elements of the apparent Mediterranean structural rigidity. Third, and also linked to the structural transformations, the chapter analyses three aspects of the organisation of farming in the South: the diverse footprint that the rural exodus left in the shaping of farm family networks, the diffusion of outsourcing and agricultural contractors and the dynamics of farm labour, with special attention to role played by foreign migration. Fourth, we focus on two other key components of productivist modernisation: farm intensification and specialisation. Fifth, the diffusion of multifunctional dynamics (quality-oriented production and non agricultural diversification) is addressed, in order to introduce some reflections about their meaning and scope in the Mediterranean regions. Since most of these analyses are based on aggregated national data, in order to illustrate the agricultural diversity of Southern countries, sixth section introduces a straightforward typology of farming systems. Finally, the concluding section goes back to discuss the two theses established above.

NORTH-SOUTH DIFFERENCES: THE STRUCTURAL TRAITS

A first step of this analysis is to check to what extent the stereotype about the structural characteristics differentiating Northern and Southern Europe is still valid. Table 1 shows a set of indicators from EUROSTAT, for 2007,[2] which allow quantifying the key elements of such stereotype.

Table 1. Farm Structural Indicators (2007).

	UAA/ Holding (ha)	LU/ Holding	Economic Size (ESU)/ Holding	% Full Time Holders	% Holders Aged >65	% Permanent Crops
Portugal	12.6	10.2	6.6	26.6	48.3	17.2
Spain	23.8	46.9	20.6	22.2	36.6	17.5
Italy	7.6	32.0	14.9	21.1	44.5	18.2
Greece	4.7	7.0	7.2	12.8	37.4	27.6
Belgium	28.6	105.0	70.3	69.1	21.2	1.5
Denmark	59.7	161.2	80.1	40.0	20.3	0.4
France	52.1	66.0	53.6	56.7	15.4	3.9
Germany	45.7	67.6	49.5	45.7	7.5	1.2
Ireland	32.3	48.6	19.4	55.0	24.9	0.0
Luxembourg	56.9	91.4	51.8	30.3	15.9	1.2
Netherlands	24.9	120.6	111.3	62.9	18.2	1.8
UK	53.8	77.0	31.4	43.2	32.6	0.2

Source: Authors' elaboration from EUROSTAT Farm Structure Surveys.

North-South differences keep well marked in the three size-focused indicators: physical (in both hectares of UAA and livestock units) and economic (in European Size Units, based on Standard Gross Margins). The Spanish farms' average superficial dimension is similar to the ones in Belgium and Netherlands. However, this hides the dual character of Spanish agriculture, with an outstanding weigh of small and very large farms, and with scarce presence of medium holdings, precisely the ones leading the processes of agricultural modernisation in the North of the continent.[3]

The other exception of the North-South dichotomy is the low economic dimension of Irish farms (similar to the Southern ones), which confirms its peripheral character compared to the Central model of European agriculture that could also, in some aspects, be attributed to Ireland.

On the other hand, the table also shows that less than a quarter of holdings in the South provides full-time employment to the holders, whereas in the North this share is notably higher. This fact does not respond to more pluriactivity or off-farm employment of holders in Mediterranean,[4] but to the limited capacity of small holding to provide employment. A considerable proportion of these non full-time farmers is, precisely, made up with the large group of aged holders shown in the same table.

Finally, the relative balance of permanent/annual crops also shows notable differences in the productive orientation of farms in the two groups of countries.

THE PROCESS OF STRUCTURAL ADJUSTMENT IN SOUTHERN AGRICULTURES: HISTORICAL RIGIDITY AND RECENT CHANGES

The other component of the stereotype about the North-South disparities deals with the dynamics of farm structures, much more rigid in the Mediterranean, whereas Northern Europe shows a more rapid adjustment (reduction of the number of holdings and increase of the size of remaining ones). This concentration trend of agricultural production in fewer and larger production units constitutes ones on the major traits of the so-called productivist modernisation.

EUROSTAT publishes from the 1960s homogeneous series on the evolution of farm structures of the countries that progressively access the EU, allowing in this way the comparison of their respective paces of adjustment (EUROSTAT, 2000). These data show a steady rhythm of concentration, around 3% of annual growth of farms' average size (has UAA) between 1966/1967 (first year with data) and late 1980s in most of Northern countries.[5] On the other hand, Italy, the unique Southern country belonging in that period to the European Community, showed a virtually frozen agrarian structure (0.3% reduction of the average size of holdings between 1966/1967 and 1990). The three other Southern countries, according to national data series (not comparable with EUROSTAT ones), showed a slow pace of adjustment, with farms' sizes growing at 1% of annual rate (Arnalte, 1992).

For the most recent period, Fig. 1 shows the global results of the structural adjustment for the 12 countries analysed in the last two decades (already with EUROSTAT homogeneous series). Data highlight the core of extensive agriculture of North-western countries, with a uniform and high rhythm of adjustment for these 20 years. This corroborates that farms concentration still goes on in these farming systems. The three exceptions already mentioned in the 'historical period' (Netherlands, UK and Ireland) also appear well delimited.

Regarding Southern countries, we notice some changes in their performance when compared with previous periods. This can be analysed from Table 2, which show for these four countries and for the same period (1990–2007) the evolution of the number of farms and the UAA.

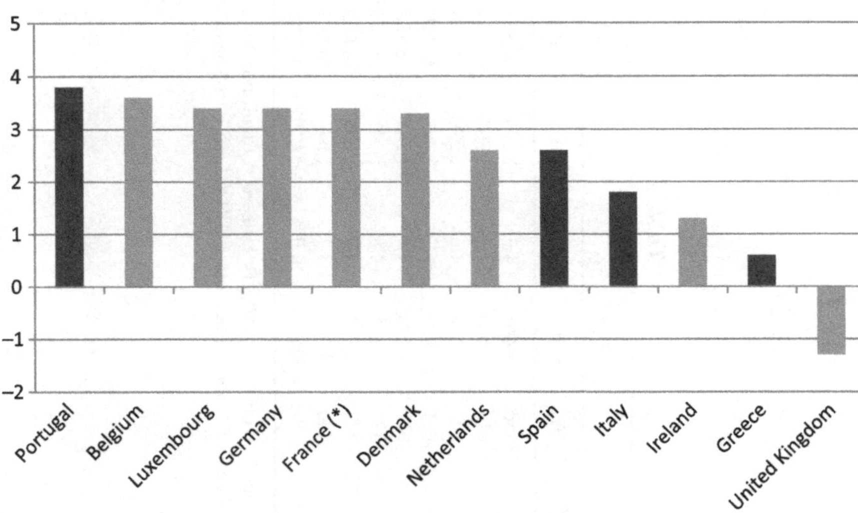

Fig. 1. Evolution UAA/Holding (Annual Variation Rates in %, 1990–2007). (*) Metropolitan France. *Source*: Authors' elaboration from EUROSTAT Farm Structure Survey.

Table 2. Evolution of Mediterranean Farm Structures.

	Greece			Spain			Italy			Portugal		
	A	B	C	A	B	C	A	B	C	A	B	C
1990	850	3,661	4.3	1,594	24,531	15.4	2,665	14,947	5.6	599	4,006	6.7
1993	819	3,539	4.3	1,384	24,714	17.9	2,488	14,736	5.9	489	3,950	8.1
1995	802	3,578	4.5	1,278	25,230	19.7	2,482	14,685	5.9	451	3,925	8.7
1997	821	3,499	4.3	1,208	25,630	21.2	2,315	14,833	6.4	417	3,822	9.2
2000	817	3,583	4.4	1,287	26,158	20.3	2,154	13,062	6.1	416	3,863	9.3
2003	824	3,968	4.8	1,141	25,175	22.1	1,964	13,116	6.7	359	3,725	10.4
2005	834	3,984	4.8	1,079	24,855	23.0	1,729	12,708	7.4	324	3,680	11.4
2007	860	4,076	4.7	1,044	24,893	23.8	1,679	12,744	7.6	275	3,473	12.6
Var. 90/07	1.2%	11.3%	10.0%	−34.5%	1.5%	54.9%	−37.0%	−14.7%	35.3%	−54.1%	−13.3%	88.7%

A: Number of Holdings (x000); B: Utilised Agricultural Area, UAA (× 000 ha); C: UAA/holding (ha).
Source: Authors' elaboration from EUROSTAT Farm Structure Surveys.

Structural adjustment underwent a strong acceleration in both Spain and Portugal after their accession to the UE in 1986, as illustrated by the rapid drop of the number of farms in the 1990s. In the 2000s, the process slowed down in Spain,[6] but continued at similar rates in Portugal, which provoked the disappearance in 17 years of more than a half of agricultural holdings in the country.

Agricultural structure in Italy left behind its historical stability, with a notable pace of adjustment, particularly in the second part of that period (27% fewer holdings between 1997 and 2007). The considerable reduction of agricultural area explains that, despite that drop, the growth of holdings' size has been moderate. Lastly, Greece still responds to the Southern stereotype during this more recent period, maintaining a persistent structural rigidity.

SOME EXPLANATORY ELEMENTS OF THE STRUCTURAL RIGIDITY

Several explanatory factors have been suggested to explain the weakness of the process of farm concentration in Southern Europe for the last decades, among them: cultural and legal reasons (heritage systems, land transfer regulations), lack of effective policies to stimulate farm modernisation, factors related to the productive orientation of Mediterranean agriculture, as well as other arguments around the characteristics of the processes of economic development and the urban–rural linkages in these countries (Arnalte & Ortiz, 2006; European Commission, 1997; Fabiani & Scarano, 1995).

We will refer at this point to two of these factors. On the one hand, we tackle a simple and quantifiable aspect: the low diffusion of land renting. This has been also a relevant issue in the debates about the most effective tools of structural policy to facilitate agricultural adjustment. On the other hand, we focus on a more general and less analysed (due to the lack of quantitative and sound information) aspect: the differences between 'official' farm structure statistics and the real performance of these agricultures.

The Diffusion of Land Renting

Specialised literature has frequently referred the constraints of farmland markets in Mediterranean countries. In particular, the low diffusion of renting in these countries has been pointed out, since this tenancy regime was the main form of land transfer among holdings in the process of adjustment, taking place in a number of North-western European and other industrialised countries (Blandford & Hill, 2005). This lower dissemination of renting in the South is also related to the relative importance of permanent crops in these countries (see Table 1); crops with productive cycle hardly adapt renting contract periods.

Data from Table 3 show the evolution of the relative weigh of farmland under renting. They confirm the high presence of this tenancy regime in Northern Europe (above 50% of UAA, and rising, in four countries). In line with other adjustment indicators, exceptions are found in Netherlands, UK and Ireland, as well as Denmark, where this regime – despite its recent increase – has had little tradition (Harrison, 1982).

Figures also confirm the low diffusion level of renting in Mediterranean countries (around, 20% of UAA in 1990), although it has increased 8–10% in Greece, Italy and Spain during the period considered. The progress of the liberalisation of rental agreements, notably in Spain and Italy is these years, would have pushed its development.

Nevertheless, it is not possible to obtain definitive conclusions in terms of simple cause–effect relationships between farmland renting diffusion and structural adjustment rates. In Portugal, the Southern country where adjustment has been more rapid, the level of renting has not changed substantially. Whereas in Greece, where farm structures seemed to be frozen, data show a considerable expansion of renting.

Table 3. Evolution % of UAA Tenant Farmed.

	1990	2000	2007
Portugal	24.6	23.2	23.3
Spain	19.9	27.0	27.3
Italy	18.4	23.4	27.9
Greece	22.2	27.7	31.8
Belgium	66.2	67.2	66.9
Denmark	19.1	25.2	29.3
France*	56.2	62.7	74.3
Germany	53.0	62.8	61.7
Ireland	12.4	18.6	18.4
Luxembourg	48.9	53.2	56.5
Netherlands	31.4	27.6	25.5
UK	38.4	33.8	31.7

*Metropolitan France.
Source: Authors' elaboration from EUROSTAT Farm Structure Surveys.

Informal Land Cessions and the 'Real' Structure of Holdings

The doubts about what is really behind the data provided by censuses and farm structure surveys have been often addressed in the Mediterranean literature. This has been the case of Italy, where a number of authors (Barbero, 1982; Sotte, 2006) have wondered about 'how many' were in fact the agricultural holdings in the country, and what sort of economic performance is hidden under that image of agricultural activity atomised in many thousands of very small farms. The problem has always been the lack of empirical evidences about the undercurrent of official data.

A recent study focused on a Northern Greece rural area and precisely aimed to obtain 'an authentic picture of the reality of family farming in modern Greece that often lies concealed behind official figures and myths' and to explain how Greek agriculture 'remains in the productive system', provides illuminating data on this issue (Koutsou, Partalidou, & Petrou, 2011, p. 404). In spite of its local character (Kilkis prefecture), we consider the 'real' structure of farm holders of this study to be very illustrative of this situation. Fig. 2 summarises its main findings:

This classification of farm heads interrelates to two informal forms of land cession described in the study:

- Inter-family cessions within enlarged families. In that case, the holding of 'real' farmers are made up of farmland belonging to several 'official' holders – e.g. farmer's own land, his wife's or sons' land, or that of other relatives who migrated. This situation is a consequence of the egalitarian systems of heritage and the maintenance of family linkages after the rural exodus, as well as because parents use to register the land in the name of the sons living in the city, so that they could maintain their roots in their original villages and the agrarian family identity – which land ownership symbolises. Evidently, land is considered a 'family asset', rather than a component of 'the farming business' (Koutsou et al., 2011, pp. 409–410, 416).
- Informal land cessions to 'professional farmers', who are the real operators. The study illustrates how most of these cessions do not imply an economic remuneration; rather, owners only demand the land to be maintained 'in good agricultural condition, in accordance with CAP obligations'. In this way, the 'official' holder gets CAP subsidies. According to the study's informants, roughly 60% of 'non farmers' (holders living elsewhere, pensioners or widows) participate in this way of land transfer (Koutsou et al., 2011, pp. 411–412).

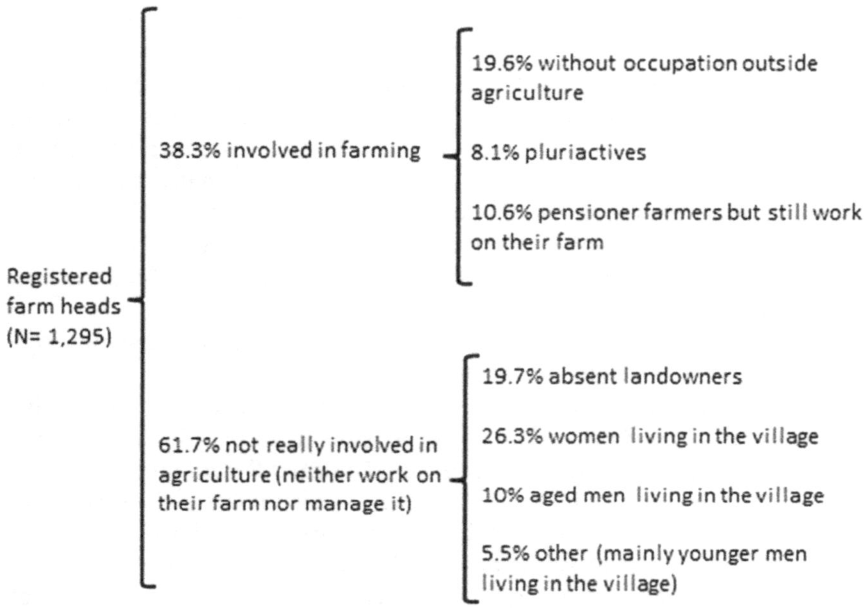

Fig. 2. Farm Heads Classification in Kilkis Prefecture (Greece). *Source*: Elaboration from Koutsou et al. (2011, pp. 411–415).

As a consequence of these informal flows of land, the real size of holdings is notably larger than the official one, and more likely to be economically feasible. The study concludes that 'the fragmentation of farmland in the area is in essence a fiction and has nothing to do with actual land use or land management practices'. It also underlines that, besides economic factors, social and cultural driving forces also lead to these types of land cessions and this organisation of farming (p. 410).

Koutsou et al. (2011, pp. 415–416) also insist on the informal and temporary character of these cessions, which contributes to somewhat instability, although they acknowledge this has lasted for decades, even 'before the CAP'. These authors also reject this constitutes a simple response to the 'policy framework' and a way to get subsidies, although they also recognise the capacity of this model of organisation to adapt to changing circumstances.

Another case study located in the other extreme of the Mediterranean (Spain) focuses on the organisation of small-scale farming with some similarities to the former one. Indeed, Moragues' (2011) research in Alto

Palancia – an inland rural area in the Region of Valencia well connected to urban zones, analyses the typology of agrarian holders members of the cooperatives of the area. She identifies a small group (3–4%) of full-time farmers (also here referred by the own field informants as 'real farmers'), while the rest distributes fifty-fifty between retired people and farmers with off-farm employment. In this case, pluriactivity is considerably more frequent than that of the case study of Northern Greece (Moragues, 2011, pp. 169–170).

This analysis depicts a continuum from 'real farmers' who 'manage an important amount of land (…) and in many cases are agricultural contractors' (p. 182), to the opposite extreme, that of those holders 'dissociated from farming'. In the middle, Moragues finds a range of levels of involvement in farming operations and farm management, as well as of levels of outsourcing – i.e. amount and type of tasks that are outsourced to either agricultural contractors or cooperatives' services (Moragues, 2011, pp. 182–186).

Our point is that these forms of agricultural structures allow, at least for the time being, for some stability of small-scale farming and could explain the weak concentration of holdings shown by the official statistics. Evidently, these two studies are not representative of the whole Mediterranean, which also include areas of more professionalised farming systems and where the processes of farm concentration and differentiation adopt patterns more similar to the Northern European ones.

However, these two studies bring forward a number of elements and processes that are often found in Mediterranean agricultures, and which can be considered as characteristics of their models of organisation. In this regard, the following section deepens into these informal forms of land cessions.

FARMING ORGANISATION IN THE SOUTH

Rural Exodus and Farm Family Networks

The model of inter-family land cessions within enlarged families described in the Greek region of Kilkis is not so frequent in other Southern areas, like for instance rural Spain. Possibly this would be the consequence of some differences of the intense process of rural exodus these two countries underwent in the 1960s–1970s of the 20th century.

In Greece, some authors insist on the idea that the intense process of out-migration did not break solidarity family networks. In this sense, Kasimis and Papadopoulos (2001, p. 205) argue that the rural exodus was part of a larger family strategy, 'the exodus of some family members is planned'. For Damianakos, 'the peasant family never emigrates as a unit: one or two members always stay behind in the village and take care of the farm' (1997, p. 203).

Out-rural migration did not produce, therefore, the disintegration of rural society; so that the modalities of in-family land cessions currently at play show the way farm organisation and management respond to a strategic family project, in the framework of the emerging relationships between the community of out-migrants and those who remain in the villages (Goussios and Duquenne, 2003, p. 46).

The process was different in Spain. In many rural areas (particularly mountain and dryland interior plains) society did not resist the intensity of the process. Many villages and counties emptied 'from one day to the next' (Camarero, 1997, p. 230). Thus, entire families migrate 'without looking behind'.[7] In these regions, only aged people remained, which provoked deep demographic transformations and seriously affected the dynamic of farms: the two first agricultural censuses carried out in Spain, in 1962 and 1972, reported the disappearance of 365,000 holdings, 12.5% of the total.

A similar situation of massive out-migration and collapse of many areas is depicted for the Portuguese rural exodus in the same period. For instance, Baptista (1996, reprinted in 2001, p. 36) describes 'the flight of hundreds of thousands of men and women who, once open the door to leave, escaped from the arduous life and work conditions they had in the places and villages of rural society'. This massive flight was also a rejection to the 'praise to rural life' still frequent among the elites of traditional rural societies.

At present, although there are still family relationships around land tenancy both in Spain and Portugal (see for instance Chapter 7 in Baptista, 2010), they are neither so widespread nor so intense than the ones the Greek literature points out.

Outsourcing and Agricultural Contractors

The two previous case studies (Kilkis in Greece and Alto Palancia in Spain) also illustrate the presence of a figure that, in diverse forms and more or less diffused, can be found in most of Southern European agricultures. This is the outsourcing of specific farming tasks, or even of the complete management of the holding, carried out by agricultural contractors (with different levels of formalisation as service firms or cooperatives) without the farm 'owner' losing his/her position as holder. Agricultural contractors manage and carry out farm works using their own machinery and equipment and, if necessary, resorting to hired labour.

This modality of farm organisation is found, as shown above, in Greece,[8] as well as in Italy, where a number of authors made in the 1980s interesting conceptual contributions to this phenomenon (De Filippis, 1985; Pugliese & Ceriani-Sebregondi, 1981; Vellante, 1981). In Spain, outsourcing has been also analysed nationally and at the level of several farming systems (Arnalte, 1989, 2002; Gallego Bono, 2010; Langreo, 2002; Ortiz, Arnalte, Moragues, & Doñate, 2011). In Portugal, Canadas (1998) analysed in depth farm outsourcing in viticulture specialised regions.

It has to be acknowledged that this phenomenon is not exclusive from Southern Europe; rather several authors have described similar processes of agricultural labour reorganisation leading to a more frequent utilisation of external services both in UK (Ball, 1987; Errington & Gasson, 1996) and France (Harff & Lamarche, 1998). However, which clearly differentiates Northern and Southern developments of this phenomenon is its degree of formalisation. Thus, for example, while we find in France a diversity of legal entities giving formalisation to outsourcing – e.g. CUMA (cooperatives of agricultural equipment utilisation) or ETA (farm work firms), in Southern Europe the majority of these services are carried out informally and without a clear regulatory framework.

This also hinders official statistics to capture properly the phenomenon. Indeed, less of 1% of Southern agrarian holders declares they provide contracted farm services using his/her own machinery (a question included in the EUROSTAT's Farm Structure Survey). This figure exceeded 10% in UK and Scandinavian countries (Ortiz et al., 2011 based on data from 2007).[9]

This also means that the characteristics of providers of agricultural services are not well known, so it is necessary to resort to specific researches on this topic (e.g. Fanfani & Pecci, 1991 for the Po Valley in Italy; Ortiz et al., 2011 for three productive systems in Mediterranean regions of Spain; and the already cited Koutsou et al., 2011 and Moragues, 2011). In summary, these studies show that (despite the existence of specialised agricultural service firms) most of the services are provided by farmers having own machinery. These farmers range from large holders having an important fleet of own – and often specialised (e.g. harvesters) – machinery, to medium size holders for whom service provision means an important share of family income. Moreover, authors also highlight this is a clear growth strategy for these farmers.

In addition, what is has been empirically shown in the South is that the relationships between demanders and suppliers of farm services often inscribe and intertwine in the context of territorialised social networks. For instance, Gallego Bono (2010) argues, in his analysis of Spanish Mediterranean citric areas, that part-time farming and outsourcing belong to and evolve within an organisational-institutional network which is defined by both formal rules and shared values and conventions. Moragues' (2011) analysis of small-scale olive production in the Spanish region of Valencia, demonstrates the correlation between the degree of social integration of farmers (the higher) and the modalities of services contracted (the more informal and cheaper agreements).

All this implies that small-scale holdings outsourcing certain farm operations, can take advantage of technical progress and innovation, as well as of economies of scale. This allows for some theoretical considerations. On the one hand, we notice that precisely economies of scale associated to technical progress keeps being a key element even within a context of farm structures seemingly 'frozen' (Arnalte, 2006). On the other hand, the individual holding is no longer the basic unit of agricultural production, which challenges some traditional conceptualisation of farm structure analysis. At higher levels of outsourcing, the farm becomes the mere physical location of a set of operations externally decided and managed, so that the performance of the agricultural productive process takes place at another relevant level, dissociated from the structure of individual farms (Arnalte, 1989; De Filippis, 1985).

In short, outsourcing is one of the several 'heterodox' forms of agricultural organisation in the Mediterranean Europe, through the setting up of several and intense interdependences among different types of farms and actors (between service providers and service takers, between out-migrants and relatives in charge of managing the holdings within family cessions). This model differentiates from that of the Danish model (on the basis of well-sized and professional farms) which dominates Northern Europe. However, the economic rationale of this heterodox way of restructuring has also allowed substantial technical progress in this type of agriculture.

Farm Labour Market and the Role of Migrant Work

The utilisation of a considerable amount of salaried labour, basically provided by foreign migrants, in the Mediterranean areas of intensive farming systems is a well-known feature of these agricultures. In the 1980s,

Berlan (1986, 1987) alerted about the risks of Mediterranean agriculture to become a 'European California' based on the overexploitation of foreign working class, and leading to both social tensions and the 'sacrifice of an important part of family-based farms' (1986: 22; 1987: 244). This contrast between this model of agriculture that was emerging in the South (particularly in Spain, Southern France and Italy) and the 'virtues' of family farming still dominating Northern Europe, was attracting the attention of a part of literature. Other studies focused on the underlying rationale of the substitution of local labour force with foreign migrants in the process of consolidation of intensive farming systems (see Hoggart & Mendoza, 1999, on migrants in Northern littoral in Catalonia).

The current situation of foreign agricultural workers is addressed in other chapters of this volume (Moreno, Laurent and Kasimis and Papadopoulos). However, it is useful to contextualise the situation of Southern countries and the role played by incoming foreign population in the framework of the evolution of agricultural labour markets in Europe during the last decades.

For this purpose, Fig. 3 shows the evolution of salaried labour within agricultural occupied population in selected EU countries. Data from 1980 show that only UK – historically different from the continental one – could be considered to have a non-family agriculture from the point of view of employment. In the rest of the countries agriculture could be said to maintain its family character, although Spain and Italy already had slightly higher levels of salaried employment.

This situation radically changes in the two following decades. The most striking case is that of Dutch agriculture, which passed from 22% of salaried in 1980 to 49% in 2000. A rapid rise is also found in the more extensive French agriculture. Behind this evolution it is the well-known process of breaking up of the family labour group (Blanc, 1987) leading to a gradual individualisation of family farming, i.e. the holder is the unique member involved in farm work (or management) and the rest of family members follow differentiated off-farm labour paths. In any case, the effects of the reduction of work family availability are different according to the characteristics of farming: in more extensive agricultures, where mechanisation is easier, this rise has been slower (as the French case illustrates); in intensive agricultures, where the substitution of labour with mechanisation is more constrained, salaried labour increases rapidly (Netherlands).[10]

It is precisely this later correlation (intensive specialisation and salaried labour) which explains the trend in Southern countries. As Fig. 3 shows, salaried labour increases in Spain the period considered. Also in Italy and

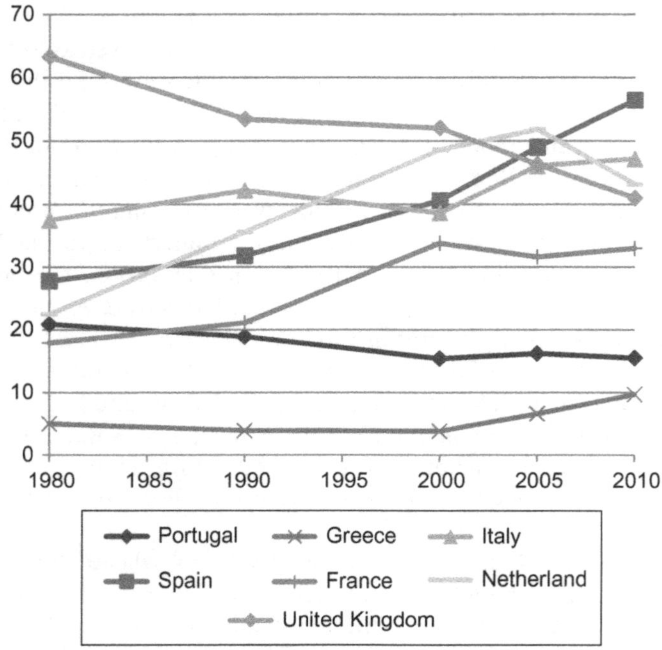

Fig. 3. Evolution of Salaried Employment in Agriculture (% Out of the Total Agricultural Employment). *Source*: European Commission. *Agriculture in the European Union. Statistical and Economic Information. Several years.* Available in http://ec.europa.eu/agriculture/statistics/agricultural/index_en.htm. Accessed January 2013.

Greece (the most 'family' agriculture) salaried labour increases from 2000. The only exception is Portugal, where this variable even drops slightly. In addition, this growing demand of salaried labour has been satisfied, almost exclusively, with foreign workers. In order to analyse in-depth that recent process we will refer specifically to the Spanish case.

Two circumstances converge in late 1980s and early 1990s. On the one hand, once finished the transitional period after the accession to the EU (1986), European markets opened, which increased dramatically vegetable exports to the North. On the other hand, the growth of Spanish economy since middle 1980s was absorbing the labour stock that remained in agriculture during the crisis of the 1970s and early 1980s. As a result, intensive agriculture expanded in many Mediterranean regions supported on the massive arrival of foreign migrants. What Pedreño (1999) calls 'vegetable factories' consolidated.

Several studies have tackled the organisation of farming and the resulting social structure in these intensive systems (particularly in South-eastern Spain, Gómez López, 1993; Pedreño, 1999, 2012). The massive and flexible availability of foreign labour force – together with the development of new forms of market organisation and supply concentration – has been a key factor of competitiveness and consolidation of intensive horticulture. The demand of work comes from two types of farms: on the one hand large commercial firms and on the other hand family farms adopting a range of modalities of entrepreneurial character (see Moreno, this volume) and utilising an important share of salaried labour. Pedreño's (1999) analysis of horticultural areas in Murcia shows the diversity of labour sub-contracting modalities – either by means of temporary work agencies or mere informal labour arrangements, allowing non-direct labour relationships between employees and the holdings where they work. This outsourcing of labour management facilitates a high degree of deregulation and informality, which constitutes a characterising trait in the Mediterranean, giving rise to a social construction based upon a vulnerable and flexible work force.

This close correlation between intensive agriculture and labour flexibility is corroborated by the weight of flexible agricultural work in European countries (see Table 4). Indeed, the highest percentages of Annual Work Units resulting from the addition of non-regular salaried plus workers not

Table 4. Distribution of Annual Work Units in EU Agriculture (2010).

	Family	Non-Family		
		Regular	Non-Regular	No Directly Contracted
Portugal	80.1	11.3	7.5	1.1
Spain	61.1	17.1	18.2	3.6
Italy	79.1	8.8	11.6	0.5
Greece	81.7	4.2	13.1	1.0
Belgium	74.6	18.2	6.7	0.5
Denmark	57.3	39.8	2.9	0.0
Germany	62.8	25.4	10.1	1.7
Ireland	90.6	5.8	1.8	1.8
France	43.0	44.4	11.0	1.5
Luxembourg	74.8	20.1	4.0	0.8
Netherlands	55.3	26.6	11.7	6.4
United Kingdom	64.6	23.8	7.0	4.6

Source: Authors' elaboration from EUROSTAT Agricultural Censuses.

directly contracted by the holder are found in Spain (21.8%), Netherlands (18.1%), Greece (14.1%), France (12.5%) and Italy (12.1%), all of them with an outstanding presence of intensive agriculture. These data are useful to frame the several references that some of the chapter of this volume make about labour.

And last but not least, the entry of foreign migrants – that reaches 24% of total occupied in Spanish agriculture (MAGRAMA, 2012a) – is not exclusive of intensive littoral farming areas. Gradually, it has taken place also in inland regions where family farming has 'individualised' (as in other European regions), both as itinerant employees for harvesting in several systems (grapes, olives, tobacco) and fixed workers in intensive livestock. Moreover, migrant population has grown in Spanish rural areas (from 2.8% out of total in 2001 to 6.7% in 2007, Camarero et al., 2009, pp. 136–142), and its works in construction and service industry, contributing in this way to the demographic renewal of many rural areas severely affected by the exodus of previous decades. For Greece, Kasimis and Papadopoulos (2005) also point out the role played by migrants in the economic revitalisation of rural areas.

OTHER PRODUCTIVIST TRAITS: INTENSIFICATION AND SPECIALISATION

Together with the concentration of production, literature points out farm intensification and specialisation as the two other identification traits of productivist modernisation. This section reviews, upon the basis of some straightforward indicators, the evolution of these traits for Southern Europe and its differences with that of Northern countries.

The thesis that Mediterranean agricultures have reached lower levels of *intensification* than Northern ones is widely supported. For instance Lamarche (1996) showed that difference in his analysis of the evolution of a set of productive indicators (mechanisation, consumption of chemical fertilisers, livestock intensification) between the 1980s and early 1990s. However, he also noticed a slight convergence due to the slowing down of these indicators in the North, whereas they went on in the South. This author concluded that, in this period, this would be showing that Southern farmers were fully involved in farming system intensifications, whereas Northern ones were beginning to question, or at least moderate, that trend (Lamarche, 1996, p. 90). Also Caraveli (2000) corroborated these differences, but also showed local intensification processes in some areas of Mediterranean countries, parallel to ongoing processes of extensification and even farmland abandonment in mountain areas or less favoured areas.

In order to update the situation and analyse these changes in the beginning of this century, we utilise two indicators (Table 5). The first one is the most used indicator on physical intensification of farming: chemical fertiliser consumption per hectare of UAA, mainly linked to yields increase. The second one is an economic and more global indicator: the ratio between Intermediate Consumptions (i.e. farms' expenditure in external inputs throughout the productive cycle) and the total value of agricultural production (Output of the Agricultural Industry).

Data show that, in the nineties, fertilisation levels dropped in Northern countries, a similar trend than the observed in Greece and Portugal. On the contrary, Italy and Spain still showed slight increases in fertilisation consumption per hectare in that period, although they began to decrease in the first years of the 21th century. As a result, North-South fertilisation gap remains (being Italy the unique exception). These lower levels are related to the Mediterranean climatic conditions (lower precipitations, periodic droughts) which limit the use of fertilisers, even if the notable expansion of irrigated perimeters in these countries has mitigated these constraints (Caraveli, 2000).

The other indicator of Table 5 (IC/OAI) led to similar conclusions: analogous evolutions in the first decade of the century in both Northern and Southern European countries, so that the existing gap (lower levels of intermediate consumption of external inputs in the Mediterranean) remains. In any case, some factors affect this indicator. On the one hand, we cannot forget that this indicator is strongly associated to the importance of livestock production (and therefore fodder expenditure, a major component of IC) in the North. On the other hand, the general increase of the ratio is explained by the relative evolution on paid input prices by farmers and received output prices (the so called 'price squeeze'), which deteriorates economic margins in all European farming systems.

Together with concentration and intensification, the third trait associated to the productivist modernisation of agriculture is the higher specialisation of farms, which would progressively abandon productive diversity that characterises traditional farming, to specialise its activity in fewer different products.

A way to quantify specialisation from official statistics is to resort to the classification of farm types carried out within the farm structure surveys, upon the basis of the relative contribution of the different products to each

Table 5. Indicators of Agricultural Intensification.

	Total Consumption of Chemical Fertilisers kg/ha, UAA			Intermediate Consumption (IC)/Output of the Agricultural Industry (OAI) (%)	
	1990	2000	2007	2000	2010
Portugal	88.1	75.0	57.0	50.0	61.9
Spain	103.0	106.6	101.4	37.1	44.6
Italy	166.1	177.2	137.3	37.4	46.6
Greece	217.3	153.4	97.2	33.7	47.9
Belgium	351.9*	249.5*	258.5*	61.3	68.8
Denmark	287.8	169.1	149.4	62.9	71.9
France	278.8	191.3	170.2	51.9	59.0
Germany	250.1	195.2	188.8	59.7	69.7
Ireland	203.4	172.1	133.0	53.6	76.0
Luxembourg				52.7	73.7
Netherlands	332.5	248.3	206.4	53.2	65.1
UK	181.7	134.8	120.2	58.7	66.1

*Data for Belgium and Luxembourg together.
Source: EUROSTAT.

holding's Standard Gross Margin (SGM). A holding is considered as specialised in a certain product (or group of products) when it contributes at least 2/3 of the SGM. Table 6 shows, for several European countries, the relative weight of specialised farms.[11]

Data show a notable level of agricultural specialisation in both Northern and Southern Europe. More than 70% of farms (except Portugal) are specialised. Moreover, the level of specialisation increased slightly between 2000 and 2007, with the exception of Greece and UK. Logically, this generalised specialisation differs among countries. In the North, bovine livestock predominates: 92% of Irish farms, 60% in Luxembourg, 50% in Netherlands and 49% in UK are specialised in *Grazing livestock* type. Also in the North, 54% of Danish and 23% of French holdings specialise in *Field Crops*.

In the Mediterranean, holdings show a different specialisation profile, since most of them are included into the *Permanent Crops* type (52% of Greek and Spanish farms, 47% in Italy and 36% in Portugal). More concretely, most of these farms are specialised in olive production (34% of Greek holdings, 21% in Italy and 20% in Spain), in fruit and citric trees (18% of Spanish farms) and vineyards (11% of Portuguese and 10% of Italian farms).

Table 6. Farm Specialisation.

	% Holdings in Specialised Types (Types 1 to 5, Except Subtype 34)*	
	2000	2007
Portugal	50.6	56.3
Spain	77.3	80.6
Italy	74.1	75.7
Greece	75.9	72.0
Belgium	77.4	78.5
Denmark	77.5	80.5
France	77.8	80.0
Germany	71.7	78.6
Ireland	95.9	97.0
Luxembourg	81.6	86.1
Netherlands	86.1	86.4
UK	87.7	64.7**

*See endnote 11.
**For UK, the drop is associated to an increase of 'non-classified' farms. The percentage of 'mixed farms' remains for the whole period between 5% and 7%.
Source: EUROSTAT Farm Structure Survey, 2007.

In any case, it is necessary to highlight that this Southern specialisation in not the result of a recent process of productivist modernisation, as it could be perhaps in the North. Rather, the Mediterranean specialisation in export-oriented commercial crops dates back to centuries ago. For instance, Damianakos (1997, pp. 194–196) describes the high specialisation in raisin production in Ionic Islands and Peloponeso during the second half of the 19 century. In the Spanish Mediterranean, Piqueras (1985) analyses the several commercial crops that have been succeeding from the XVIII century. In both cases, these historical processes led to large monocrop areas in these regions. This leads to outline the necessity of a longer historical perspective to analyse some facets of agrarian change.

SOME (INTERNALLY COMPLEX) MULTIFUNCTIONAL DYNAMICS

Previous sections have focused on the dynamics of the structural characteristics of holdings, which relate with the physical traits of farms, as well as with the modalities of management and organisation of farm work. Besides them, other major transformations, which would be related to post-productivist or multifunctional models of agricultural development have taken place. In this regard, we will refer here to two of these changes: the expansion of food quality-oriented practices in the holdings and the adoption of non-agricultural diversification strategies. The aim of these sections is not to present a detailed analysis for the four countries we are dealing with, but to highlight some elements we consider to be relevant for our discussion.

The 'Boom' of Quality-Oriented Dynamics in Southern Europe

Marsden and Sonnino (2006) argued that food quality differentiation based upon the territorial linkages of production has prevailed in Mediterranean countries. Rather, in Northern Europe, quality attributes have been more related to aspects like public health, hygiene or the environmental implications of production. According to these authors, this difference would be in part due to the distinctive legal frameworks, more oriented towards promoting private labels in the North, whereas in the South national policy makers (with the support of EU Regulations) have strongly promoted the development of territorial labels.

Indeed, food territorial labels – namely Protected Designation of Origin (PDO) and Protected Geographical Indication (PGI) – expanded rapidly in the last 15–20 years in the Mediterranean. In 2008, more than 60% of EU food (wine excluded) PDO/PGI and 47% of their production (in value) were concentrated in Portugal, Spain, Italy and Greece, leading, together with France, EU statistics.[12]

This is clearly illustrated by Spain, which passed from 74 (PDO and PGI) in 1998 to 160 in 2010.[13] Among them, olive oil production (an emblematic Mediterranean product) had in 1998 only 6 PDOs covering 218,804 ha. Ten years later, they were 28 olive oil PDOs with almost a million hectares of olive groves within the protected perimeters.

Our point is that, although these quality-oriented practices have been usually considered as an advance of non-productivist farming models, their growth has been often fully parallel to strong productivist transformations. This can be illustrated again by Spanish olive oil production, which, at the same time it was leading PDO expansion, was simultaneously witnessing an outstanding increase of the land devoted to this crop (leading to mono-cropping in many areas and transforming and homogenising traditional landscapes) and a rise of yields thanks to intensive practices (new irrigation districts, new and more dense plantations) (Scheidel & Krausmann, 2011). Similar processes can be found in vineyard production (see Moreno-Pérez, this volume). In this way, some PDO/PGIs have resulted in a rising pressure over natural resources (e.g. water). In addition, the regulation of some territorial labels located in high nature value areas would be presenting limitations when it comes time to integrate environmental considerations – as illustrated by Beaufoy (2009) for the jam PDO of 'dehesas', the Spanish version of Portuguese montados (see Pinto-Correia and Godinho, this volume).

The other outstanding process regarding quality orientation is the diffusion of organic farming practices. Contrary to territorial approaches, this is more a single farmer decision level, although collective action can also play a decisive role (Ortiz-Miranda, Moreno-Pérez, & Moragues-Faus, 2010). EUROSTAT data show how the four EU Mediterranean countries have experienced rapid expansions of organic certification, particularly in the first years of the 21th century (see Table 7). This has led these countries to surpass most non-Mediterranean ones, which could be reflecting an

Table 7. Share of Total Organic Crop Area Out of Total UAA.

	2000	2010
Austria	13.8	17.2
Sweden	5.9	14.3
Italy	6.7	8.6
Greece	0.7	8.4
Finland	6.7	7.4
Spain	1.5	6.7
Denmark	5.9	6.1
Germany	3.2	5.9
Portugal	1.2	5.8
United Kingdom	3.3	4.1
Belgium	1.5	3.6
France	1.2	2.9
Luxembourg	0.8	2.8
Netherlands	1.6	2.5
Ireland	0.6	1.1
EU-27	-	5.1
EU-15	3.0	6.4

Source: EUROSTAT website http://epp.eurostat.ec.europa.eu/portal/page/portal/agriculture/data/database. Accessed on January 2013.

already mature organic sector in Northern and Central Europe and a certain Southern delay in the adoption of these practices. However, the extensive character of many Mediterranean farming systems (due to their agro-ecological conditions) would be also facilitating organic diffusion, which could also explain this comparative trend. The Spanish organic expansion (more than 1.8 million ha in 2011) corroborates this. Two thirds of that figure correspond to area of forests, grasslands, and fallow land. Organic crops occupied 'only' 610,000 ha. It is not to say this is a minor change; however, a more detailed analysis is needed to assess the driving forces and real implications of this massive organic conversion.

In addition, this expansion has been more remarkable in terms of area than in terms of number of holdings. The reason is that, and this seems to be a distinctive trait of Mediterranean (as well as other EU-12 exceptions), holdings adopting organic practices use to be the larger ones (in physical size) (see Table 8). The reasons of why the existing economies of scale in the adoption of organic practices seem to be playing more importantly in some countries (among them the Mediterranean) than others, remain unclear.

Finally, the expansion of organic production in the South contrasts with the lower level of consumption of organic food in these countries. This would be reflecting, in spite of the lack of specific statistics (European Commission, 2010), that an important share of Southern organic production

Table 8. Percentage of Organic Holdings Out of Total Holdings per UAA Strata (2007).

	UAA of holdings (ha)			
	<20	20–<50	50–<100	>=100
Portugal	0.4	1.4	2.8	5.4
Spain	1.2	2.3	3.9	4.5
Italy	2.1	7.6	9.7	14.2
Greece	3.5	7.7	13.5	20.3
Belgium	0.5	1.3	2.2	2.7
Denmark	3.9	5.5	4.9	5.6
France	1.4	2.1	2.4	1.6
Germany	3.1	4.7	3.9	5.2
Ireland	0.4	0.6	0.6	0.5
Luxembourg	1.7	2.9	3.0	2.2
Netherlands	1.0	1.9	2.4	4.0
UK	0.7	1.7	2.1	3.2

Source: Farm Structure Survey EUROSTAT.

is participating in the intra-EU trade that is feeding the higher levels of organic consumption in the North. For instance, Italy and Spain are among the three first organic exporters (in value) in the world (MAGRAMA, 2012b, from IFOAM statistics). On the opposite, Germany, France and UK are among the four main world organic importers.

In-between farm diversification and product differentiation, other holdings' strategies relate to their connections with the final food consumer (a set of practices that literature has considered, together with quality production, as deepening, see Chapter 2, this volume). The clearest example is direct selling. Here again, we find how this traditional modality has been reassessed through the lens of new paradigms of agrarian development. The re-gained recognition of these practices has led, also here, to include them into official agricultural statistics. Thus, EUROSTAT statistics identify, for some countries (basically Mediterranean and Eastern ones), the number of sole holdings producing mainly for direct selling. The Farm Structure Survey of 2007 shows that 24% of Greek sole holdings producing mainly for direct selling, 17.4% in Italy, 6.4% in Portugal and only 0.1% in Spain.

There are not historical records to assess the evolution of this modality of marketing. It is true that this, as well as other forms of short supply chains, are now more visible thanks to the renewed attention of statistics and researchers. Nevertheless, we wonder if they might have been present in the Mediterranean for long, interwoven with the dense and complex rural-urban, agrarian-non agrarian linkages we have discussed in this chapter and that constitute a historical distinctive trait of Southern countries.

Non-Agricultural Farm Diversification

Another major trait of multifunctional agriculture expresses when other farm-related activities become a business for farmers. This information is collected into EUROSTAT's Farm Structure Surveys from early this century, which reflects again how these practices (historically associated to marginal farming) have become to be considered more relevant under the discourse of multifunctionality. Hence, the diffusion of (statistically identified) diversification practices shows a clear North-South different profile, again with some exceptions (Fig. 4).

Here again, diversification is closely related to farm size (Table 9): the larger the holding, the more prone to diversify, due to a combination, among other, of economies of scale, capital availability, available underutilised land, etc. Contrary to organic farming, which showed some North-South

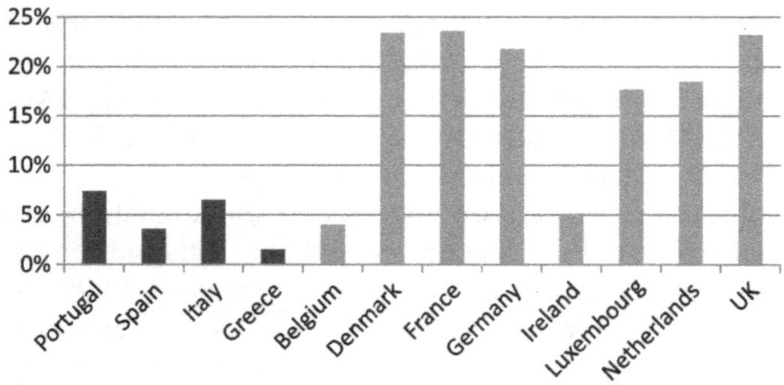

Fig. 4. Percentage of Holdings with Another Gainful Activity (2007).
Source: EUROSTAT Farm Structure Surveys.

Table 9. Percentage of Diversified Holdings Out of Total Holdings per UAA Strata (2007).

	UAA of holdings (ha)			
	<20	20–<50	50–<100	>=100
Portugal	8.3	8.0	9.7	12.5
Spain	3.1	5.1	6.8	8.1
Italy	6.4	14.4	16.7	19.6
Greece	1.4	6.3	8.3	18.0
Belgium	3.2	4.6	5.0	6.8
Denmark	19.2	22.5	26.0	32.3
France	27.5	23.8	20.7	24.3
Germany	19.2	24.6	26.6	29.7
Ireland	3.5	5.4	8.1	11.1
Luxembourg	9.8	12.2	24.6	24.8
Netherlands	11.9	25.5	30.9	33.9
UK	22.7	25.4	27.8	36.1

Source: EUROSTAT Farm Structure Surveys.

differences, in this case, diversification-size correlation seems to be the general tone in all of the countries considered (excepting France). This correlation would also explain the lower diffusion of diversification in Mediterranean countries having smaller holdings.

Needless to say that, under these national aggregated data, it is possible to find regions and areas where agri-tourism, on farm processing activities, the provision of farm and non-farm services with own machinery and other modalities of diversification constitute frequent practices, and play a key role in underpinning farming systems. Several studies illustrate these cases, see for instance Alonso Mielgo, Sevilla Guzmán, Jiménez Romera, and Guzmán Casado (2001) for Los Pedroches in Spain, or several examples from Italy: Brunori and Rossi (2000), Sonnino (2004) or Kanemasu and Sonnino (2009) for Tuscany, Ventura and Milone (2000) for Umbria.

Finally, much has been written about the relationships between farm diversification and the provision of environmental services, a key element of the multifunctional paradigm. Moreover, this linkage would seem to offer more opportunities in Southern countries. Indeed, data show unequivocally the deep integration between farming systems and high nature value (HNV) areas in the Mediterranean. According to EUROSTAT's IRENA Indicators, 53% of Greek UAA is within HNV areas, 37% in Spain, 34% in Portugal and 21% in Italy. From EU-15, only two other countries have similar figures: 27% in UK (mainly uplands) and 24% in Ireland. This contrasts with the higher dissociation between farmland and natural areas in countries like France (15%), Germany (3%), Denmark (3%) or Netherlands (2%).

The possibilities to turn the provision of environmental services into income require both complex institutional arrangements and an adequate regulatory framework. By the time being these modalities both privately funded (through diverse forms of private contracts, see for instance Hodge & Adams, 2012) and as part of public policies (whose clearest example is agri-environmental payments included into the Pillar II of the CAP) are much more developed and diffused in Northern Europe than in the Mediterranean (Rosell, Viladomiu, & Correa, 2010).

A STRAIGHTFORWARD APPROXIMATION TO MEDITERRANEAN AGRICULTURAL DIVERSITY

The pages above have addressed the main common traits of agriculture in four Mediterranean countries and its differences from Northern Europe. The analysis has been based on national data, so that the outstanding internal differences among Southern farming systems – in terms of productive orientations, farm structures, processes of intensification and the like – remained concealed. The studies included in the following chapters of this volume will precisely illustrate this internal agricultural and rural diversity in the Mediterranean. However, in the global picture of this chapter, we consider useful to outline a straightforward typology as a first approximation to that diversity.

Needless to say that this is not an exhaustive scheme including all the existing Southern agricultures, rather it aims to cover a sort of easily identifiable ideal-types (with several intermediate situations) in these countries. This is a typology of farming systems upon the characteristics of farms (with different territorial linkages) predominating in each area, and it also illustrates the degree of advance of the productivist modernisation in several territories.

A *first type* is made up with farming systems of consolidated family-based, modern and professionalised holdings. Some examples are found in the Po Valley (Italy), where farming is closely connected to the strong regional agri-food (livestock oriented) district. In Spain we also find several examples: intensive horticulture in Southeast, intensive olive groves in Andalucía or Castilla's cereal-oriented plateaus where rural outmigration allowed farm concentration. We have also similar cases in Portugal (e.g. vineyard holdings in the Douro region) and Greece (intensive farming in irrigated plateaus). They are the most similar situations to professional and competitive farming in the North, although still far from the agricultural characteristics of Dutch Flevoland or Paris Bassin, emblematic examples of specialised agricultural areas.[14] In addition, this type of agriculture is more vulnerable in the Mediterranean due to climatic constraints, since they use to be irrigated systems highly dependent on – more and more uncertain – water resources.

A *second type* corresponds to areas where there is a stable system of small holdings, well integrated with local labour markets, either in close urban areas or within diversified and dynamic rural contexts. They are about systems with a majority of pluriactive holders, frequently resorting to outsourcing. Saraceno (1994) described these characteristics for the agriculture of the province of Udine (North-eastern Italy), although it might be also present in Central Italy. This is also the type of structure that can be found in Northern littoral of Portugal (Baptista, 2001, 2002) and in several areas of Spanish Mediterranean coast. They are systems where land concentration has not taken place (because pluriactivity opportunities), but where productive intensification is going on. In addition, the difficulties to enlarge farm size have led professional farmers to grow by the way of becoming agricultural contractors.

A *third type* can be found in many mountain areas and depressed inland regions in the four Mediterranean countries. They are deep rural areas poorly connected to urban ones. When these agricultures are dominated by small farms (a common situation) outmigration does not result in land concentration, but in farmland abandonment. Here, retirement pensions and CAP subsidies become the main sources of income. Some examples can be found in the Portuguese region of Tras os Montes (Rodrigues, 2000), the Spanish Cantabrian Mountains, large extensions of Italian Mezzogiorno or in the many Greek mountains. In most cases, they are about high nature value areas where opportunities for environmental valorisation are open, but they remain uncertain.

Finally, a *fourth type* of agrarian reality can be identified. It is not so general in the Mediterranean, but it has a clearly defined presence in Southern Spain (Andalucia and Extremadura) and Portugal (Alentejo). This is the *latifundio* model. They are about very large agricultural properties, owned by holders following often conservative strategies, based on the perception of CAP subsidies and maintaining extensive agriculture, except when the development of irrigation districts has allowed diverse degrees of intensification.

CONCLUDING REMARKS

This chapter aimed to be a sort of introductory and historical backdrop for the following chapters of this volume, which will enrich with deeper and more detailed studies many of the elements that have been introduced here. Therefore, the conclusions we present in this section are provisional and will be developed and improved along the several case studies. These conclusions revolve around the two working thesis we departed from: one the one hand, that of Mediterranean farms' structural delay and rigidity; one the other hand, that precisely this delay in the modernisation path becomes an opportunity for adopting new practices and strategies in line with the model of multifunctional agricultural development.

Regarding the first argument (the progress of the productivist modernisation), official statistics show unequivocally that the 'stereotype' persists. Average holdings are smaller (physical and particularly economically). However, also clear sights of growing structural dynamism are found in the two last decades, although it does not mean to reach the pace of core countries of the 'Northern model'.

But under the picture drawn by the official statistics, the rich Mediterranean literature (which not always has being able to permeate theoretical paradigms on agrarian change) also reveals the several and diverse flows of informal land cessions leading to the enlargement of 'real' production units, i.e. a heterodox (but effective) way of concentration. In addition, Southern countries have been adopting (from early 90s) some patterns of transformation typical of modern agricultures, which parallel them with Northern dynamics. Among them:

- Drop of some indicators of intensification like the level of fertiliser consumption.
- Trend towards the individualisation of family farms associated to the decrease of family labour availability in modernised holdings.

However, even if the progress of modernisation or productivism or the adoption of Northern evolution are indeed taking place, they do in a more informal and unregulated way in the South. And this has consequences in terms of:

- Weaker institutional stability of holdings, which differentiates from the Danish model of well-sized holdings upon the basis of a clear institutional support.
- But more territorially and socially embedded, interwoven in fluid forms of rural-urban, agricultural-non agricultural networks and interdependent linkages among different types of farms and actors.

The second thesis we have discussed was related to the development of the multifunctional model of agrarian change in the Mediterranean. The data we have provided show the growing diffusion of these attributes, in part taking advantage of the agro-ecologic conditions. However, some clarifications are needed at this point. First, these developments have shown not to be 'pure' forms as depicted within the paradigm of sustainable rural development. On the one hand, we find parallel processes of orientation towards quality production and productivist developments, particularly in terms of intensification. And it is not just that intensification and quality orientation can coexist in a same area or holding, or that these two trends are 'related'. Our point is that precisely quality orientation has pushed intensive paths. On the other hand, the export orientation of an important share of the growing Southern organic food production is another hybrid situation: organic products sent thousands of kilometers away. The growth of organic demand in the North is a magnificent market stimulus for farming systems accustomed to participate in international markets. And we should not underestimate the capacity of Mediterranean intensive agricultural districts (the 'European California') to adapt to organic standards.

Second, the clear correlations between farm size and adoption of these practices raise doubts regarding their capacity to be a 'solution' for the mass of small holdings spread over the whole Mediterranean Europe. The adoption of typical multifunctional practices frequently requires well-sized and entrepreneurial farms.

Finally, as we have insisted, a longer historical approach is needed to soundly assess and interpret national and local agrarian dynamics, a perspective even beyond the post-World War II period – when the debate of paradigms centres. European Mediterranean agriculture has had several commercial specialisations, which have developed and depressed again and again as a consequence of both markets' evolution and, sometimes the exhaustion of the natural resource base – a key Achilles heel of these agricultures. So, we could assert that, all in all, their historic and current evolutions are simply forms and phases of the capitalist development in agriculture, obviously adapted to the peripheral conditions of these economies in Europe, as well as to the new demands (culture, origin, environment, leisure) these productive systems are confronted to.

NOTES

1. Although some authors also dissent about the existence of a common model for these four countries' agricultures. Damianakos (1997, p. 190) argue that 'the Greek agrarian system has little in common with that of Italy, Spain or Portugal'.
2. On December 2012, EUROSTAT provided data from the 2009/2010 Agricultural Censuses from all the European countries (see *Structure of Agricultural Holdings 2010* at http://epp.eurostat.ec.europa.eu/portal/page/portal/statistics/search_database, accessed on December 2012). However, these data are not yet homogeneous due to the important changes the Censuses introduce regarding the statistical universes in some countries (see http://epp.eurostat.ec.europa.eu/cache/ITY_SDDS/EN/ef_esms.htm#comparability, accessed on December 2012). Therefore they cannot be compared with the former data series. Consequently, it is recommendable to base the comparative analysis upon 2007 data. However, some of the national studies included in this volume use data from 2009/2010 censuses, based on national sources that do allow time comparisons.
3. The Spain-Netherlands comparison is, at this point, very illustrative. They show a very similar average dimension (ha of UAA) of their farms. However, 77.9% of Spanish holdings are below 20 ha of UAA (58% in Netherlands), 5.2% are above 100 ha (2.6 in Netherlands), whereas in Spain only 16.9% are between 20 and 100 ha, being 39.4% in Netherlands.
4. Actually, in 2007 the percentages of holders with Other (main) Gainful Activity are higher in some Northern countries like Sweden (49.4), Germany (43.5) or Denmark (39.8) than in Southern ones like Spain (27.1), Italy (22.7), Portugal (21.8) or Greece (18.3).
5. Three exceptions, with lower rates of size growth, are found: (i) Netherlands, where holdings are engaged in trajectories of intensification (leading to a rapid

increase of their economic dimension) and less in terms of physical growth; (ii) UK, whose already large holdings (around 60 ha of SAU in 1970, tripling French, Danish of German ones) were less pressed to grow in the search of economies of scale; and (iii) Ireland, whose lower rate of adjustment would confirm its peripheral character in the domain of European agriculture.

6. Arnalte, Ortiz, and Moreno (2008) show how farm growth slowed down in the last decade in terms of physical size, but it accelerated in economic size due to intensification, particularly in some productive orientations.

7. These verses from the Spanish song-writer Jose Antonio Labordeta reflect the uprooting feelings of the people leaving inland areas of Aragon (one of the regions where the exodus was more intense): 'If you find in some way people carrying their home on their shoulders, do not ask them about their land, they will look at you furiously' (song 'Todos repiten lo mismo', from the disc 'Cantar y callar', 1974).

8. Damianakos (1997, pp. 200, 203) also refers to the development of farming with hired machinery or the existence of sub-contracting relationships within farms.

9. Some estimations of the demand of external services, using national statistics, show the widespread diffusion of this phenomenon. In Spain, from data from the Agricultural Census of 1999, 37% of farms used tractors belonging to other holdings, cooperatives or service firms; and 24% also did with other specialised machinery (particularly harvesters) (Ortiz et al., 2011).

10. See Blanc, Brun, Delord, and Lacombe (1990) about the maintenance (or not) of the family character of French agriculture; or the more general analysis of Gasson et al. (1988) on family farming transformations.

11. There are 8 basic (1 digit) types: 5 specialised (*field crops, horticulture, permanent crops, grazing livestock* and *granivores*) and 3 mixed types (mixed cropping; mixed livestock; mixed crops-livestock), as well as a type of non-classifiable holdings. We consider as specialised the 5 first groups (with the exception of the sub-type *various permanent crops combined*). This later type (code 34) is a common sort of farm diversification in the Mediterranean (more than 2/3 of SGM comes from mixed permanent crops).

12. European Commission, http://ec.europa.eu/agriculture/quality/schemes/index_en.htm, accessed January 2013.

13. Spanish Ministry of Agriculture, http://www.magrama.gob.es/es/alimentacion/temas/calidad-agroalimentaria/calidad-diferenciada/dop/htm/cifrasydatos.aspx, accessed May 2012.

14. See the preliminary typology of rural regions established in the EU Project ETUDE (Van der Ploeg & Marsden, 2008).

REFERENCES

Alonso Mielgo, A. M., Sevilla Guzmán, E., Jiménez Romera, M., & Guzmán Casado, G. (2001). Rural development and ecological management of endogenous resources: The case of mountain olive groves in Los Pedroches comarca (Spain). *Journal of Environmental Policy and Planning*, *3*(2), 163–175.

Arnalte, E. (1989). Estructura de las explotaciones agrarias y externalización del proceso productivo. *Información Comercial Española, 666*, 101–117.

Arnalte, E. (1992). Agriculturas del sur de Europa y reforma de la Política Agrícola Común. *Revista de Economía, 12*, 91–95.

Arnalte, E. (2002). Ajuste estructural y cambios en los modelos productivos de la agricultura española. In C. Gómez Benito & J. J. González (Coord.), *Agricultura y Sociedad en el cambio de Siglo* (pp. 391–426). Madrid: McGraw-Hill-UNED.

Arnalte, E. (2006). Economía política del proceso de ajuste estructural en la agricultura de los países desarrollados. In E. Arnalte (Coord.), *Políticas agrarias y ajuste estructural en la agricultura española* (pp. 17–54). Madrid: Ministerio de Agricultura Pesca y Alimentación.

Arnalte, E., & Ortiz, D. (2006). Lectura comparada de los procesos de ajuste: factores determinantes e incidencia de las políticas. In E. Arnalte, (Dir.), *Políticas agrarias y ajuste estructural en la agricultura española* (pp. 353–381). Madrid: Ministerio de Agricultura Pesca y Alimentación.

Arnalte, E., Ortiz, D., & Moreno, O. (2008). Cambio estructural en la agricultura española: Un nuevo modelo de ajuste en el inicio del siglo XXI. *Papeles de Economía Española, 117*, 59–73.

Ball, R. M. (1987). Agricultural contractors: Some survey findings. *Journal of Agricultural Economics, 38*(3), 481–488.

Baptista, F. (1996). Declínio de um Tempo Longo. Reprinted in Baptista, F. (2001) *Agriculturas e Territórios* (pp. 9–37). Oeiras: Celta Editora.

Baptista, F. (2001). *Agricultura e capitalismo na Europa do Sul*. In Baptista, F. Agricultura e Territórios (pp. 63–81).

Baptista, F. (2002). Los caminos de la agricultura familiar. In C. Gómez Benito & J. J. González (Coord.), *Agricultura y Sociedad en el cambio de Siglo* (pp. 377–390). Madrid: McGraw-Hill-UNED.

Baptista, F. (2010). *O Espaço Rural. Declínio da Agricultura*. Oeiras: Celta Editora.

Barbero, G. (1982). Quante sono le aziende agricole italiane? *Rivista di Economia Agraria, 2*, 329–366.

Beaufoy, G. (2009). Environmental considerations of pig production in the dehesa – are they incorporated in the PDO label Dehesa de Extremadura? In K. Biala (Ed.), *Environmental labelling and certification initiatives in the agri-food sector – a way of marketing agricultural sustainability*. Proceedings of the JRC Summer Expert Meeting Ranco, 1–3 July 2008.

Beopoulos, N. (2003). Mediterranean agriculture in the light and shadows of the CAP. In Ch. Kasimis & G. Stathakis (Eds.), *The reform of the CAP and rural development in Southern Europe* (pp. 39–51). Aldershot: Ashgate.

Berlan, J. P. (1986). Agriculture et migrations. *Revue européenne de migrations internationales, 2*(2–3), 9–32.

Berlan, J. P. (1987). La agricultura "mediterránea" y el mercado de trabajo: ¿una California para Europa? *Agricultura y Sociedad, 42*, 233–245.

Blanc, M. (1987). Family and employment in agriculture: Recent changes in France. *Journal of Agricultural Economics, 38*(2), 289–301.

Blanc, M., Brun, A., Delord, B., & Lacombe, P. (1990). L'agriculture française, est-elle encore familiale? In P. Coulomb, H. Delorme, B. Hervieu, M. Jollivet & P. Lacombe (Eds.), *Les agriculteurs et la politique* (pp. 310–327). Paris: Presses de la Fondation National des Sciences Politiques.

Blandford, D., & Hill, B. (2005). *Facilitating farm-level adjustment to the reform of trade and agricultural policies*. Trade Policy Issues Paper no. 4. International Agricultural Trade Research Consortium.

Brunori, G., & Rossi, A. (2000). Synergy and coherence through collective action: Some insights from wine routes in Tuscany. *Sociologia Ruralis*, *40*(4), 409–423.

Buller, H. (2001). Is this the European model? In H. Buller & K. Hoggart (Eds.), *Agricultural transformation, food and environment* (pp. 1–8). Aldershot: Ashgate.

Camarero, L. A. (1997). Pautas demográficas y espaciales de las transformaciones del medio rural: Ruralidad y agricultura. In C. Gómez Benito & J. González (Eds.), *Agricultura y Sociedad en la España contemporánea* (pp. 225–246). Madrid: MAPA-CIS.

Camarero, L. A. (Coord.), Cruz, F., González, M., Del Pino, J. A., Oliva, J., & Sampedro, R. (2009). *La población rural de España: de los desequilibrios a la sostenibilidad social*. Fundación La Caixa, Barcelona.

Canadas, M. J. (1998). Trabalho, território e tecnologia: Transformação e situação actual na viticultura. PhD Dissertation. Universidade Técnica de Lisboa.

Caraveli, H. (2000). A comparative analysis on intensification and extensification in Mediterranean agriculture: Dilemmas for LFAs policy. *Journal of Rural Studies*, *16*(2), 231–242.

Damianakos, S. (1997). The ongoing quest for a model of greek agriculture. *Sociologia Ruralis*, *37*(2), 190–208.

De Filippis, F. (1985). Il part-time nel dibattito sulla stratificaziones aziendale dell'agricoltura italiana. *La Questione Agraria*, *18*, 3–16.

Delorme, H. (2000). *La 'cuestión' del agricultor profesional en Francia*. In Agricultura Familiar en España 2000 (pp. 65–75). Madrid: Fundación de Estudios Rurales.

Errington, A., & Gasson, R. (1996). The increasing flexibility of the farm and horticulture workforce in England and Wales. *Journal of Rural Studies*, *12*(23), 127–141.

European Commission. (1997). *Rural developments*, Working document, CAP 2000 Situation and Outlooks, DG VI.

European Commission. (2010). *An analysis of the EU organic sector*. Directorate-General for Agriculture and Rural Development, Brussels.

EUROSTAT. (2000). *Structure des exploitations agricoles. Résultats historiques – Enquêtes de 1966/67 à 1997*. Luxembourg.

Fabiani, G., & Scarano, G. (1995). Una stratificazione socioeconomica delle azienda agricole: Pluralismo funzionale e sviluppo territoriale. *La Questione Agraria*, 59, 27–91.

Fanfani, R., & Pecci, F. (1991). Innovazione e servizi nell'agricoltura italiana. Il caso del contoterzismo. *La Questione Agraria*, *42*, 79–121.

Fonte, M. (2001). Reconstructing rurality in Mediterranean Italy. In L. Granberg, I. Kovacs & H. Tovey (Eds.), *Europe's green ring* (pp. 263–282). Aldershot: Ashgate.

Gallego Bono, J. M. (2010). La agricultura a tiempo parcial y la externalización de servicios agrarios como vehículo del cambio estructural. *Revista Española de Estudios Agrosociales y Pesqueros*, *225*, 13–45.

Gasson, R., Crow, G., Errington, A., Hutson, J., Marsden, T., & Winter, D. M. (1988). The farm as a family business: A review. *Journal of Agricultural Economics*, *39*(1), 1–41.

Gómez López, J. D. (1993). *Cultivos de invernadero en la fachada Sureste Peninsular ante el ingreso en la C.E.* Madrid: Ministerio de Agricultura Pesca y Alimentación.

Goussios, D., & Duquenne, M. N. (2003). L'exploitation agricole à distance en Grèce: mobilité, pluriactivité et ruralisation. *Méditerranée*, *100*, 45–48.
Granberg, L., Kovách, I., & Tovey, H. (2001). Afterword. In L. Granberg, I. Kovách & H. Tovey (Eds.), *Europe's green ring* (pp. 359–366). Aldershot: Ashgate.
Harff, Y., & Lamarche, H. (1998). Le travail en agriculture: nouvelles demandes, nouveaux enjeux. *Économie Rurale*, *224*, 3–11.
Harrison, A. (1982). *Factors influencing ownership, tenancy, mobility and use of farmland in the member states of the European Community*. Information on Agriculture, 86. Luxembourg.
Hodge, I. D., & Adams, W. M. (2012). Neoliberalisation, rural land trusts and institutional blending. *Geoforum*, *43*(3), 472–482.
Hoggart, K., Buller, H., & Black, R. (1995). *Rural Europe. Identity and change*. London: Arnold.
Hoggart, K., & Mendoza, C. (1999). African in-migrant workers in Spanish agriculture. *Sociologia Ruralis*, *4*, 538–562.
Kanemasu, Y., & Sonnino, R. (2009). Dynamics of power and cooperation in rural development: The case of chianina beef production in Italy. *International Journal of Sociology of Agriculture & Food*, *16*(2), 36–53.
Kasimis, C., & Papadopoulos, A. G. (2001). The de-agriculturalisation of the Greek countryside: The changing characteristics of an ongoing socio-economic transformation. In L. Granberg, I. Kovacs & H. Tovey (Eds.), *Europe's green ring* (pp. 179–218). Aldershot: Ashgate.
Kasimis, C., & Papadopoulos, A. G. (2005). The multifunctional role of migrants in the Greek countryside: Implications for the rural economy and society. *Journal of Ethnic and Migration Studies*, *31*(1), 99–127.
Koutsou, S., Partalidou, M., & Petrou, M. (2011). Present or absent farm heads? A contemporary reading of family farming in Greece. *Sociologia Ruralis*, *51*(4), 404–419.
Lamarche, H. (1996). Europe du Nord, Europe du Sud: le chassé-croisé de la course à l'intensification. In M. Jollivet & N. Eizner (Dir.), *L'Europe et ses campagnes* (pp. 77–98). Paris: Presses de la Fondation National des Sciences Politiques.
Langreo, A. (2002). La externalización del trabajo agrario y las empresas de servicios a la agricultura. *Economía Agraria y Recursos Naturales*, *2*(1), 45–67.
Laurent, C. (1998). Activités agricoles et territoires ruraux. Nouvelles questions de recherche. In Delorme, H. (Dir.), *Les territoires ruraux en Europe*. Questions de recherche (pp. 80–92). Paris: CERI-CNRS.
MAGRAMA (2012a). Empleo en el sector agroalimentario. *Análisis y Prospectiva*, 7. Ministerio de Agricultura Pesca y Alimentación, Madrid.
MAGRAMA (2012b). *Caracterización del sector de la producción ecológica española en términos de valor, volumen y mercado*. Madrid.
Mahé, L. P. (2001, Spring). Can the European model be negotiable in the WTO? *EuroChoices*, 10–16.
Marsden, T. (2003). The condition of rural sustainability: Issues in the governance of rural space in Europe. In C. Kasimis & G. Stathakis (Eds.), *The reform of the CAP and rural development in Southern Europe* (pp. 19–37). Aldershot: Ashgate.
Marsden, T., & Sonnino, R. (2006). Beyond the divide: Rethinking relationships between alternative and conventional food networks in Europe. *Journal of Economic Geography*, *6*, 181–199.

Moragues, A. (2011). *How is agriculture sustained? Following farmers from the field to the market. A case study of olive oil in a Mediterranean region.* PhD Dissertation Universitat Politècnica de Valencia.

Ortiz, D., Arnalte, E., Moragues, A., & Doñate, S. (2011). Las explotaciones agrarias de servicios: la configuración de una vía de gestión de la agricultura. *VIII Congreso de Economía Agraria*. Madrid.

Ortiz-Miranda, D., Moreno-Pérez, O. M., & Moragues-Faus, A. M. (2010). Innovative strategies of agricultural cooperatives in the framework of the new rural development paradigms: The case of the region of Valencia (Spain). *Environment and Planning A*, *42*(3), 661–677.

Pedreño, A. (1999). *Del jornalero agrícola al obrero de las factorías vegetales: estrategias familiares y nomadismo laboral en la ruralidad murciana.* Madrid: Ministerio de Agricultura Pesca y Alimentación.

Pedreño, A. (2012). Trabajadores y agriculturas mediterráneas en la globalización. *Regiones, Suplemento de Antropología*, *47*, 16–20.

Piqueras, J. (1985). *La agricultura valenciana de exportación y su formación histórica.* Madrid: Ministerio de Agricultura Pesca y Alimentación.

Pugliese, E., & Ceriani-Sebregondi, F. (1981). Destrutturazione Aziendale, famiglia e classi sociali in agricoltura. *Agricoltura e Societá*, *4*, 19–31.

Rodrigues, O. (2000). Utilização do Território e Propriedade Fundiária. PhD Dissertation. Universidade Técnica de Lisboa.

Rosell, J., Viladomiu, L., & Correa, M. (2010). Mejora del medio ambiente y nivel de desarrollo: las opciones de los Programas de Desarrollo Rural (2007–2013) de la Unión Europea. *Revista Española de Estudios Agrosociales y Pesqueros*, *226*, 13–37.

Saraceno, E. (1994). The modern functions of small farm systems: An Italian experience. *Sociologia Ruralis*, *34*(4), 308–328.

Scheidel, A., & Krausmann, F. (2011). Diet, trade and land use: A socio-ecological analysis of the transformation of the olive oil system. *Land Use Policy*, *28*(1), 47–56.

Sivignon, M. (1996). Les systèmes agraires européens: Héritages, mutations, frontières. In M. Jollivetet & N. Eizner (Dir.), *L'Europe et ses campagnes* (pp. 37–55). Paris: Presses de la Fondation National des Sciences Politiques.

Sonnino, R. (2004). For a 'piece of bread'? Interpreting sustainable development through agritourism in Southern Tuscany. *Sociologia Ruralis*, *44*(3), 285–300.

Sotte, F. (2006). Quante sono le imprese agricole in Italia? *Agriregionieuropa*, *2*(5). Available at http://www.agriregionieuropa.univpm.it/dettart.php?id_articolo=111. Accessed on January 2013.

Tracy, M. (1989). *Government and agriculture in Western Europe 1880–1988* (3rd ed.). New York: New York University Press.

Van der Ploeg, J. D., & Marsden, T. (2008). *Unfolding webs.* Assen: Van Gorcum.

Vellante, S. (1981). Innovazioni tecnologiche, forme di produzione emergenti e organizzazione aziendale. *La Questione Agraria*, *4*, 67–108.

Ventura, F., & Milone, P. (2000). Theory and practice of multi-product farms: Farm butcheries in Umbria. *Sociologia Ruralis*, *40*(4), 452–465.

CHAPTER 4

CHANGING AGRICULTURE – CHANGING LANDSCAPES: WHAT IS GOING ON IN THE HIGH VALUED *MONTADO* LANDSCAPES OF SOUTHERN PORTUGAL?

Teresa Pinto-Correia and Sérgio Godinho

ABSTRACT

The Portuguese montado is a particular land use system, characterized by the combination, in the same area, of the forestry and the grazing components interrelating with each other, in large-scale farm units. Mostly, this system is acknowledged due to its specific landscape character, in a savanna-like phisionomy, with changing densities along a continuous tree cover of holm and cork oaks and grazing in the under cover. The montado is a production system, and its extensive character and particular pattern makes it possible to support a multitude of ecosystem goods and services nowadays valued by society. Nevertheless the system is threatened and the resulting landscape is under strong reduction in the last decades. This paper shows the dimension of the ongoing reduction, for the whole region of Alentejo, since 1960 and up to now. And furthermore, based on a survey to land managers of montado

Agriculture in Mediterranean Europe: Between Old and New Paradigms
Research in Rural Sociology and Development, Volume 19, 75–90

ISSN: 1057-1922/doi:10.1108/S1057-1922(2013)0000019006

in a Natura 2000 site, it shows how the land management options for the most are still focusing on production and productivist ideals, even when keeping a multifunctional system. These orientations do not result in a radical replacement of the system, and therefore the illusion is kept that the multifunctionality is mantained – but progressively the system loses its balance and the tree cover decays in such a way that the montado disappears. This unique landscape is thus under severe threat. The paper ends with a discussion on the urgent need for integrated policy goals and tools for the montado as a system, and for much more colaboration with the land managers in order to strength the multifunctionalty relevance and support a novel attitude replacing the productivist concept of farming, misleading in the context of this system.

Keywords: *Montado*; agro-silvo pastoral system; savanna-like landscape; land management; productivism and post-productivism; decay

THE *MONTADO* SYSTEM AND LANDSCAPE

The Portuguese *montado* is an agro-silvo pastoral system quite similar to the *dehesa* in Spain, and covering in Portugal most of the Southern region, Alentejo (Fig. 1).

Even if the different data sources do not show the same exact dimension, there is agreement that the total area covered is over one million hectares (Pinto-Correia, Ribeiro, & Sá-Sousa, 2011) (Fig. 2).

The most singular characteristic of the *montado* is its savanna-like physiognomy, spread throughout a large-scale mosaic, in changing densities, of cork (*Quercus suber*) and holm oak (*Quercus ilex rotundifolia*) trees (Fig. 3). Opposite to what characterize most well-known Mediterranean land use systems, the *montado* can only be found on large-scale properties, in a *latifundia* structure, where each farm unit can be from 100 to 1,000 hectares, and sometimes even larger. The highly extensive use of this system, developed along centuries of progressive human adaptation of the natural vegetation (Aronson, Santos-Pereira, & Pausas, 2009), explains why it is not manageable in small-scale farms.

Montados are complex agroforestry systems, in which complexity increases with the conjunction of production activities that share the

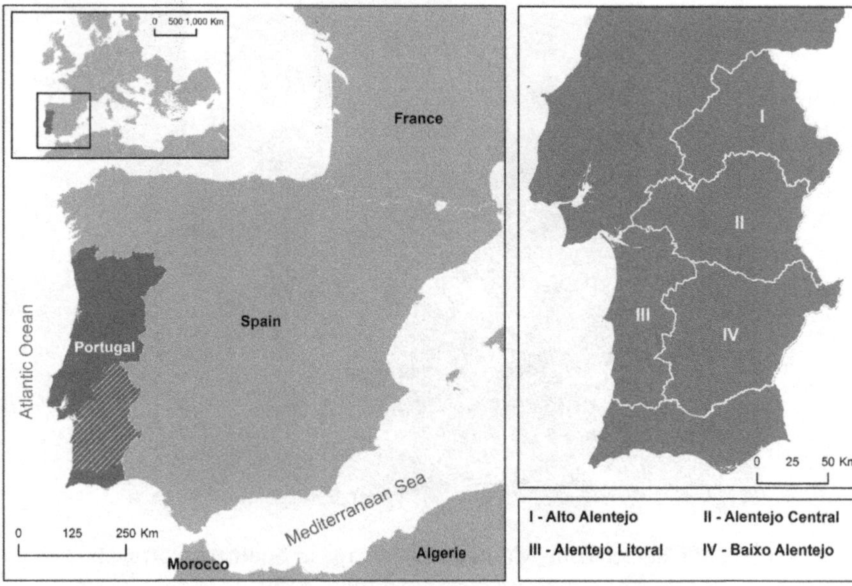

Fig. 1. Study Area and NUTS III Boundaries.

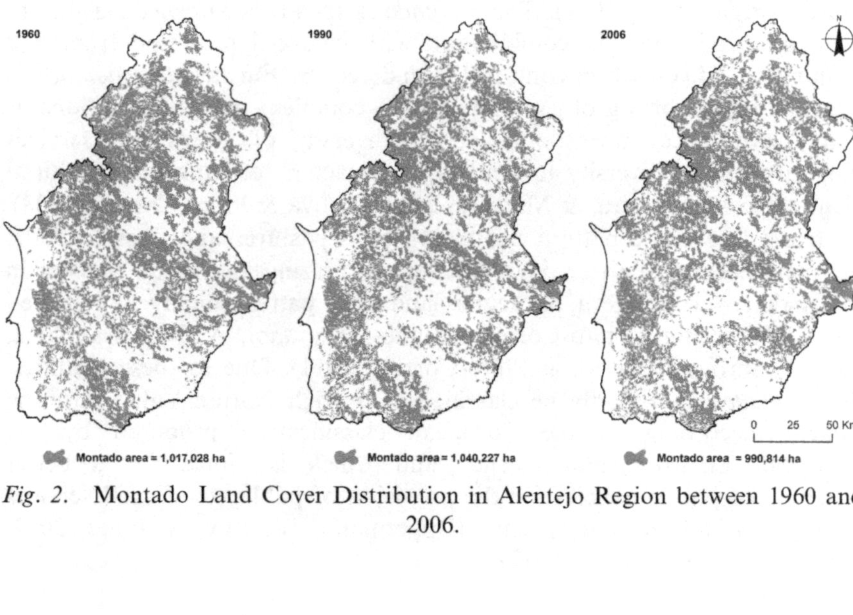

Fig. 2. Montado Land Cover Distribution in Alentejo Region between 1960 and 2006.

Fig. 3. Characteristic Montado Landscape in Southern Portugal.

same growing space in a region characterized by its site variability, especially at the climate, soil and topography levels (Pinto-Correia, 1993; Pinto-Correia et al., 2011). The *montado* is mostly acknowledged due to the cork production, in combination with charcoal produced from tree pruning, and livestock grazing in the undercover. Further, the *montado* is known for supporting other multiple and complementary productions as honey, aromatics, mushrooms, and a diversity of goods and services ranging from biodiversity to landscape character, recreation and cultural identity (Costa, Pereira, & Madeira, 2009; Surová & Pinto-Correia, 2008). Both the large-scale pattern combining both pastures and trees, and the fuzziness both of the tree cover and of the grazing and shrub patches in the undercover, create a particular landscape pattern highly appreciated today (Paracchini, Pinto-Correia, Ramos, Capitani, & Madeira, 2012; Surová, Surový, Ribeiro, & Pinto-Correia, 2011). Due to these qualities, the *montado* can broadly be classified as a High Nature Value Farming System, according to the European classification proposed by the European Environmental Agency and which is aimed for a better targeting of public policies for the provision of public goods and services through agriculture and forestry (Oppermann, Beaufoy, & Jones, 2012; Pinto-Correia & Ribeiro, 2012).

Fig. 4. Extensive Livestock Grazing in Montado Landscape.

Production activities in the undercover are extensive livestock grazing of cattle, sheep, goats, cattle and the Iberian pig (Fig. 4). These were combined in former days with cereal crops cultivated in long rotations with fallow, but cultivation is seldom today (Gaspar, Mesías, Escribano, Rodriguez de Ledesma, & Pulido, 2007; Plieninger, 2007). The open tree cover is maintained through natural regeneration and trees are seldom planted. Nevertheless, trees have a direct value as fodder crop, providing acorns and leafy branches in autumn and winter, when the herbage production is low, and an indirect value as shelter against cold in the winter and heat in summer (Cañellas, Roig, Poblaciones, Gea-Izquierdo, & Olea, 2007; Moreno-Marcos et al., 2007). Furthermore, the trees create the ecological characteristics that are fundamental to the sustainability of all activities occurring at stand level (Ribeiro, Surový, & Oliveira, 2006). The sustainable management of the combination sets of production activities requires a detailed knowledge of the resilience and elasticity of the forest components, in each particular conditions of climate, soil and topography (Ribeiro et al., 2004).

Despite its many qualities and the general acknowledgement of their value, the balance of the *montado* is threatened today by a combination of different factors (Costa et al., 2009; Eichhorn et al., 2006; Pinto-Correia et al., 2011).

The most acknowledged threats are, on one hand, the over-exploitation of the tree cover, as cork harvest and pruning for charcoal production, in non-balanced terms, which will harm and weaken the trees (Cañellas et al., 2007). On the other, intensification of the activities in the undercover, such as overgrazing and mechanized ploughing, may hinder tree regeneration, so that the long-term regeneration of the tree cover is not guaranteed (Plieninger, 2007). Natural regeneration can thus become unreliable. Artificial planting is seldom successful, as the development of the young trees is severely constrained by the hard natural conditions, when the root system and the canopy protection of adult trees is absent. The reduction in trees and the lack of regeneration is a severe threat to the future maintenance of the system. Mechanized and deep ploughing also affects the root system and weakens the trees. Overgrazing may result in compaction of the soil and higher erosion risks.

Finally, in more peripheric and more fragile areas, there has been a trend of extensification or abandonment, that allows an invasion of shrubs and other oaks increasing the competition (reducing cork production), resulting in shrub encroachment and the risk of forest fires (Pinto-Correia & Mascarenhas, 1999). Furthermore, ongoing climatic change may presumably induce more repeated and severe droughts, while the spread of wildfires may force the turnover of *montado* mosaic into large and persistent shrublands (Acácio, Holmgren, Rego, Moreir, & Mohren, 2009).

From a landscape perspective, the impact of these changes in management will lead to a simplification, with the progressive reduction of the *montado* area, through an extreme opening of the tree cover or, on the opposite, shrub encroachment and thus a turn into a dense maquis or forest formation.

RECENT CHANGES IN THE *MONTADO* PATTERN

Considering the region of Alentejo, the distribution shown in Fig. 2 is clear in showing how the *montado* is dominant in the landscape pattern of the region. In the area of *montado* shown in the maps, all different types of pattern are included: with the tree cover dominated by *Quercus suber*, by *Quercus rotundifolia*, a few with *Quercus pyrenaica* (in northern Alentejo), and many areas also with mixed tree cover. Further, the area of *montado* includes areas with more dense tree cover, also eventually more shrub, and a more forestry oriented use, and areas with a more sparse tree cover and a more relevant role of agriculture. The *montado* is an extremely resilient land use system, where the different components can be maintained under a dynamic balance, even with different weights along time, resulting from different management options. Therefore, the local land cover pattern can be variable, within what is classified as *montado*.

This also means that cartographic delimitation of *montado* areas is a difficult task and the extent of these areas varies according to the criteria followed for example for tree density or for the use under the tree cover. The areas of *montado* can be classified as agricultural areas, under pastures or even sometimes cultivation, but they can also be classified as forest, or also shrub or maquis areas, and for example only part of the *montado* areas fall within what is classified in CORINE land cover maps as agro-silvo pastoral areas. For the maps shown in Fig. 2, data has been gathered from different cartographic sources specifically to produce a map of the *montado* distribution in the region of Alentejo, and its changes along recent decades (Godinho et al., submitted). The dates of the maps, 1960, 1990 and 2006 corresponds to the years from which the existing information is available.

What has been happening so far? The most striking change is that while the area of *montado* has been improving in the decades 1960 to 1990, in recent years it has been strongly reducing. In global terms, between 1960 and 1990 there has been an increase of 23,199 ha, while from 1990 to 2006 a decrease of 49,413 ha. This means an average rate of expansion of 773 ha year^{-1} in the first period, and then from 1990 to 2006 an average rate of reduction of 3,088 ha year^{-1}. The global patterns of *montado* change observed in 1960–1990 and 1990–2006 periods are the result of the magnitude of the loss and gain spatial processes all over the region. Despite the observed *montado* total area increase between 1960 and 1990, several areas in the study region registered a loss of *montado* in this period, mainly in the Baixo Alentejo where the holm oak *montado* is dominant. On the contrary, for the period 1990–2006, results of the spatial analysis showed there was only an insignificant *montado* area increase, in a few areas. And mainly, as shown above, the global balance of this 16-year period put in evidence there is a long-term declining trends of the *montado*. The figures are alarming, as they show a continuous decrease. And this alarm is strengthened by other data on case-study areas in the region, which show how this trend has continued in the years from 2006 until now.

The *montado* has been developed along centuries, and results from the progressive adaptation of the natural ecosystem into a productive land use system that could deal with the natural constrains of the region and, through the integration of the agro-silvo and pastoral components, profit from all the scares resources (Aronson et al., 2009; Pinto-Correia & Fonseca, 2009). Nevertheless, this integration requires multiple levels of human intervention, and a careful intervention in the system at each level. Most work has been done until the 1960s–1970s with manual labour, what was made possible due to the property and social structure in the region of Alentejo. With the progressive increase in labour and other production costs, the general replacement of manual by mechanized work, and after the change in the Portuguese political regime and the integration in the European Union as well as the globalization of markets, the former balance in the management of the *montado* was disentangled. The changes registered reflect the disappearance of the former balance and the quest for new paradigms and management pratices.

The areas most affected by this reduction are the areas of holm oak, more dominant in the in-land, in areas with poor soils and harsh dry Mediterranean climate, and formerly valorized to a large extent by the grazing of Iberian pig. The propagation of the African swine fever in the sixties and the prohibition of grazing pigs, have stimulated the destructions of significant areas of holm oak *montado* to promote other land-use types with better economic returns (Crespo, 2005). The holm and especially the cork oaks have been protected by law since the past, and are since 1977, and again 1997, under sever legislation controlling their use. They cannot be withdrawn without permission from the competent authorities – but a land use damaging the tree balance and impeding natural regeneration leads in the same way to the tree cover depletion, in the medium term. In a first stage, in the middle of the 20th century, many of these holm oak areas were used for intensive cereal cultivation with heavy machinery that led to soil degradation and depletion, and also in the medium term, to the decay in the trees, with no regeneration, and thus a progressive disappearance of the tree cover. More recently (1986–2006) the afforestation with faster growing trees (*Pinus* and *Eucalyptus* spp.) in the depleted soils and generally in soils with low agricultural capacity, emerged as a viable alternative to the economy of the agroforestry holdings. Eucalyptus have been planted since the 1970s in the whole Portugal, as a response to agricultural abandonment and the demand of fast growing timber by the paper industry. Pine trees have been planted in the Alentejo region since the middle of the 1990s, as a direct consequence of the afforestation measures within the CAP (Common Agricultural Policy), after the reform of 1992 (Pinto-Correia & Breman, 2009). Forestry measures have been used mostly in the *montado* areas which were already under decay, in poor or depleted soils, steep slopes or extreme peripheral locations, as a strategy to secure a regular income in plots where the maintenance of the agro-silvo pastoral system was seen as difficult. On another side, on the better soils and also including cork oak *montado*, mainly since two decades ago, intensification in livestock production in the undercover is the major driver of change. Again here the major driver is the CAP measures, in particular the livestock premium. Payments according to the livestock headage lead to intensification in grazing. Further, as payments for cattle have so far been considerably superior to payments for sheep, many *montado* areas formerly grazed by sheep and goats have changed into cattle grazing, even if sheep or goats, with a much lower grazing impact, were better adapted to the preservation of the *montado* balance (Bugalho, Caldeira, Pereira, Aronson, & Pausas, 2011; Pinto-Correia & Breman, 2009). The high intensity of grazing and the use of heavy breeds leads often to soil compaction, depletion and the decay of the tree cover by lack of natural regeneration.

In this way, even if there is still much *montado* in the region of Alentejo, and even if this system is broadly acknowledged due to its specificity and the multiple goods and services provided, some drivers are affecting its management so strongly that the result is a decay in an extremely strong pace. There is no particular pressure caused by urban growth or infrastructure development. Besides the large Alqueva dam, where many hectares of *montado* have disappeared, mostly these pressures affect the small-scale farming and land cover mosaic. Changes in *montado* distribution are mainly resulting from changes in its management. The areas with more marginal conditions, poor soils, steep slopes and peripheral locations, are mostly under extensification and have been replaced by shrub encroachment or new forestations. In other areas, and especially in the more fertile soils, the changing trend is mostly intensification by cattle grazing and mechanization, leading to the decay in tree cover and progressive opening of the *montado*.

LAND MANAGERS OPTIONS IN FACE OF CONFLICTING PRESSURES

The maintenance of this land use system and related landscape is a challenge today, both for policy makers as well as for land owners. The increased acknowledgement of the *montado* as a highly multifunctional land use system opens up multiple possibilities for new market driven activities as well as for an increased support through public policies. But for these new possibilities to support new management orientations, a paradigm shift is required, not only in policy targeting, but mainly in the land owners goals and strategies (Marsden & Sonnino, 2008; Wilson, 2009).

Recent literature shows how land managers are now following different pathways. The main divide described from the Northern and Western European context is between those continuing the productivist strategy and those opting for a more post-productivist orientation (Holmes, 2006; Wilson, 2007, 2008). It is also described how those last land managers are often struggling to transmit this orientation in their practices, and may opt to adapt their management so that spatially there is a divide between productive areas and marginal non-used areas, and thus continue a dominant productivist practice (Marsden & Sonnino, 2008; Sutherland, 2010; Wilson, 2008, 2009).

The analysis of the options taken by land managers in the *montado* show a quite different arrangement between the attitudes expressed and the options followed in practice. The multifunctional character of the *montado* as it appears and is valued today creates the impression that its management aims for multifunctionality. A survey in the Natura 2000 site of Monfurado, an area of 24,000 ha in the centre of Alentejo, with large-scale holdings and a land use dominated by a dense *montado* has been applied in 2011 (Pinto-Correia, Menezes, & Barroso, 2012). 70 land managers have been interviewed, about their management practices and the relation to their expressed attitudes concerning farming. Land managers are here all those who take the decisions on management orientations for the farm unit; they include land owners, even if they are absent, not living in the farm but taking major decisions that are set to practice by an employee; they include also in some cases others which are not the owners but are responsible for management, as leasers.

This survey reveals how, even if the system is kept extensive, integrating multiple components, and supporting multiple ecosystems goods and services, this does not necessarily mean a post-productivist strategy from the land manager side. Production keeps on being the main driver of the land manager options, and the productivist ideal is strongly embedded in the farmers self-concept. A so far hidden tension is made clear. Previous surveys (Bruckmeier & Tovey, 2009; Rodrigo & Veiga, 2009) have showed how the conservationist behaviour by managers of these extensive systems may be grounded in reasons such as the awareness of the farm environmental constrains, sensitivity concerning heritage values, lack of entrepreneurship and the prevailing property structure. All these factors have been there for long. They do not reflect a transition process from productivism to post-productivism or a quest for a multifunctional landscape.

In the Monfurado survey, only 20% of the land managers inquired have expressed a conviction that their land should be managed for multi-functionality, and also in practice build their everyday management strategy based on this conviction. They keep an extensive management and promote or support other non-productive land–based activities in their land, and are much aware of their role in preserving a delicate balance for the long-term sustainability of the *montado*. The remaining expresses clear productivist goals. From these, a group of 25% keep the conventional extensive practice, with low intensity of grazing and a specific care for the natural regeneration of the tree cover, but express strong production oriented goals. This is the opposite to what often has been described from other regions of Europe, where some farmers describe themselves as non-productivist but opt in their everyday decisions for keeping a productivist orientation (Sutherland, 2010; Wilson, 2008). The remaining 55% in the inquired land managers express themselves and their convictions as agri-business farmers, and manage their land with a main focus on intensification and increase in production levels. This is mostly done through an increase in cattle density, and sometimes a change in cattle breeds to higher meat productive breeds. Such intensification in livestock production is reflected in a need to intensity cultivation in order to produce fodder, hay or forage, generally creating irrigated areas in the farm. Otherwise fodder to supplement the pastures needs to be bought outside the farm, increasing farm dependency from the exterior. In the Monfurado site, conditions for agriculture are relatively good and this is the most common trend. Extensification or abandonment trends are not significant.

The design of the Common Agricultural Policy tools so far surely has a role in the intensification strategies, as the payments for livestock production in Portugal have been kept fully coupled to production, and thus income from subsidies in the *montado* farms is strongly related to the number of livestock units in each farm. Further, cattle payments are considerably higher than sheep payments, what has led to a replacement of the later by the first, even in areas where cattle is far too damaging for the balance of the system.

Furthermore, previous studies on Alentejo land managers have shown how failing to observe the productivist ideal can have a significant and direct impact on the social position of farmers within the agricultural community and strong influence their ability to perceive themselves as good farmers (Rodrigo & Veiga, 2009). In order to keep their position in the farming community, land managers want to keep and be seen as producers, and better, successful producers. In the region of Alentejo, besides the role of the

CAP and its economic incentives in the last two decades, several historical conditions may contribute to explain the resistance of the productivist paradigm and the weakness of true transitions: fragility of the rural institutional framework; absence of capacity building networks; lack of social fabric open to new developments. The innovative and skilled rural entrepreneur faces a difficult context. Therefore probably, the transition into a multifunctional management model is still weaker than what can be expected from the analysis of the landscape multifunctional capacity.

As far as the landscape is concerned, the *montado* structure in each farm unit reflects the options of the landowner, and this production oriented paradigm has consequences. The cork and holm oaks are protected through specific legislation. They cannot be cut or withdrawn, or even pruned, without a specific permission by the competent authorities. But grazing intensity affects soil compaction and tree regeneration. And soil mobilization may increase erosion risks and affects the root system and also the survival of the young tree shoots (Ribeiro et al., 2004). Therefore, impacts of the options taken can only be observed to a large extent, in the long term. When land owners are not explicitly paying attention to the interlinked components of the *montado*, there is a lower attention to the tree cover and thus a progressive decay of the same, through lack of natural regeneration and weakening of the existing trees. In the long term, the result of intensification in livestock production is the progressive opening of the tree cover, with no younger trees to replace the old ones, and thus a replacement of former *montado* areas by open patches which will be occupied by shrub, if grazing or cultivation are not maintained.

The observed fast reduction in the *montado* area, shown in the maps from Fig. 2 and discussed in the previous section, demonstrates how the recent practices have affected the system and thus also the underlying fragility of this system.

DISCUSSION: *MONTADO*, A THREATHENED LANDSCAPE?

The fast reducing area of *montado* in Southern Portugal is shown clearly in the maps in Fig. 2. This is an evidence, and this reduction is much stronger than any change in the *montado* area observed before. This is nevertheless not a perceived change, nor by the public nor by the public authorities, as the extension of *montado* is still largely dominant in the land cover in the region, and decay and disappearance of the *montado*, as well as oaks mortality, are seen as local processes. Nevertheless, understanding the land managers attitudes and options reveals the deepness of the threats affecting the *montado*. The heritage value that most land managers consider in the *montado*, and the respect they also express for this system and for its richness and specific character is being overwhelmed by the economic constraints they are facing, and mostly, their farmer self-concept.

The revealed conflict between the practices and the attitudes in the analysed land managers is quite the opposite to what has been found in other places in Europe. It has been described from different European examples how a post-productivist attitude and discourse is in opposition to the more productivist management, explained by land managers as needed in relation to economic survival (Marsden & Sonnino, 2008; Sutherland, 2010). Or also, in other examples, how different practices in the same farm, by the land owner as a producer or as a landscape manager, reflect different and complementary views on the farm (Primdahl & Kristensen, 2011). What this Mediterranean example shows is another combination of attitudes and practices. Many land managers keep a so far multifunctional system but many of them are aiming for a production oriented, and if possible a more specialized farm. Only a few opt for a management strategy aiming for multifunctionality and for valuing the multiple goods and services provided by the *montado*.

The question then, when aiming for the future preservation of the *montado* system and landscape, is what will be the future pathways taken by most land managers, and if it is possible to progress towards a renewed balanced management of this system.

The design of the Common Agricultural Policy and of the options for its implementation in Portugal, have so far resulted in conflicting messages to the land manager in such extensive and high nature value land use systems. The strength of CAP payments has created an increase dependency of land owners from its measures and schemes, and a deviation from long-term reasoning on the land use system. From one side, the coupled livestock payments and the higher premium for cattle lead to the replacement of sheep by cattle, in many areas where the pressure by cattle grazing is unsustainable, and further, it lead to a progressive increase in cattle grazing intensity. On the other, agri-environmental measures and other environment supporting schemes, including the afforestation support measures, aim for the opposite, e.g., extensive grazing and the maintenance of a semi-natural vegetation and natural regeneration of the tree cover. The new focus on land-based payments as discussed for the next coming CAP period may create a new reasoning. But also, even if policy mechanisms deriving from the CAP will change in the next future, the global economic crisis and evolving context may lead to a higher pressure for food production in these Southern European regions and systems.

On one side, policy goals and design are crucial in this issue. If multifunctional landscape assets are to be exploited and contribute to the higher resilience of the *montado* in face of changing global context, then there is an urgent need to clarify policy objectives and avoid placing the land manager in face of conflicting instruments and options. On the other side, the need emerges to work more comprehensively with land managers for a broader understanding of their options and an integrative support and guidelines for a management oriented to multifunctionality, what could undoubtedly be strengthened by the construction of new land manager identities.

If land managers would value more explicitly the landscape qualities of their *montado*, focus on their landscape potential, they would be prone to enroll in a more complex and variable mix of production, consumption and protection goals in their farm management (Pinto-Correia & Breman, 2009). These changes would require a deepening, a re-grounding or a broadening in farm activities (Van der Ploeg & Marsden, 2008), similar to what has been done before in many *montado* farm units. The need for a joint strategy, playing at all levels of decision affecting the *montado*, is obvious. Only such a strategy can lead to a change the threatening trends observed today.

REFERENCES

Acácio, V., Holmgren, M., Rego, F., Moreir, F., & Mohren, G. M. J. (2009). Are drought and wildfires turning Mediterranean cork oak forests into persistent shrublands? *Agroforestry Systems*, *76*, 389–400. doi:10.1007/s10457-008-9165-y

Aronson, J., Santos-Pereira, J., & Pausas, J. G. (2009). General introduction. In J. Aronson, J. Santos-Pereira & J. G. Pausas (Eds.), *Cork oak woodlands on the edge: Ecology, adaptive management, and restoration* (pp. 1–10). Washington, DC: Island Press.

Bruckmeier, K., & Tovey, H. (Eds.). (2009). *Rural sustainable development in the knowledge society*. England: Ashgate.

Bugalho, M., Caldeira, M. C., Pereira, J. S., Aronson, J., & Pausas, J. M. (2011). Mediterranean cork oak savannas require human use to sustain biodiversity and ecosystem services. *Frontiers Ecological Environment*, *9*(5), 278–286. doi:10.1890/100084

Cañellas, I., Roig, S., Poblaciones, M. J., Gea-Izquierdo, G., & Olea, L. (2007). An approach to acorn production in Iberian dehesas. *Agroforestry Systems*, *70*, 3–9. doi:10.1007/s10457-007-9034-0

Costa, A., Pereira, H., & Madeira, M. (2009). Landscape dynamics in endangered cork oak woodlands in Southwestern Portugal (1958–2005). *Agroforestry Systems*, *77*, 83–96. doi:10.1007/s10457-009-9212-3

Crespo, D. G. (2005). The role of pasture improvement in the rehabilitation of the "montado/dehesa" system and in developing its traditional products. In *Animal products from the Mediterranean area*, 25–27 September 2005, Santarém, Portugal. EAAP Publication n° 119, 2006, 185–195.

Eichhorn, M. P., Paris, P., Herzog, F., Incoll, L. D., Liagre, F., Mantzanas, K., … Dupraz, C. (2006). Silvoarable systems in Europe: Past, present and future prospects. *Agroforestry Systems*, *67*, 29–50. doi:10.1007/s10457-005-1111-7

Gaspar, P., Mesías, F. J., Escribano, M., Rodriguez de Ledesma, A., & Pulido, F. (2007). Economic and management characterization of dehesa farms: Implications for their sustainability. *Agroforestry Systems*, *71*, 151–162. doi:10.1007/s10457-007-9081-6

Godinho, S., Guiomar, N., Machado, R., Freire, M., Santos, P., Neves, N., Sá-Sousa, P., Pinto-Correia, T. (submitted). Spatial and temporal analysis of *montado* land cover in southern Portugal: A 100 year case study on major landscape drivers and dynamics. *Annals of Forest Science*.

Holmes, J. (2006). Impulses towards a multifunctional transition in rural Australia: Gaps in the research agenda. *Journal of Rural Studies*, *22*, 142–160.

Marsden, T., & Sonnino, R. (2008). Rural development and the regional state: Denying the multifunctional agriculture in the UK. *Journal of Rural Studies*, *24*, 422–431.

Moreno-Marcos, G., Obrador, J. J., García, E., Cubera, E., Montero, M. J., Pulido, F., & Dupraz, C. (2007). Driving competitive and facilitative interactions in oak dehesas through management practices. *Agroforestry Systems*, *70*, 25–40. doi:10.1007/s10457-007-9036-y

Oppermann, R., Beaufoy, G., & Jones G. (Eds.) (2012). High Nature Value Farming in Europe. 35 European countries – experiences and perspectives. *Verlag Regionalkultur*, Ubstadt-Weiher, 544.

Paracchini M. L., Pinto-Correia, T., Ramos, I. L., Capitani, C., & Madeira, L. (2012). Progress in indicators to assess agricultural landscape valuation: How and what is measured at different levels of governance. *Landscape Research* (submitted 2nd version, October 2012).

Pinto-Correia, T. (1993). Threatened landscape in Alentejo, Portugal: The montado and other "agro-silvo-pastoral" systems. *Landscape and Urban Planning*, *24*, 43–48. doi:10.1016/0169-2046(93)90081-N

Pinto-Correia, T., & Breman, B. (2009). The new roles of farming in a differentiated European countryside: Contribution to a typology of rural areas according to their multi-functionality. Application to Portugal. *Regional Environmental Change*, *3*(9), 143–152.

Pinto-Correia, T., & Fonseca, A. (2009). Historical perspective of montados: The example of Évora. In J. Aronson, J. Santos Pereira & J. G. Pausas (Eds.), *Cork oak woodlands on the edge: Ecology, adaptive management, and restoration* (pp. 49–54). Washington, DC: Island Press.

Pinto-Correia, T., & Mascarenhas, J. (1999). Contribution for the extensification/intensification debate: What is happening to the Portuguese Montado? *Lands Urban Plan*, *46*, 125–131. doi:10.1016/S0169-2046(99)00036-5

Pinto-Correia, T., Menezes, H., & Barroso, F. (2012). The landscape as an asset in Southern European fragile agricultural systems: contrasts and contradictions in land managers attitudes and practices. *Landscape Research* (accepted for publication Sept. 2012).

Pinto-Correia, T., & Ribeiro, S. (2012). HNV in 35 countries in Europe: Portugal. In R. Oppermann, G. Beaufoy, & G. Jones (Eds.), High Nature Value Farming in Europe. 35 European countries – experiences and perspectives. *Verlag Regionalkultur*, Ubstadt-Weiher, 336–345.

Pinto-Correia, T., Ribeiro, N., & Sá-Sousa, P. (2011). Introducing the montado, the cork and holm oak agroforestry system of Southern Portugal. *Agroforestry Systems*, *82*(2), 99–104. doi:10.1007/s10457-011-9388-1

Plieninger, T. (2007). Compatibility of livestock grazing with stand regeneration in Mediterranean holm oak parklands. *Journal for Nature Conservation*, *15*, 1–9. doi:10.1016/j.jnc.2005.09.002

Primdahl, J., & Kristensen, L. (2011). The farmer as a land manager: Management rules and change patterns in a Danish region. *Geografisk Tidsskrift-Danish Journal of Geography*, *111*(2), 107–116.

Ribeiro, N. A., Dias, S., Surový, P., Gonçalves, A. C., Ferreira, A. G., & Oliveira, A. C. (2004). The importance of crown cover on the sustainability of cork oak stands: A simulation approach. *Advances in Geoecology*, *37*, 275–286.

Ribeiro, N. A., Surový, P., & Oliveira, A. C. (2006). Modeling cork oak production in Portugal. In H. Hasenauer (Ed.), *Sustainable forest management: Growth models for Europe* (pp. 285–313). Berlin Heidelberg: Springer-Verlag.

Rodrigo, I., & Veiga, J. F. (2009). Portugal: Natural resources, sustainability and rural development. In K. Bruckmeier & H. Tovey (Eds.), *Rural sustainable development in the knowledge society*. England: Ashgate.

Surová, D., & Pinto-Correia, T. (2008). Landscape preferences in the cork oak Montado region of Alentejo, southern Portugal: Searching for valuable landscape characteristics for different user group. *Landscape Research*, *33*(3), 311–330.

Surová, D., Surový, P., Ribeiro, N., & Pinto-Correia, T. (2011). Integration of landscape preferences to support the multifunctional management of the Montado system. *Agroforestry Systems*, *82*(2), 225–237. doi:10.1007/s10457-011-9373-8

Sutherland, L. A. (2010). Environmental grants and regulations in strategic farm business decision-making: A case study of attitudinal behavior in Scotland. *Land Use Policy*, *27*, 415–423.

Van der Ploeg, J. D., & Marsden, T. (2008). Unfolding webs. The dynamics of regional rural development. Van Gorcum, 262pp.

Wilson, G. A. (2007). *Multifunctional agriculture – A transition theory perspective*. Trowbridge: CABI, Cromwell Press.

Wilson, G. A. (2008). From 'weak' to 'strong' multifunctionality: Conceptualizing farm-level multifunctional transitional pathways. *Journal of Rural Studies*, *24*, 367–383.

Wilson, G. A. (2009). The spatiality of multifunctional agriculture: A human geography perspective. *Geoforum*, *40*, 269–280.

CHAPTER 5

PROCESSES OF FARMLAND ABANDONMENT: LAND USE CHANGE AND STRUCTURAL ADJUSTMENT IN GALICIA (SPAIN)

Edelmiro López-Iglesias, Francisco Sineiro-García and Roberto Lorenzana-Fernández

ABSTRACT

The objective of this chapter is to provide an approach to the farmland abandonment problem in Galicia, the Spain's north-western region. We describe the land use pattern that characterized the traditional agricultural system, and analyze the process of structural adjustment and changes in land use recorded in the last 50 years. The empirical basis is provided mainly by an original elaboration of agricultural census data for the period 1982–2009. The results show that in the last five decades the area devoted to crops and pastures was constrained to a small portion of the territory (just over 20%), while the agro-livestock uses of hill land which were very important up to the mid-twentieth century disappeared. All this led to a remarkable expansion of abandoned land, which currently occupies at least 20% of the regional area. The drivers of this farmland abandonment are diverse and vary from one zone to another. But among

Agriculture in Mediterranean Europe: Between Old and New Paradigms
Research in Rural Sociology and Development, Volume 19, 91–120

ISSN: 1057-1922/doi:10.1108/S1057-1922(2013)0000019007

them the conditioners derived from the structure of land ownership must be emphasized, coupled with the poor functioning of the land market and leasing. Land abandonment has had a major impact on the dynamics of the agricultural sector, limiting the size of farms and causing an increasing intensification in a small portion of the territory. This has also led to severe environmental problems, especially forest fires. Consequently, improving mobility and land use should be a priority of agricultural and rural development policies in this region.

Keywords: Land use change; farmland abandonment; Galicia; structural adjustment; land mobility; livestock intensification

INTRODUCTION

The farmland abandonment and subsequent reduction in the utilized agricultural area (UAA) are major issues of concern in many regions and countries of the European Union. However, the analysis of the phenomenon is faced with a fundamental problem of absence of the definition of land abandonment, both in the scientific literature and in the legal regulations (Keenleyside & Tucker, 2010; Moravec & Zemeckis, 2007; Pointereau et al., 2008). Therefore, it is necessary to distinguish the abandonment of agricultural activity from the disappearance of any use or exploitation of land (Baudry & Asselin, 1991). The farmland abandonment can refer to the state of the land at any given time (static approach) or to a process that develops gradually over time (Keenleyside & Tucker, 2010), including situations of semi-abandonment or underutilization. In any case, the operating criteria and the information sources to quantify the abandonment are markedly different according to the approach adopted: administrative, economic, ecological or agronomic (Pointereau et al., 2008).

In this chapter we discuss the farmland abandonment, which can be simply defined as 'the cessation of agricultural activities on a given surface of land' (Pointereau et al., 2008, p. 10). It is estimated that this phenomenon in Europe has reached some 770,000 hectares per year between 1961 and 2003, although with widely varying intensities among countries (Pointereau & Coulon, 2009). It has mainly affected surfaces of low agricultural productivity, especially in mountain areas, but has also occurred in other areas. It is especially remarkable in the countries of Eastern Europe, where abandonment has been linked to the reforms and the transition of their economies in the period after the collapse of communist regimes (Moravec & Zemeckis, 2007; Pointereau et al., 2008).

The abandonment is often a gradual process, starting with a reduction in production intensity and culminating in the complete cessation of agricultural use of the land. This process first results in the accumulation of herbaceous and shrub growth, and finishes with the formation of a forest ecosystem in the richest soils after a variable period, which depends on the weather and soil conditions (Keenleyside & Tucker, 2010). When the process has made little progress abandonment is relatively easy to reverse with proper management, while greater efforts and investments are needed when it is in more advanced stages. Being a complex and gradual process, there may be situations of semi-abandonment and various forms of temporary or permanent abandonment. Specifically, in the context of the current Common Agricultural Policy (CAP), the land can be not formally abandoned but in a state of semi-abandonment, being subject to some form of management that allows it to meet the requirements of cross-compliance, but with very extensive modalities (Pointereau et al., 2008). All this makes the precise quantification of the phenomenon very difficult.

The complexity of the process makes its causes diverse and markedly different from one region to another. In principle, the abandonment is largely the result of a decline in the viability of extensive farming systems and those based on small farms (Baldock, Beaufoy, Brouwer, & Godeschalk, 1996; Baudry & Asselin, 1991; Keenleyside & Tucker, 2010; Mac Donald et al., 2000). But it can also be linked to other causes as in zones close to urban areas, where agricultural land may be abandoned against the alternative of greater potential earnings in the cities (Moravec & Zemeckis, 2007). In an overview, Pointereau et al. (2008) classify the factors conditioning the risk of farmland abandonment into various types: geographic (steep slope, distance from the farm to the parcels, difficult accessibility), agro-ecological (fertility and other conditions), demographic (decreasing number of farmers, general dynamics of the population), socioeconomic (cost and profitability of production, farm size, land prices, other institutional factors) and the effect of agricultural policies.

The main effects of abandonment occur at the economic, social and environmental levels. From the economic point of view, land abandonment involves a loss of an agricultural asset, to which must be added the possible negative impact on the surroundings farms. The loss of an agricultural asset also includes the accompanying deterioration of productive infrastructure such as fences and irrigation systems. In social terms the land abandonment contributes to lower farm incomes, loss of employment and socioeconomic regression of rural areas. Finally, from an environmental perspective, the land abandonment causes the accumulation of spontaneous vegetation, which may result in increased vulnerability to fire, while the effect on biodiversity depends on the type of abandoned land (Rey Benayas, Martins, Nicolau, & Schulz, 2007; Pinto Correia & Breman, 2008).

In short, the abandonment of agricultural land is a difficult phenomenon to measure and study in practice. Both the explanatory factors and their consequences are complex, changing over time and also very dependent on local circumstances.

OBJECTIVES AND CONTENT OF THE STUDY

To study the farmland abandonment in Galicia, a region situated in the northwest of Spain (Fig. 1), we focus on the analysis of the process of structural adjustment in agriculture and on the changes in land use since the middle of the twentieth century.

Galicia has been characterized in contemporary times by comparatively low levels of economic development in the Spanish and European context, which is reflected in the fact that it remains among the convergence objective regions of the European Union during the period 2007–2013. In terms of agriculture, at the middle of the twentieth century the traditional system still remained in force defined by a high fragmentation of property, the small size of farms and a productive model characterized by its labour intensity and a

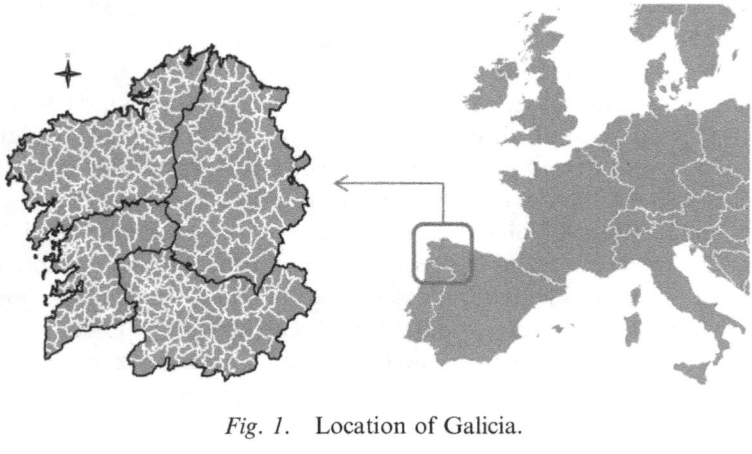

Fig. 1. Location of Galicia.

high degree of utilization of the territory. This situation has changed substantially in the last 60 years. On one hand, there was a late but very intense reduction in the agricultural labour.[1] It was accompanied by drastic changes in technology and production, which led to the establishment of an agricultural sector specialized in livestock production (mainly milk) and highly dependent on industrial inputs (López Iglesias, 2005; Sineiro, López, Lorenzana, & Valdés, 2006). However, the result of these changes has been largely conditioned by the persistence of significant deficiencies in the structure of farms and land use. Among the problems that this has caused, two should be noted: the maintenance of a very small agricultural area per farm, which hinders its economic viability, and the proliferation over recent decades of forest fires, which has high economic and environmental costs (López Iglesias, 2005; Sineiro García, 2006).

The overall objective of the chapter is to study the farmland abandonment and its relationship to technological and structural changes of the Galician agriculture in recent decades, also paying attention to the institutional constraints related to land ownership. The first step is to examine the evolution of the number of farms and changes in land use since the mid-twentieth century and then to estimate the abandonment of agricultural land. The statistical base is essentially made by the six agricultural censuses conducted in Spain (1962, 1972, 1982, 1989, 1999 and 2009). In this respect, we note two methodological changes that affect data homogeneity:

- The most important methodological change concerns the farm population. The agricultural censuses conducted in Spain until 1999 were based on a very broad definition of 'farm', including all the units with either more than 0.1 hectare of the total area or more than two livestock units. This led to a large number of very small farms and others of greater extension but with little or no agro-livestock activity being surveyed. Instead the last census (that of 2009) has used a more restrictive definition of 'farm', applying the criteria of the farm structure surveys of the European Union.[2] This has had a strong effect on the data, so they are not comparable with previous censuses.
- Another relevant change for our purpose is that the first two agricultural censuses (1962 and 1972) provided information on surfaces covered by spontaneous vegetation and even forestry land used by grazing, which were part of a broadly defined concept of 'utilized agricultural area'. By contrast, this information is no longer collected from the 1982 census onwards.

Based on these conditioning factors, we divide the analysis of the evolution of the structure of farms and land use into two periods:

- First, we examine the changes in the period 1962–1982, which corresponds to the decades prior to European integration, taking here as the basis the original data of the first agricultural censuses (with a broad concept of 'farm').
- Second, we perform a more detailed analysis of the trends in the last three decades (1982–2009), that is, roughly the period after joining the European Union (which happened in 1986). For this we use an original elaboration of the data of the last four agricultural censuses, carried out from the microdata (individual information at farm level). This original elaboration allowed us to obtain a homogeneous series adapted to the criteria of the Agricultural Census of 2009 and also down to the evolution at a municipal level (LAU2 in the nomenclature of territorial units of the European Union).

The next section provides an overview of the current land use, trying to quantify the amount of abandoned land. In the fourth section, we point out some factors contributing to land abandonment in Galicia. The fifth section is devoted to the main production and environmental consequences of this phenomenon. In the last section, we finish highlighting the main conclusions.

LAND USES IN GALICIAN AGRICULTURE: SUMMARY OF PREVIOUS TRENDS TO EUROPEAN INTEGRATION

Initial Situation: Land Use in the Traditional Agricultural System

The modernization process of agriculture has developed in Galicia with a marked delay compared to other Spanish regions and especially in comparison with the more developed countries of Central and Northern Europe. Thus, in the early 1960s the sector still retained the essential features of the traditional agricultural system based on mixed crops and livestock with a high labour intensity and low utilization of industrial inputs. This was accompanied by a share of agricultural employment that exceeded 65% of total occupied (Fernández Leiceaga & López Iglesias, 2000).

Under the traditional agricultural system what we can consider utilized agricultural area (UAA) in a strict sense (crops and pastures) occupied a small proportion (about 25%) of the territory. However, the remaining 75% of land played an important role in the survival of farms as a space for occasional cropping, for livestock grazing, for fuel wood and timber, and for supplying the plant material for obtaining compost that maintained the fertility of arable land (Bouhier, 1979, 1984; Sineiro, 1983; Sineiro & Díaz, 1984).

These uses were the basis to include much of the uncultivated land in a broad concept of UAA, because it was subjected to agricultural and livestock utilization. The first agricultural census conducted in Spain, in 1962, gives a good picture of this situation. Specifically, according to the 1962 census, in Galicia 21% of the geographical area was devoted to crops and meadows, and 32% of the area indicated rough grazing by the livestock. So the 'UAA in a broad sense', that is the space that was the object of some agricultural or livestock utilization, covered just over half of Galicia, although it encompassed two very different types of utilization.

With respect to the territorial base of farms, the consolidated historical structures in effect in the mid-twentieth century were characterized by two basic features: (1) the small size of farms and (2) the division of that surface into a large number of very small plots. Thus, according to the agricultural census of 1962 the average farm size was limited to 5.75 hectares, and it was fragmented into 22 plots with an average size of 0.25 hectares. These structures did not raise problems in the traditional agricultural system, as they were consistent with the other parts of that system, particularly with two parts: (1) the high density of labour and (2) the simplicity of the tools. Agricultural structures were rational for the existing technology (Bouhier, 1979; López Iglesias, 1996).

The Changes in the Period 1962–1982

Both agricultural structures and land use are going to experience great changes from the 1960s, driven by the sharp drop in labour and transformations in technology and production systems, which are summarized as an accelerated substitution of labour by capital (Fernández Leiceaga & López Iglesias, 2000).

Thereby in the two decades, 1962–1982, there is already an important process of disappearance and concentration of farms, which was hampered by the deficient functioning of the mechanisms of land mobility (both purchases and leasing). The sharp reduction of the population engaged in agriculture led to the disappearance of a substantial number of farms, thereby causing the release of an increasing volume of farmland. However, much of this land was not sold or leased to other farms, but was abandoned or put to non-agricultural uses (forest, urban). The result was that the disappearance of farms largely resulted in a decline of the area occupied by farms (López Iglesias, 1996).

Specifically, according to the agricultural census, in the period 1962–1982 the number of farms fell by 14.5%, but this was associated with a decrease in the total area surveyed by 7.5% so that the increase in average farm size was limited to 8.2% in 20 years, from 5.8 to 6.2 hectares (Table 1).

The beginning of the process of disappearance and concentration of farms was accompanied by major changes in land use, which can be summarized as follows:

- A reduction in the UAA, especially by the progressive disappearance of agro-livestock utilization of rough grazing

Table 1. Evolution of Agricultural Holdings in Galicia during 1962–1982 (Thousands of Farms and Thousands of Hectares).

	Number of Farms	Total Area of Agricultural Holdings	UAA1[a]	UAA2[a]	Crops	Pastures	Other Grazing Areas[b]
1962	421.3	2424.0	629.7	1570.1	476.8	152.9	940.4
1972	385.8	2456.6	687.8	1263.4	444.1	243.7	575.6
1982	360.4	2242.0	646.0	646.0	330.8	315.2	–
Change during 1962–1982							
Thousands of farms and thousands of hectares	−60.9	−182.0	16.3	−924.1	−146.0	162.3	−940.4
% change	−14.5%	−7.5%	2.6%	−58.9%	−30.6%	106.1%	−100.0%
Annual variation rate	−0.8%	−0.4%	0.1%	−4.3%	−1.8%	3.7%	–

Source: INE; agricultural censuses.

[a]Utilized agricultural area (UAA). UAA1, crops + pastures. UAA2, crops + pastures + other grazing areas.

[b]Rough grazing (consisting of grass and shrub species, such as gorse and heather; in the case of being slightly grazed or abandoned evolves into a scrub vegetation). This information stopped being collected from the 1982 agricultural census. On the evolution of these surfaces in recent decades, see the text and the footnote number 3.

- A remarkable expansion of wooded area, both on privately owned land and on communal property, which occupies more than 20% of the territory
- A sharp increase in abandoned land, by the convergence of two processes: the abandonment of some cropland and pastures belonging to farms which had disappeared, and especially the crisis of the traditional uses of rough grazing, which caused much of this land to become unused

Focusing on the latter process, from the 1960s there is a strong reduction in the occasional cropping and use of scrub to obtain compost, due to the decrease in labour and the availability of mineral fertilizers. We also noted a gradual abandonment of grazing, linked to changes in livestock production models: loss of beef cows and sheep, growing orientation towards dairy and intensive livestock production (poultry, pigs). The result was a sharp decline in the use of rough land as a temporary space for cropping and especially for grazing. The areas that were not devoted to forest plantations were abandoned and progressively occupied by scrub, although some of them end up growing trees spontaneously and without an orderly forest exploitation (Bouhier, 1979, 1984; Sineiro, 1983; Sineiro & Díaz, 1984). These changes are clearly reflected in the agricultural censuses: land used for rough grazing fell substantially in the decade 1962–1972 (Table 1). Although this information stopped being collected from the 1982 census, available studies allow to affirm that the decline was accelerated in the last four decades. Therefore, the land used for rough grazing by livestock remains actually in only a very small area.[3]

As a consequence the agricultural area was being restricted to what we call 'UAA in strict sense', that is croplands and grasslands. This is reflected in agricultural censuses with a sharp contraction of 'UAA in broad sense', which changed from 53% of the territory in 1962 down to 22% in 1982, due to the abandonment of agro-livestock use of the hill land. With respect to the 'UAA in strict sense', two facts should be noted:

- The UAA's overall volume remained stable or increased slightly, which can be explained because the abandonment of some croplands and meadows was offset by the conversion of rough grazing into permanent pasture made by many farms.
- This global stability was accompanied by significant changes in land use linked to an increased livestock specialization: a sharp decline in cultivated area in parallel to expansion of grasslands (Table 1).

EVOLUTION OF THE FARMLAND DURING 1982–2009

To analyze the evolution during the period 1982–2009, we will use a homogenized database of the last four agricultural censuses, based on the universe of farms of the 2009 Census and farm structure surveys. This delimitation has excluded from the original data of the censuses of 1982, 1989 and 1999 those farms with less than 1 hectare of UAA that have under a minimum of livestock units.

The effect of that clearance can be seen in Table 2, where the original figures of the census and those resulting from applying the new definition are compared for the period 1982–1999. The differences are very important with regard to the number and total area of farms, while its magnitude is small for the UAA.

All these indications show that the excluded farms are very small units engaged in subsistence or larger fields with little or no farming activity, composed mainly of rough grazing. In this sense, the broad definition of 'farm' applied in Spanish censuses until 1999 led to continue accounting many such units, overestimating the real area linked to farms.

Global Trends

With the new database, the dynamics in the last three decades (1982–2009) appear defined by four notes (Tables 3 and 4):

Table 2. Agricultural Holdings in Galicia during 1982–1999 (Thousands of Farms and Hectares).

	Number of Farms	Total Area of Agricultural Holdings	UAA
A. Original data from agricultural censuses			
1982	360.4	2242.0	655.0
1989	358.9	2217.1	675.0
1999	269.0	2041.8	696.7
B. Data referred to the universe of farm structure surveys			
1982	233.4	1364.7	616.8
1989	170.8	1247.4	623.8
1999	118.7	1154.6	657.3
B/A (%)			
1982	64.8	60.9	94.2
1989	47.6	56.3	92.4
1999	44.1	56.5	94.3

Source: Own calculations from agricultural censuses' microdata.

Table 3. Changes in the Agricultural Holdings in Galicia during 1982–2009.

	Number of Farms ('000)	Total Area ('000 ha)	UAA ('000 ha)	Total Area per Farm (ha)	UAA per Farm (ha)
1982	233.4	1364.7	616.8	5.8	2.6
1989	170.8	1247.4	623.8	7.3	3.7
1999	118.7	1154.6	657.3	9.7	5.5
2009	81.2	914.9	647.6	11.3	8.0
% change during 1982–2009	−65.2	−33.0	5.0	92.7	201.8
Annual variation rate (%)					
1982–1989	−4.4	−1.3	0.2		
1989–1999	−3.6	−0.8	0.5		
1999–2009	−3.7	−2.3	−0.1		
1982–2009	−3.8	−1.5	0.2		

Source: Own calculations from agricultural censuses' microdata.

Table 4. Change in the Number of Farms, Total Area of Agricultural Holdings and UAA by Farm Size Levels, Galicia, 1982–2009.

	Change During 1982–2009		
Farm Size (UAA)	Number of Farms ('000)	Total Area ('000 ha)	UAA ('000 ha)
<10 ha	−164.0	−801.3	−334.0
>=10 ha	11.8	351.4	364.8
Total	−152.2	−449.9	30.8

Source: Own calculations from agricultural censuses' microdata.

1. Strong acceleration of the disappearance of farms, with a reduction of two thirds of farms in the period 1982–2009. This loss is concentrated in small units (less than 10 hectares of UAA), although the 'survival threshold' was rising with the passage of time, up to 20 hectares in the period 1999–2009.
2. Of the total land released by farms which had disappeared (including crops, pasture and rough grazing), more than half was abandoned or was put to non-agricultural uses (forest, urban), which resulted in a sharp decline in the total area occupied by farms (Table 4).

Specifically, the disappearance of the two thirds of the farms was accompanied by a reduction in one third of the area linked to agricultural production units (Table 3).

3. The overall volume of UAA remained, and even rose slightly, which corresponds to the fact that the UAA gained by farms over 10 hectares exceeded the loss of those under that size. This meant that, while the total area per farm increased by 93% in the three decades (from 5.8 to 11.3 hectares), the UAA per farm tripled (from 2.6 to 8.0 hectares).
4. Chronologically, the disappearance of farms reached a peak in 1982–1989 (4.4% annual), remaining at rates of 3.6% or 3.7% in the last two decades. With respect to the land, this has shown a continuous reduction in the total area of farms, which accelerated in the last decade to cause a slight contraction of the UAA (Table 3).

The explanation of this evolution in agricultural land use (sharp contraction of the total area integrated into farms, stability or slight increase in UAA) must be sought in the first instance in two phenomena:

- In the land left vacant by the disappeared farms, the portion of cropland and pastures which was transferred to other farmers is much higher than that of rough grazing, so the abandonment mainly affected the latter type of surfaces.
- In addition, as it had been happening in previous decades, there were farmers who expanded their UAA by converting rough grazing into pastures.

The result of these processes is summarized in Table 5. In 1982 farms managed 46% of the geographical area of Galicia, but of that area less than

Table 5. Percent of Territory Occupied by Total Area of Agricultural Holdings and UAA during 1982–2009.

	Galicia		Spain	European Union-15
	1982	2009	2009	2007
Total area of agricultural holdings/Geographical area	46.1%	30.9%	60.5%	48.1%
UAA/Total area of agricultural holdings	45.2%	70.8%	77.6%	80.0%
UAA/Geographical area	20.9%	21.9%	46.9%	38.5%

Source: Own calculations from INE, Agricultural Census 2009; microdata of Agricultural Census 1982; EUROSTAT, Farm Structure Survey 2007.

half was covered by crop or permanent pasture (the remaining were covered by land for rough grazing, with scrub or wooded, and unproductive land). Starting from this time, in the last three decades there has been a sharp decline in the total area managed by farms, from 46% to 31% of the territory, while the overall volume of UAA rose slightly. Therefore, farms manage a decreasing part of the territory. But the percentage of that area exploited for crops and pastures is increasing, reducing the area of rough land integrated into farms (Table 5).

These data become more meaningful when we place them in a comparative perspective (Table 5):

- The percentage of the geographical area occupied by the UAA, used for crop and livestock production, is limited in Galicia in 2009 to 22%, less than half of the Spanish average and just over half of that achieved in the EU-15.
- This narrowness of UAA corresponds mainly with the low percentage of the territory operated by farms.

Territorial Differences: Evolution of UAA at the Municipal Level

The overall stability of the UAA in the last three decades hides very different dynamics in Galicia, which are reflected in the data at the municipal level. For this analysis we started to classify municipalities into three groups, depending on the density and population dynamics: urban, rural intermediate and rural in severe recession.[4]

Urban municipalities are composed of major cities and their zones of influence as well as some intermediate towns; geographically they are concentrated in the coastal areas of the western third. Occupying 13% of the regional territory and accounting for almost 70% of the population, in recent decades they have experienced a moderate increase in population (Tables 6 and 7).

Intermediate rural municipalities comprise the majority of the interior of Galicia, plus the western end of the coastal strip, being the most dynamic areas of livestock. They cover almost two thirds of the geographical area but only 29% of the population, after having lost almost a quarter of it in the last three decades. Finally, rural municipalities in severe recession correspond with mountain zones, with higher natural constraints to agricultural production and under strong demographic decline. They occupy one quarter

Table 6. Types of Municipalities: Relative Weight in the Main Variables, Galicia, 2009.

	Absolute Figures ('000)	Distribution by Types of Municipalities (%)			
		Urban	Rural Intermediate	Rural in Severe Recession	Total
Population	2795.4	68.7	28.5	2.8	100.0
Geographical area (ha)	2957.8	13.4	62.6	24.0	100.0
Total area of agricultural holdings (ha)	914.9	7.1	74.0	18.9	100.0
UAA (ha)	647.6	6.3	73.7	19.9	100.0
Number of farms	81.2	13.6	73.0	13.4	100.0
Basic ratios on agricultural land use (%)					
Total area of farms/Geographical area		16.4	36.5	24.3	30.9
UAA/Geographical area		10.3	25.8	18.2	21.9

Source: Own calculations from INE, microdata of Agricultural Census 2009 and Municipal Population Census.

Table 7. Change in Population, Number of Farms, Agricultural Area and Livestock by Types of Municipalities, Galicia, 1982–2009 (% Change).

	Types of Municipalities			
	Urban	Rural Intermediate	Rural in Severe Recession	Total
Population	19.6	−22.9	−54.8	−0.6
Number of farms	−77.0	−63.0	−66.3	−66.3
Total area of agricultural holdings	−51.9	−29.6	−35.4	−33.0
UAA	−38.2	11.9	4.2	5.0
Cultivated area	−53.7	−22.8	−61.4	−34.4
Grassland and pastures area	−8.4	43.9	59.2	43.4
Livestock	−53.2	21.0	−8.6	7.6

Source: Own calculations from INE, microdata of Agricultural Census 2009 and Municipal Population Census.

of the territory but have only 2.4% of the population. After losing more than half of their inhabitants in the period 1981–2011, their average population density is limited to about 10 inhabitants per square kilometres (Tables 6 and 7).

Based on this characterization, we can conclude as follows on the evolution of the total area and the UAA of farms in the period 1982–2009 (Table 7):

- The total area occupied by farms fell sharply in all three types of municipalities, showing that the retreat of the 'agricultural frontier' was a widespread phenomenon throughout the territory. However, this process reached its greatest intensity in urban municipalities (−52%), with more moderate magnitude in the rural in severe recession (−35%) and in rural intermediate (−30%). These differences are largely correlated with those in the rate of disappearance of farms, but the differences in mobility of land released (the extent to which they were transferred to other farmers) also affect. In any case, the evolution of the farmland appears clearly related to two factors: the agricultural, especially livestock, dynamism of municipalities (positive effect), and the pressure of the non-agricultural uses of land (negative effect).
- With respect to the UAA, the slight increase in the whole of Galicia is the net result of a sharp decline in urban municipalities (−38%), a stability in the rural in severe recession (+4%) and a moderate expansion in intermediate rural (12%) (Table 7).

In short, the farmland abandonment affected the three types of municipalities unevenly: whether we measure this process by the change in UAA or by the decline in the total area of farms, the abandonment reaches its greatest intensity in urban municipalities. In these municipalities, there was a loss of agricultural land due not only to its 'artificialization' but also to the abandonment induced by the greater earning potential in non-agricultural jobs, which Moravec and Zemeckis (2007) note as abandonment by social causes. In addition, the blocking of the sale and lease of land for agricultural uses derived from the expectation of future conversion to urban uses also acts (López Iglesias, 1996). In rural municipalities the worst dynamics are observed in the mountain areas. Although the exact conclusion depends on the type of data that we look at: if we take the UAA, this remained globally stable even in those municipalities so that we can speak of abandonment only if we identify it with the decrease of the total area of farms.

One additional note to emphasize is the change in the uses of the agricultural area. Continuing the trend observed since the 1960s, the slight increase in UAA in the period 1982–2009 is the result of a sharp decline in cultivated area (35% of the initial amount), compensated by the expansion of grasslands (an increase of 43%). The withdrawal of crops is a general trend in all the three types of municipalities, although less intense in rural intermediate, while the expansion of the area devoted to grassland reaches a similar relative magnitude in the two rural groups in contrast to the contraction in the urban (Table 7). These changes must be put in relation to the increasing livestock specialization, which has continued in recent decades (Sineiro et al., 2006). It is noteworthy that according to the agricultural census the surface of meadows and pastures records a greater percentage increase in mountain municipalities, where livestock was reduced, than in rural intermediate, which tends to concentrate the livestock more and more (Table 7). This fact points to a growing contrast between extensive and intensive livestock models that characterize one and the other municipalities, respectively.

CURRENT USES OF LAND AND ABANDONMENT

Global Data

The changes summarized above, over the last five decades (1962–2009), caused the area devoted to crops and meadows to be constrained to the small portion of the territory it occupied in the traditional agricultural system. At the same time agro-livestock use of rough grazing, so important up until the mid-twentieth century, almost completely disappeared.

The result is that at present, according to the Agricultural Census 2009, the UAA is limited to 22% of the territory of Galicia. This figure varies according to other statistical sources, depending mainly on the more or less broad definition of the grazed area (Corbelle and Crecente, 2009). Thus, official statistics on land use in 2011 raise the UAA up to 23.4% of the territory (Table 8). In any case, we can conclude that the UAA occupies a very low percentage of geographical area compared to both the Spanish and European average.

The rest of the region is divided into three main groups (Table 8):

- The wooded forest area occupies about 45% of the territory. But only about 5–7% of which is managed for forestry production (OSE, 2011), while most is not subject to any productive care, much being abandoned for practical purposes.
- 'Other land', which includes internal waters and both artificial and unproductive lands, accounts for 11% of the total.

Table 8. Current Land Use in Galicia, Compared with Potential Land Use.

Current use (2011)	Land Use ('000 ha)	%	Use Capability	%
Cultivated area	388.3	13.1	Continuous cultivation	34.7
Permanent grassland	303.5	10.3	Occasional cultivation	15.7
Utilized agricultural area (UAA)	**691.8**	**23.4**		
Scrubland	607.8	20.5		
Wooded area	1,337.6	45.2	No cultivation	36.1
Other land	320.6	10.8	Unproductive or low use	13.5
Total	2,957.8	100.0	Total	100.0

Source: Own calculations from data of Consellería do Medio Rural and Díaz-Fierros and Gil Sotres (1982).

- Finally, there is a minimum of 21% of the territory occupied by scrub. This land is not currently subject to any economic use, so it can be assimilated with land completely or almost completely abandoned.

Overall, the problem of abandonment is focused on the areas of scrub. But to this we must add the situation of semi-abandonment or pronounced underutilization affecting much of the woodland and extensive grazing.

These current uses contrast sharply with those which might enable the productive capability of the soils. Particularly on one point, the UAA is well below 50% of the surface that could be used for crops and pasture according to their potential, being suitable for occasional or continuous cultivation. This difference between the actual and potential UAA roughly corresponds to the volume of abandoned land, occupied by scrub (Table 8).

Geographic Disparities

This overall picture hides huge differences in agricultural use of the territory of individual counties. Beginning with the typology of municipalities previously defined, in Table 6 we can see that

- the rural intermediate group accounts in 2009 for three quarters of both farms and UAA, registering a ratio of UAA and geographical area (26%) well above the regional average,
- in mountain areas, where most of the rural in severe recession group is found, this ratio drops to 18%, and
- in the urban areas it is limited to 10%.

Table 9. Distribution of Municipalities According to the Percentage of UAA on the Geographical Area, Galicia, 2009.

UAA/ Geographical Area (%)	Number of Municipalities	UAA/Geographical Area (Average % in each group)	Geographical Area ('000 ha)	UAA ('000 ha)
<10	122	5.5	832.4	46.2
10–20	80	15.3	670.9	102.5
20–30	51	25.4	616.1	156.7
≥30	62	40.8	838.4	342.2
Total	315	21.9	2957.8	647.6

Source: Own calculations from INE and microdata of Agricultural Census 2009.

If we classify the municipalities according to the percentage of UAA in the geographical area, we get the picture summarized in Table 9:

- In two thirds of the municipalities the percentage of the territory dedicated to crops and pastures is below 20%.
- Even within this set there is still a large group (40% of total) in which the UAA does not reach 10% of the area. These are in counties where the agro-livestock use of territory has fallen sharply in recent decades and is today almost marginal. This corresponds to two situations: urban and coastal areas with high population densities, under significant pressure from urban land uses; and the rural and mountainous areas of the south of the region with a marked demographic decline.
- In the opposite group they are one fifth of them where the ratio of UAA on geographical area is over 30%, reaching an average value (41%) similar to the European one. These zones correspond almost exactly with the counties which in recent decades have been more dynamic in livestock production, mainly linked to dairy (Sineiro et al., 2006).

In conclusion, we can say that if land abandonment is now a widespread phenomenon, its intensity varies greatly from one county to another. However, micro-level studies with more accurate data show that, even in counties with a relatively high utilization of the territory, there is a significant presence of high potential agriculture lands that are abandoned. In this regard, the cartographic analysis of the abandonment is interesting, and also the consequent potential supply of land for agricultural uses, from parish-level data for four regions of Galicia, which is offered in Corbelle, Díaz Manso, Crecente, and Martínez Rivas (2011).

EXPLANATORY FACTORS OF FARMLAND ABANDONMENT IN GALICIA

The explanatory factors of the high level of farmland abandonment that is registered in Galicia are very complex and are also influenced by local circumstances, which explain the geographic variability of the phenomenon (Corbelle, Crecente, & Santé, 2011). Among them two must be emphasized, which have operated continuously during the last five decades: (1) the ownership structure and (2) the poor performance of the mechanisms of land mobility. In addition, other factors should be considered, which are mainly related to the dynamics of agricultural production, the changes in production systems and the impact of different public policies.

The Constraints Derived from the Ownership Structure

A key element that has been contributing to the farmland abandonment, and that hinders the productive mobilization of the farmland already abandoned, is the ownership structure. This is characterized by a pronounced division and the existence of an increasing number of owners disconnected from exploitation of the land. Specifically, according to land registry data for 2011, the rural area of Galicia (2.85 million hectares) is divided between 1.6 million owners. Given that the region's population is around 2.8 million people, this means that the vast majority of the adult population are landowners, while the number of persons employed in agriculture is less than 65,000. Furthermore, that area is fragmented into a total of 11.4 million plots. Thus, the average area per owner is limited to 1.8 hectares, composed of seven plots (Table 10).

These aggregate data require important qualifications. Strictly speaking, from the point of view of ownership we distinguish three types of surfaces,

Table 10. Land Ownership Structure, Galicia, 2011.

Rural area ('000 ha)	2,854.3
Number of landowners ('000)	1,611.4
Number of plots ('000)	11,397.2
Area per landowner (ha)	1.8
Plots per landowner	7.1
Area per plot (ha)	0.3

Source: General Directorate of Land Registry.

which pose different problems in terms of their use and the risk of abandonment:

1. *Private lands integrated into individual farms.* Accounting for only 31% of the territory (22% of UAA and 9% of forest land), these surfaces, although belonging mostly to farmers and their families, are divided among hundreds of thousands of proprietors. And, except in areas where land consolidation is completed, they exhibit a high fragmentation and deficiencies in access to plots, which hinders their proper utilization.
2. *Forest and abandoned land held by non-farmer owners, which cover 39% of the territory.* They are spread over hundreds of thousands of owners with each one having a very small area, suffering from a similar or greater fragmentation than the previous ones and with problems of access to plots. Furthermore, to this we must add that the vast majority of these owners are completely or almost completely detached from the use of their land, all of which contributes to the semi-abandonment or complete abandonment of an important part of them.
3. *Common lands.* They occupy 23% of the territory, belonging to more than 3,000 neighbourhood communities. These units generally are large (average of 206 hectares) without problems of fragmentation into plots, but they face other constraints contributing to a high level of abandonment or underutilization. These include the sharp demographic decline in most proprietary communities, little or no involvement of community members in the use of these lands, and the inadequacy of the legal framework and public policies applied (Fernández Leiceaga et al., 2006).

Although land abandonment is present in all the three types of properties, its magnitude is reduced in the first while achieving high levels in both the second and the third.

Deficiencies in Land Mobility

Starting from this ownership structure, the farmland abandonment has been driven by the deficient operation of the mechanisms of land mobility. Existing studies show that the land market is characterized in Galicia throughout the period analyzed by low mobility, the existence of a significant demand for non-agricultural buyers and high prices. All these elements had a role limiting the increase in size of the farms. To this is added a diffusion very scarce of land renting until recently (López Iglesias, 1996, 2005). This land transfer pathway has only to acquire a significant presence during the last decade, especially on larger farms. The consequence was that a large part of the space left vacant by the disappearance of farms in the last five decades was abandoned, or was transferred to urban uses or forestry.

The factors behind this lack of mobility, and resulting in the abandonment or conversion of land to non-agricultural uses, are different. But among them may be noted the following (López Iglesias, 1996):

1. Limitations in the demand for land by farmers in some counties, mainly due to the limited resources of most farms and to demographic factors (aged holders).
2. *The strong competition with forest use, an issue fostered by the absence of a land use planning and certain public policies.* In this regard it should be noted that the priority to the afforestation of rough land, instead of the possible conversion to pasture, has characterized the actions of both Spanish and Galician governments in the last decades. This prioritization of forestry option is a constant since the 1940s, and it is reflected both in public incentives to investments by private owners and in the intervention of the Forest Administration in the direct management of many common lands (first central government and since the 1980s the autonomous government of Galicia). In recent decades the concurrence from forest uses has also been driven by the support for the afforestation of agricultural land implemented with European funds from 1992 onwards.
3. *The competition with urban uses.* The scattered urban growth that characterizes Galicia, together with the absence of effective management plans, not only caused the artificialization of major extensions but also a phenomenon of much greater magnitude: the abandonment of farmland in the expectation of a possible future urbanization.

   In this sense, the urban–rural connections have largely been focused in the land market, favouring a strong decline of farmland close to urban areas. In addition growing links have also developed through the labour market, with an expansion of pluriactivity. But these links have not stimulated the maintenance of agricultural use of the land in nearby rural areas, given the difficulties in reconciling pluriactivity with livestock production, especially with dairying. Finally, proximity to urban markets has only had some carryover effects on agricultural production in certain specialized and intensive crops, particularly horticultural glasshouse, with little impact on land occupation.

4. *The small size and poor access of many plots, which hinder the mechanization of work and favours abandonment.* This problem has been corrected only in a small part of the territory, given the slowness of land consolidation policy implemented from the 1950s onwards.
5. *The absence of policies to stimulate the mobility of the land and especially to promote renting, through adequate regulation and the establishment of economic incentives and/or taxes on abandoned farmland.* The only action in this direction was the creation in late 2007 of the Land Bank of Galicia, an interesting instrument that so far has had very limited effect, due inter alia to the non-implementation of one of its central elements: the penalties for owners of abandoned lands.

Dynamics of Agricultural Output and Production Systems

Given these background structural constraints, the evolution of agricultural land uses has strong links with the dynamics of the main productions and trends of production systems. In the decades prior to European integration, there is the breakdown of traditional agricultural system that used an integrated method of cultivating and grazing more than half of the territory. The highlight was the rapid abandonment of the agro-livestock use of rough land, without traditional uses being replaced by a new model of livestock use of these areas, as the new technology available would have allowed (Bouhier, 1979; Sineiro & Díaz, 1984).

This was accompanied by significant changes in production. Livestock specialization led to a strong fall in the area of cropland, especially that devoted to cereals. The increasing specialization in dairy farming and the growth of landless productions of pigs and poultry came at the expense of a decline in beef cows and sheep. The progressive abandonment of rough land caused the withdrawal of the farms into the croplands and grasslands, which remained almost unchanged at 20% of the territory. This limitation on available land, especially in the dairy sector, led to intensification as the main way to increase the economic size of the farms.

These trends have continued subsequent to integration into the European Union (in 1986), but with some new features. Adaptation to the European framework and the successive CAP reforms contributed to accelerate the rhythm of disappearance of farms, especially in the dairy sector by the quota policy (López Iglesias, 2006). The application of quotas did not prevent production from continuing to undergo a notable increase. However, its implementation accelerated the restructuring of the sector, favoured by a 'free regionalized market' of quotas which was in force in Spain for most of the period, and was complemented by abandonment of production and early-retirement plans. That intense restructuring continued to be accompanied by a growing intensification, a phenomenon stimulated by the favourable performance until recent years of the price ratio of milk to animal feed. Finally, we note the increasing concentration of milk production in the zones with better conditions, while at the same time the process initiated previously of virtual disappearance of dairying in areas of medium and high mountain was culminating (Sineiro et al., 2006).

Together with the consolidation of dairy specialization, the dynamics of major productions in these decades show two relevant developments: a recovery in beef cattle, which take refuge in many farms leaving the dairy sector, and a greater dynamism of certain specialized and intensive crops (wine, horticulture). Data from the Agricultural Census 2009 allowed us to estimate the actual share of the main types of farming both in terms of agricultural output and in the UAA. Globally, livestock farms occupy two thirds of the UAA and account for 80% of output. We should note here three segments: dairy farms, which manage one third of UAA and generate 44% of the production value; those specialized in beef cattle, with a similar share of the occupation of the territory (28% of UAA) but, given their extensive nature, generating only 10% of the output; and pig and poultry, where the reverse situation applies (one fourth of the output but only 1% of the UAA). Farms specialized in crop production have a very limited share of the management of the territory, occupying 6% of UAA (specifically viticulture and horticulture account for only 2%). Although the role of unspecialized farms is much more important, as they manage almost 30% of croplands and pastures, they only contribute 12% of the output.

At the global level, the core dynamics of Galician agriculture continues to be marked in recent decades by the features of the 'productivist model': increasing farm size, specialization and intensification. This latter process is closely linked to the lack of land mobility and abandonment. Of the other possible ways for the maintenance of farms, and subsequent land use, crop diversification has been severely limited by the small size of farms and its high fragmentation into small plots. With respect to differentiation by way of quality, this strategy has been consolidated in the wine industry with a remarkable development of the five Protected Designations of Origin existing in Galicia. While this production is very important in some counties, where it allowed slow land abandonment, it has a limited weight as a whole in the occupation of territory. More important from this point of view has been the consolidation of the Protected Geographical Indication 'Galician Veal' in beef. The valorization of this production through this official quality scheme has contributed to the maintenance of farms, and thus slowed the abandonment of land that would have occurred in its absence, in certain mid-mountain areas. However, this effect has been limited by the fact that a significant part of the production comes from feedlots. Finally, it should be noted that this strategy of obtaining differentiated quality products has very low impact in the dairy sector, as illustrated by one fact: the four cheeses with Protected Designations of Origin existing in the region represented only 2.2% of milk production.

THE CONSEQUENCES OF FARMLAND ABANDONMENT: PRODUCTIVE AND ENVIRONMENTAL IMPACTS

Farmland abandonment has important effects in Galicia, both for agriculture and for the dynamics of rural areas. We will focus on two fields: (1) the impact on agriculture production and (2) the environmental consequences.

Productive Impacts: Limitation of Farm Size and Intensification

From the point of view of the productive dynamics of agriculture and the economic performance of farms, the fact that the UAA was restricted to a small part of the territory, and the parallel process of farmland abandonment due to the factors mentioned above, had two consequences:

1. It contributed to keep a small UAA per farm, limiting the increase in labour productivity and in economic output.
2. It caused the production growth to be achieved largely by means of an increasing intensification. Given the difficulties in expanding the agricultural area, most dynamic farms had to orient their strategy towards increased output per unit area. This resulted in an increase in the stocking capacity of farms, especially dairy, associated with a strong dependence on purchases of animal feed (Sineiro et al., 2006).

The latter has important disadvantages (López Iglesias, 2005; Sineiro, Santiso, Calcedo, & Lorenzana, 2010):

- At the individual level it raises the costs of farms, reducing net income. It also makes them more vulnerable to fluctuations in international markets. This has been highlighted in recent years, with the sharp rise and volatility in the prices of raw materials for animal feed that characterizes global markets since 2007.
- From the perspective of the regional economy, livestock intensification caused a growing dependence on imports of animal feed, which is a large factor in the deficit in the agrifood trade balance of Galicia. At the same time, hundreds of thousands of hectares of land with high agricultural potential are currently abandoned and thus wasted.

The two effects are illustrated by the statistics. As it is well known, the economic size of farms, the standard gross margin (SGM) per farm, depends on two elements: SGM per hectare of UAA, and UAA per farm. This fulfils the following equation: SGM per farm = SGM per hectare of UAA × UAA per farm. Comparing these indices in Galicia with the EU-15, using data from the Farm Structure Survey 2007, shows that the SGM per farm is limited in Galicia to 42.7% of the EU average; this is due entirely to a low UAA per farm; instead, the SGM per hectare of UAA reaches a value higher than the European average (110.2%).

These trends also tended to be accentuated in recent decades, which registered a limited convergence with the EU in the UAA per farm, despite the much higher intensity in the disappearance of farms in Galicia, while the growth of output per hectare far exceeded the European average.

Environmental Problems

In environmental terms, the evolution of farmland use in recent decades causes major problems resulting from the coexistence of two opposing phenomena:

- Increasing intensification of production in a limited part of the territory, particularly in areas of dairy specialization, which causes risk of contamination of water and soil.
- Abandonment or semi-abandonment of large areas, especially in certain counties. This, among other things, has contributed to increasing the risk of forest fires.

Starting with the latter, the accumulation of plant material, of biomass, derived from abandoned land makes them more vulnerable to fires. This is a problem with a strong impact on Galicia in recent decades: in the last two decades about 543,000 hectares, 18% of the territory, have burned, two thirds of which corresponds to scrub lands (the areas that can be identified to a greater degree with abandoned land) (Sineiro García, 2006).

The analysis of municipal data shows that the counties where a greater percentage of the area was affected by fires correspond broadly with those that suffered a major contraction of the UAA in recent decades and that currently register a smaller proportion of UAA over the total area. More precisely, the fires are mainly concentrated on two types of zones: (1) the coastal areas of the western third, corresponding to urban or peri-urban municipalities with high population densities and a strong predominance of wooded land and (2) the rural and mountain counties characterized by a sharp demographic decline and an increased presence of scrub. At the other extreme, the municipalities less affected by fires coincide fairly roughly with the main livestock production counties of the northern half in which there is a greater agricultural land use (Sineiro García, 2006).

The other side of the coin is the high livestock intensification found in the most dynamic farming municipalities. The limited UAA that even these municipalities have, linked to the abandonment of farmland which is also occurring in them (Corbelle, Díaz Manso et al., 2011), causes a high density of animals per hectare of UAA, which is still higher per forage area. Specifically, 51% of cattle is now concentrated in 22% of the territory of Galicia, so that the stocking density per hectare of UAA in these municipalities exceeds 1.5 livestock units (LU), being above 2.2 LU per hectare of forage area. That intensification causes significant environmental problems, mainly related to waste management and the risks of soil and water pollution (Pérez Fra, García Arias, & Docío Rodríguez, 2006).

CONCLUSIONS

In this chapter we have discussed an approach to the farmland abandonment in Galicia, by putting this in relation to the process of structural adjustment and changes in land use recorded in the last 50 years.

In the middle of the twentieth century a traditional agricultural system continued in force at this region defined by a high fragmentation of property, small size of farms and a production model characterized by labour intensity and limited recourse to industrial inputs. That system made possible a high degree of utilization of the territory.

This agricultural system has changed substantially over the past 50 years. In these decades, especially since European integration in 1986, we observed a late but very intense reduction in labour, which was accompanied by drastic production and technological changes. These led to the configuration of a sector specializing in livestock, especially dairy cattle, and highly dependent on inputs of industrial origin. These changes were associated with dramatic changes in land use, so that the space devoted to crops and pasture was constrained to a small portion of the territory, while at the same time the previous agro-livestock uses of the rough land disappeared.

The result was a sharp expansion of abandoned land that is currently covered by scrub, occupying at least 20% of the regional area, to which must be added the semi-abandonment situation affecting much of the woodland and extensive grazing. The abandonment of agricultural land is at its most intense in urban municipalities, presenting also very relevant magnitudes in the mountain areas. But this is a general phenomenon, found in a scattered manner even in those counties with the highest livestock intensification.

The drivers of this high farmland abandonment in Galicia are complex and are heavily influenced by local circumstances, which explain the geographic variability of the phenomenon. It is necessary to emphasize two of them: the ownership structure, which is characterized by a sharp subdivision, and the poor performance of mobility mechanisms of land (land market and renting), highly influenced by the concurrence of urban and forest uses and the absence of land planning policies. To these must be added the problem of communal land, occupying about 25% of the territory. Furthermore, the abandonment has been favoured by the 'productivist' orientation that continues to lead most of the farms' strategy in recent decades, particularly the most dynamic, whereas other possible pathways for the maintenance of farms, which could help improve agricultural occupation of the territory, have little presence: crop diversification and especially differentiation via quality. This latter pathway has some relevance only in the wine and beef, while it is almost completely absent in the main agricultural production of the region, the dairy sector.

Farmland abandonment has caused significant negative effects. From a production point of view it has limited the economic size of farms and fostered a growing intensification, which for farming, especially dairy, has caused a high dependence on feed purchases. With regard to environmental problems the most serious has been the increased risk of forest fires.

These negative impacts become more relevant today because of the two factors: (1) the sharp rise and volatility in prices of raw materials for animal feed and (2) the evolution towards a uniform payment per hectare of UAA that is taking shape in the Common Agricultural Policy (CAP) for the period 2014–2020. The first factor makes intensive livestock farming more vulnerable. The second may lead to a reduction of the CAP direct payments in Galicia unless a significant expansion of the UAA is achieved, since those payments, despite having a low amount per farm, reach a level per hectare clearly above the Spanish average.

In a global perspective, our analysis shows the paradoxical case of a region in which the land has historically been, and remains today, a scarce production factor in the agriculture, but at the same time it is characterized in recent decades by a very poor use of this resource, which has its clearest manifestation in the abandonment of farmland. Beyond the natural and economic constraints, this illustrates the weight of institutional factors, in our case those related to ownership structure and regulation of land mobility and land use.

NOTES

1. The population employed in agriculture fell by more than 90% in the period 1950–2010, from 70% of total employment in 1950 to rates of around 6% at present.
2. According to these criteria, it included only farms with a utilized agricultural area (UAA) over 1 hectare and those not reaching this threshold but have a minimum area on horticulture or a certain livestock and economic dimension.
3. No more than 60,000 hectares, compared to 940,000 hectares accounted by the agricultural census of 1962.
4. Those with a population density greater than 150 inhabitants per square kilometres or with a population increase of more than 20% in the last three decades are classified as urban. Rural municipalities are divided into two groups: those which lost more than 40% of the population in these decades (rural in severe recession) and the remainder (intermediate rural).

This is a typology developed by the authors, although it has some similarities to the official classification established by the Spanish Ministry of Agriculture in Sustainable Rural Development Program 2010–2014.

REFERENCES

Baldock, D., Beaufoy, G., Brouwer, F., & Godeschalk, F. (1996). *Farming at the margins: Abandonment or redeployment of agricultural land in Europe*. London/The Hague: Institute for European Environmental Policy.

Baudry, J., & Asselin, A. (1991). Effects of low grazing pressure on some ecological patterns in Normandy, France. *Options Méditerranéennes - Série Séminaires, 15*, 103–109.

Bouhier, A. (1979). *La Galice. Essai géographique d'analyse et d'interprétation d'un vieux complexe agraire*. Poitiers, France: University of Poitiers.

Bouhier, A. (1984). Las formas tradicionales de utilización del monte. Su evolución reciente. Las perspectivas de porvenir. In Seminario de Estudos Galegos, *Os usos do monte en Galicia. Cuadernos da Área de Ciencias Agrarias 5* (pp. 11–28). A Coruña, Spain: Edicións do Castro.

Corbelle, E., & Crecente, R. (2009). Evolución histórica de la Superficie Agrícola Utilizada en Galicia (1962–2006). Integración de fuentes estadísticas y cartográficas. *Economía Agraria y Recursos Naturales, 9*(2), 183–192.

Corbelle, E., Crecente, R., & Santé, I. (2011). Multi-scale assessment and spatial modelling of agricultural land abandonment in a European peripheral region: Galicia (Spain) 1956–2004. *Land Use Policy, 29*(3), 493–501.

Corbelle, E., Díaz Manso, M., Crecente, R., & Martínez Rivas, E. (Eds.). (2011). *Mercado e mobilidade de terras en Galicia. Modelos de oferta e demanda a escala parroquial.* Santiago de Compostela, Spain: Service of Publications and Scientific Exchange of the University of Santiago de Compostela.

Díaz-Fierros, F., & Gil Sotres, F. (1982). Evaluación da capacidade produtiva das terras de Galicia. *Revista galega de estudios agrarios, 7–8*, 149–172.

Fernández Leiceaga, X., & López Iglesias, E. (2000). *Estrutura económica de Galiza*. Santiago de Compostela, Spain: Edicións Laiovento.

Fernández Leiceaga, X., López Iglesias, E., Jordán, M., Besteiro, B., Viso, P., Balboa, X., … Soto, D. (2006). *Os montes veciñais en man común: o patrimonio silente. Natureza, economía, identidade e democracia na Galicia rural.* Vigo: Edicións Xerais de Galicia.

Keenleyside, C., & Tucker, G. M. (2010). *Farmland abandonment in the EU: An assessment of trends and prospects.* Report prepared for WWF. Institute for European Environmental Policy, London.

López Iglesias, E. (1996). *Movilidad de la tierra y dinámica de las estructuras agrarias en Galicia.* Madrid: Ministerio de Agricultura, Pesca y Alimentación. Secretaría General Técnica (Spanish Ministry of Agriculture).

López Iglesias, E. (2005). O papel actual da concentración parcelária na política de estruturas agrárias e desenvolvimento rural: Reflexións a partir da experiência da Galiza. In F. Oliveira Baptista (Coord.), *Terra e tecnología. Século e meio de debates e de políticas de emparcelamento* (pp. 85–114). Oeiras, Portugal: Celta Editora.

López Iglesias, E. (2006). El proceso de ajuste estructural en la agricultura española: caracterización general de las tendencias en las dos últimas décadas. In E. Arnalte Alegre (Coord.), *Políticas agrarias y ajuste estructural en la agricultura española* (pp. 55–89). Madrid: Ministerio de Agricultura, Pesca y Alimentación. Secretaría General Técnica (Spanish Ministry of Agriculture).

Mac Donald, D., Crabtree, J. R., Wiesinger, G., Dax, T., Stamou, N., Fleury, P., … Gibon, A. (2000). Agricultural abandonment in mountain areas of Europe: Environmental consequences and policy response. *Journal of Environmental Management, 59*, 47–69.

Moravec, J., & Zemeckis, R. (2007). *Cross compliance and land abandonment.* Deliverable D17 of the CC Network Project, SSPE-CT-2005-022727.

Observatorio de la Sostenibilidad en España (OSE). (2011). *Sostenibilidad en España 2011.* Madrid: OSE.

Pérez Fra, M., García Arias, A., & Docío Rodríguez, F. (2006). Efectos territoriales de la reestructuración de la ganadería bovina de la cornisa cantábrica. In E. Arnalte (Coord.), *Políticas agrarias y ajuste estructural en la agricultura española* (pp. 329–350). Madrid: Ministerio de Agricultura, Pesca y Alimentación. Secretaría General Técnica (Spanish Ministry of Agriculture).

Pinto Correia, T., & Breman, B. (2008). Understanding marginalisation in the periphery of Europe: a multidimensional process. In F. Brouwer, T. van Rheenen, S. Dhillion & A. Elgersma (Eds.), *Sustainable land management. Strategies to cope with the marginalisation of agriculture*. Cheltenham: Edward Elgar.

Pointereau, P., & Coulon, F. (2009). Abandon et artificialisation des terres agricoles. *Courrier de l'environnement de l'INRA*, *57*, 109–120.

Pointereau, P., Coulon, F., Girard, P., Lambotte, M., Stuczynski, T., Sánchez Ortega, V., & Del Rio, A. (2008). *Analysis of farmland abandonment and the extent and location of agricultural areas that are actually abandoned or are in risk to be abandoned.* Institute for Environment and Sustainability, Joint Research Centre, European Commission. Luxembourg: Office for Official Publications of the European Communities.

Rey Benayas, J. M., Martins, A., Nicolau, J. M., & Schulz, J. J. (2007). Abandonment of agricultural land: An overview of drivers and consequences. *CAB Reviews: Perspectives in Agriculture, Veterinary Science, Nutrition and Natural Resources*, *2*(57), 1–14.

Sineiro, F. (1983). Consideracións sobre os aproveitamentos da terra en Galicia e a súa evolución no período 1930–1980, con especial referencia ao uso das terras a monte. *Revista galega de estudios agrarios*, *9*, 11–34.

Sineiro, F., & Díaz, N. (1984). A produción gandeira nas terras a monte: análisis da evolución do sistema produtivo no período 1900–1980; as posibilidades de mellora e sementeira de pastos. In Seminario de Estudos Galegos, *Os usos do monte en Galicia. Cuadernos da Área de Ciencias Agrarias 5* (pp. 197–245). A Coruña, Spain: Edicións do Castro.

Sineiro García, F. (2006). As causas estructurais dos incendios forestais en Galicia. In Díaz-F. Fierros & P. Baamonde (Coord.), *Os incendios forestais en Galicia* (pp. 77–99). Santiago de Compostela: Consello da Cultura Galega.

Sineiro, F., López, E., Lorenzana, R., & Valdés, B. (2006). El proceso de ajuste en la ganadería bovina de la Cornisa Cantábrica. In E. Arnalte (Coord.), *Políticas agrarias y ajuste estructural en la agricultura española* (pp. 261–289). Madrid: Ministerio de Agricultura, Pesca y Alimentación. Secretaría General Técnica (Spanish Ministry of Agriculture).

Sineiro, F., Santiso, J., Calcedo, V., & Lorenzana, R. (2010). *El sector lácteo: Escenarios de evolución*. Córdoba: COVAP.

CHAPTER 6

REPRODUCING PRODUCTIVISM IN SPANISH AGRICULTURAL SYSTEMS

Olga M. Moreno-Pérez

ABSTRACT

The purpose of this chapter is to shed light on some outstanding patterns of change observed in Spanish agriculture over the last decades, and to discuss the ways in which the productivist rationale is reproduced in them. We start by providing an overall picture of the structural transformations of agriculture revealed by the national statistics, calling attention to the increasing importance of a "hard core" of farms progressing under a rationale of growth and modernization. Later, drawing on the literature, we comment some meaningful trajectories of farm change observed in three selected Mediterranean farming systems, namely horticulture (as an example of a long-consolidated intensive agriculture), vine growing (an orientation which has undergone stunning changes in Spain in recent times), and small ruminant production (an extensive farming system with a high conservation value). The three farming systems are advancing, to a greater or a lesser degree, along intensification, concentration, and specialization pathways. However, the introduction of new elements in the most expansionary farm strategies (such as the participation in quality schemes) will provide interesting elements of discussion of the adaptable

Agriculture in Mediterranean Europe: Between Old and New Paradigms
Research in Rural Sociology and Development, Volume 19, 121–147

ISSN: 1057-1922/doi:10.1108/S1057-1922(2013)0000019008

nature of productivism and its capacity to accommodate to external opportunities and constrains.

Keywords: Neoproductivism; Mediterranean farming systems; Farm change; Agricultural intensification; Farm concentration

INTRODUCTION

The academic debate on agricultural change in the developed world in the recent times has much revolved around the shift from a "productivist" to a "postproductivist" regime. The concept of postproductivism broke into the British agricultural and rural studies by mid-1980s (Marsden & Symes, 1987), and reached a peak in the literature over the following decade – see, for example, Ward (1993), Lowe, Murdoch, Masden, Munton, and Flynn (1993) and Shucksmith (1993). In essence, the "postproductivist transition" in agriculture encompassed a number of changes that were interpreted as a reaction of farmers to two facts. The first one was the progressive decline in the farm economic margins,[1] which, supposedly, could no longer be recouped by way of scale-based growth strategies. The second was the alteration of the prevailing model of public support of agriculture. Thus, the introduction of political tools by late 1980s aimed at containing production in Europe, in parallel with the growing emphasis in nonproductive (e.g., environmental) concerns, have been interpreted not only as driving factors of postproductivist practices, but also as a central dimension of post-productivism itself (Mather, Hill, & Nijnik, 2006).

A wide array of trajectories of agricultural change has been analyzed under the conceptual "umbrella" of postproductivism; in fact, the vagueness of the term has been widely commented in the literature (Wilson, 2007). In a much-quoted work, Ilbery and Bowler (1998) associated the postproductivist transition of agriculture with three development pathways: extensification (i.e., drop in inputs and production per unit of surface), dispersion of agricultural holdings, and diversification (i.e., implementation of activities in the farm different from conventional, mass agricultural production).[2] However, many scholars shed light shortly after on the poor empirical evidence on the progress of these trends in the field, even in the British context (Evans, Morris, & Winter, 2002; Walford, 2003).

Strong criticism was also launched at the postulations of the post-productivist transition from a theoretical viewpoint. Thus, the notion of "replacement" of a regime by another in a lineal fashion was considered "simplistic" and "naive" in subsequent workings (Ward, Jackson, Russell, & Wilkinson, 2008, p. 128). Commentators also objected that there was something of a dualistic conception of agricultural change in opposing the "postproductivist" behavior of farmers to the "productivist" one. In contrast, the idea that has prevailed in the literature is that of the *coexistence*, in the same territory and sometimes in the same holding, of characteristics that could fit in with these two ideal types (Evans et al., 2002; Wilson, 2007). Going further, some elements that had been uncritically branded as "postproductivist" before (e.g., organic farming) can be reinterpreted as adaptation strategies of productivism to external constrains or opportunities – and therefore they somehow underpin the accumulation regime (see Chapter 2 for a broader discussion of these issues).

Despite the drawbacks of postproductivism as a new paradigm of farm change, it still constitutes a useful analytical framework to formalize the investigations on this matter, and as such is found in some recent articles (Calleja, Ilbery, & Mills, 2012). However, these deliberations have been mainly led by the British academia, and had much less impact in the scientific sphere of the Mediterranean countries of Europe. In the case of Spain, the postproductivism entered in the specialized literature – concretely in the realm of rural geographers – by the second half of the 1990s.[3] It was welcomed as a theoretical reference to frame a heterogeneous array of studies on the changes underway in different agricultures and rural areas of Spain (Menor-Toribio, 2000). However, when going into the overall figures of Spanish agriculture, some analysts have shown scepticism about the progression of the "postproductivist" trends in this country (Molinero, 2006; Ortiz & Moreno, 2007).

The purpose of this chapter is to shed light on some outstanding patterns of change observed in Spanish agriculture over the last decades, and to discuss to what extent, and in which ways, the productivist rationale is reproduced in this dynamics. Emphasis will be made in the prevalence of processes such as the farm concentration, specialization, and intensification in this country, as well as on some distinctive elements that these trends show in different contexts. We will start by providing an overall picture of the changes in farm structure revealed by the national statistical sources, drawing attention on some dynamics underlying the agricultural adjustment. Later, we will focus on some outstanding trajectories of change that are being observed in three specific farming systems. The first one is horticulture in SE Spain, as an illustrative example of a long-consolidated intensive agriculture which is in continuous progress to adapt to external stimuli. The second one is vineyard farming – an orientation which has undergone stunning changes both in terms of farm structure and practices in the recent times. Finally, attention will be paid to small ruminant (i.e., sheep and goat) production, a traditional, extensive farming system with a high conservation value (Caraveli, 2000; MacDonald et al., 2000) which is also advancing, to a certain degree, along "productivist" development pathways.

STRUCTURAL CHANGE IN SPANISH AGRICULTURE: AN OUTLINE FROM NATIONAL STATISTICS

The statistical information collected by the Farm Structure Surveys[4] performed by the National Institute of Statistics of Spain (INE) reveals that agriculture in this country has been immerse in a strong process of structural adjustment in the late XX – early XXI century. As the figures of the Table 1 show, the number of farms decreased in 42% between 1987 and 2007; given that the overall Utilized Agricultural Area (UAA) experienced much slighter changes in this period, the average surface of the remaining farms underwent a spectacular increase (75%). These data contrast with the slow changes registered in the 1970s and, particularly, the 1980s – when scientific and political concerns revolved around the "rigidity" of Spanish farm structure and its negative implications for agricultural modernization (Sumpsi, 1994).

Spanish scholars tackled the causes of this structural dynamics turn in a number of studies published over the past decade. Thus, Sineiro, López, Lorenzana, and Valdés (2004), Moreno and Ortiz (2008) and the works

Table 1. Basic Data on Farm Structure in Spain.

	1987	1993	1995	1997	2003	2005	2007
No. holdings (000)	1,791.6	1,383.9	1,277.6	1,208.3	1,128.0	1,069.7	1,036.2
UAA (000 has)	24,796.5	24,713.7	25,230.3	25,630.1	25,175.3	24,855.1	24,892.5
UAA/holding (has)	13.8	18.0	19.9	21.4	22.5	23.4	24.2
SGM/holding (ESU)*	5.3	7.0	8.7	10.7	15.3	18.7	20.7

Source: Own elaboration from Farm Structure Surveys (INE).

*Standard Gross Margin, a measure of the economic size of an agricultural holding used by EUROSTAT. It is expressed in European Size Units (1 ESU corresponds to 1200 euros of SGM).

contained in Arnalte (2006a) illuminated the role of technology, demographic factors and agricultural policies as triggers of farm concentration during the 1990s. The mechanisms behind the process showed notable regional differences; however, it could be held that the structural change in this period, to a large extent, fitted in with the 'classic' model of agricultural adjustment – that is, territorial expansion combined with mechanization, in the search of economies of scale (see Arnalte, 2006b). However, a different pattern of structural change can be discerned if we go into the recent data provided by the Farm Structure Surveys. Indeed, if the annual variation rates of the farm structural data of the long period 1987–2007 are split in two decades, it becomes evident that the process slowed down in the latter (Table 2). Understanding the nature of the recent transformations requires a closer examination of the structural data in 1997 and 2007.

As happens in other developed countries, the evolution portrayed by the aggregate farm numbers is not neutral in terms of the profile of the holdings prevailing from the restructuring process. Contrarily, the disappearance of farms is concentrated in the smallest ones, whereas the segment of large holdings swells – the break-point size between falling and increasing farm numbers differing among countries (Hill, 2006). Following this pattern, the evolution of Spanish agriculture is configuring a set of farms in continuous expansion that leads the most important transformations of this sector and gains importance both in relative and absolute terms – what has been called the "hard core" of Spanish agriculture (Arnalte, Ortiz, & Moreno, 2008). Fig. 1 helps to illustrate this. It depicts the evolution of the number of farms belonging to different strata of economic dimension between 1997 and 2007, and shows the dramatic fall of the smallest holdings and the swelling of the upper strata, more noticeable for those exceeding 40 ESU.[5] Hence, this is the operative threshold that we have chosen to identify the "hard core" of Spanish agriculture. Drawing on Arnalte et al. (2008) elaborations, we will deepen in the characteristics and the recent evolution of this segment of

Table 2. Annual Variation Rates of Farm Structural Data in Spain (%).

	Annual Variation Rate 1987–1997	Annual Variation Rate 1997–2007
No. holdings	−3.9	−1.5
Total UAA	0.3	−0.3
UAA/Holding	4.4	1.3
SGM/holding	7.3	6.9

Source: Own elaboration from Farm Structure Surveys (INE).

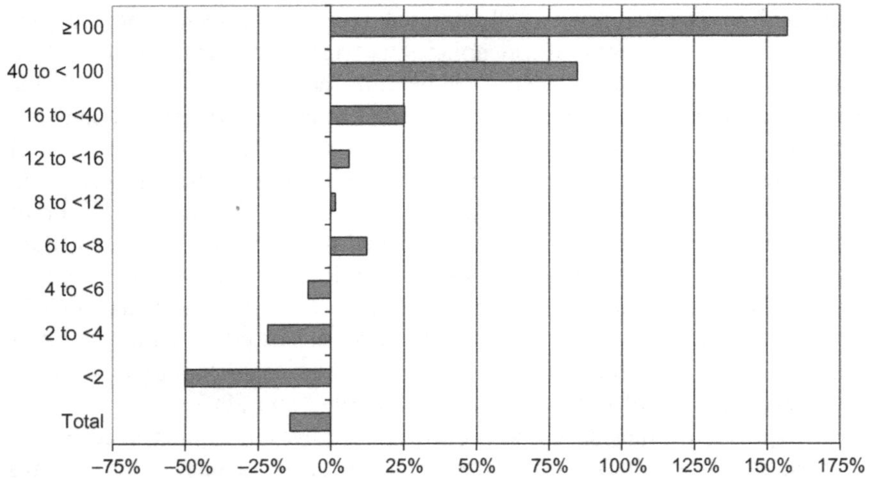

Fig. 1. Variation in the Number of Holdings between 1997 and 2007 per Economic Size Strata (ESU). *Source*: Own elaboration from Farm Structure Surveys (INE).

Table 3. Relative Weight of Holdings over 40 ESU.

	1997		2007	
	Number	% Out of All Holdings	Number	% Out of All Holdings
No. holdings	57,973	5%	116,826	11%
UAA (000 has)	9,679.4	38%	12,173.7	49%
SGM (000 ESU)	5,535.1	43%	13,618.1	63%
Farm labour (000 AWU*)	243.2	22%	348.3	36%

*Annual Work Units.

farms to further understand the recent scheme of agricultural transformations in Spain.

As Table 3 shows, 60,000 holdings joined to the "hard core" between 1997 and 2007, which means a two-fold increase in the number or productive units belonging to this group. The increasing prominence of this segment of farms can be further appreciated by observing the growing proportion of the basic magnitudes of Spanish agriculture that they jointly manage. In fact, they account for a raising proportion of the overall UAA, farm employment and, more remarkably, agricultural output.

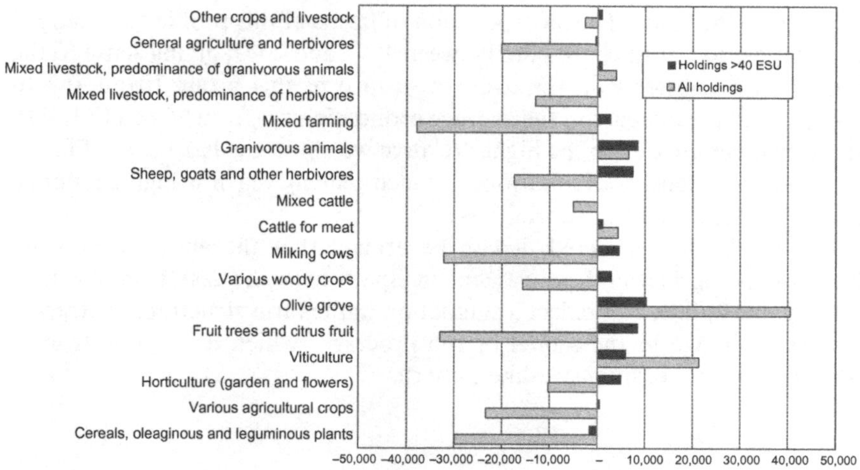

Fig. 2. Variation in the Number of Farms per Technical-Economic Orientation (1997–2007). *Source*: Own elaboration from Farm Structure Surveys (INE).

A more complete picture can be obtained if the variation in the number of farms – taken as a whole, and also segregating the "hard core" ones – is broken down by agricultural orientations (Fig. 2). A generalized drop in the number of farms is observed, with important exceptions in those specialized in viticulture, olive grove production and, to a lesser extent, certain livestock orientations.[6] In turn, the number of farms beyond 40 ESU increased in nearly all cases, including, meaningfully, a few orientations in which the overall number of effectives declined – notably horticulture, fruit trees and citrus fruit and "sheep, goats and other herbivorous." Moreover, in most of the orientations the increase in the "hard core" holdings was much more noticeable in relative than in absolute terms. The most outstanding cases are the farms specialized in olive groves (those beyond 40 ESU registered a 257% increase), sheep and goats (258%) and vine-oriented farms (for which the "hard core" registered a spectacular 789% increase).

It is important to remark that the bulk of farms incorporating to this stratum were specialized in intensive orientations (horticulture, fruit and citrus fruit, specialized livestock) or in traditional permanent crops (olive trees, vineyards) which have strongly intensified in the recent past, as literature has stated – see, for example, Cots-Folch, Martínez-Casasnovas, and Ramos (2009) and Gallardo and Ceña (2006); the next sections will go further into this issue. The incorporation of farms of this profile has changed the composition of the hard core between 1997 and 2007, giving sense to the drastic drop registered in the average surface of this group (from 167 to 104 has) and the increase in the average economic size (from 96 to 117 ESU). These changes are due to the higher relative weight in the hard core of farms belonging to intensive orientations – which usually reach a high economic dimension on a small territorial base.

All the above considered, it can be argued that the apparent slowing down of the agricultural adjustment in Spain over the 2000s (going back to the Table 2) does not reflect a stagnation in the farm structure; contrarily, it hides a change in the model of the process – which have gone from a extensive to a more intensive-like pattern.

INTENSIVE HORTICULTURE IN SOUTHERN SPAIN: THE PATHWAYS OF PROGRESS OF THE PRODUCTIVIST REGIME

Intensive horticultural production in Spain is mainly concentrated in the Southeastern coast of the country, a natural region embracing the provinces of Almería, Murcia, and the South of Alicante.[7] This is precisely the most arid region in Europe (the precipitations are fairly irregular and barely reach 300 mm per year, even less in certain zones), a serious natural limitation that has historically conditioned the agricultural and economic development in this area (Sánchez-Picón, Aznar-Sánchez, & García-Latorre, 2011). However, the introduction in the 1960s of artificial soil systems (the so-called *enarenado* technique) and plastic greenhouses in Almería allowed farmers to make a good use of notable comparative advantages for off-season vegetable production: a privileged microclimate – with an extremely high luminosity, warm temperatures in winter and very low humidity – abundant workforce and underground water resources, and low-priced land. Scholars have largely addressed the genesis of these innovations and how they detonated what is considered one of the most stunning examples of agricultural development in Europe – with profound spatial, environmental, and socioeconomic implications (see, e.g., Aznar-Sánchez, Galdeano-Gómez, & Pérez-Mesa, 2011). In these studies, commentators underline the key role played by the State in the early stages of the process, by providing with hydraulic and energy infrastructures, coupled with strong technical and financial support.

The mentioned technological and institutional changes enabled a great number of small farms to reach substantial gains in both the yields and the quality of vegetable production, based on the mobilization of family labour and an affordable setting-up investment. Following the success of the pioneers, greenhouses greatly disseminated in the "zero zone" of these changes – the "Poniente" county of Almería[8] – and also spread northwards along the SE coast of Spain, reaching the county of "Campo de Cartagena" (in the boundary between Murcia and Alicante) well into the 1970s. Other occurrences took place after this early phase that fuelled further waves of agricultural development in these territories. First, the completion in 1979 of a water transfer between the basins of the rivers Tajo and Segura alleviated the water shortage that had resulted from the spectacular farm growth of the previous decades – although it triggered further farm expansion and subsequently, induced more water demand. Another milestone in the evolution of this area was the Accession to the ECC – more concretely the end of the transitional period in 1993 – which extraordinarily boosted vegetable exportations.

The farm growth over the last decades has required a great amount of labour, since many farm operations cannot be mechanized. Farm expansion hence was fueled with the entrance in this region of immigrant workforce over the 1980s – initially from Maghrebi countries and later from Eastern Europe and Latin America – which were massively occupied in these farms (Pedreño, 1999). Indeed, before the immigrants' arrival, the limitations of the family labour and the scarcity of local hired workers were a major constrain for agricultural development. Aznar-Sánchez et al. (2011) relate how the development of auxiliary activities brought about more opportunities of employment in these areas – and thus more difficulties in finding laborers willing to work in greenhouses. As for the family labour, it gradually decreased due to the disengagement of the family members with greenhouse farming. Apart from the alternatives offered by the local labour markets in this particular case, the sociological grounds of the rupture of the family labour group – well established in the literature for long (see Blanc, 1987) – were fully applicable to this farming system. Nowadays, around two thirds of the labour performed in these farms, which oscillates between 2.4 and 2.6 AWU/greenhouse hectare, is carried out by hired workers (Gómez-Espín, López-Fernández, & Montaner-Salas, 2011; Aznar-Sánchez et al., 2011), and immigrants – mainly from Morocco, Equator, and Romania – remain providing the bulk of this labour.

A more comprehensive treatment of the multifold processes of change experienced by this farming system throughout this long period would largely exceed the scope and purpose of this chapter. However, some patterns of change observed in the recent past are worth commenting by focusing our attention in a particular area of this region. Moreno-Pérez, Arnalte-Alegre, and Ortiz-Miranda (2011) centered in a subzone of "Campo de Cartagena" where a myriad of family-based farms are specialized in vegetable production and, to a lesser extent, citrus fruits and floriculture. Following several in-depth interviews to selected informants, farm-level data were collected by conducting a 135-farmer survey dealing with the family and the farm characteristics at that moment (2009), and with the most important changes undergone by the farm from 1990 to 2009.

Informants revealed how the introduction of the *California* green pepper cultivar in early 1990s, highly appreciated in European markets, provided extraordinary benefits from exportations and boosted a forceful process of farm growth. Smallholders make a most intensive use of their little endowments of land and water, cultivating horticultural crops outdoors and, typically, in greenhouses.[9] As for open-air cultivation, several crop cycles per year are cultivated by combining winter and summer species – lettuce, artichoke, broccoli, celery, melon, watermelon-, or by obtaining two or three harvests of the same crop (in the case of lettuce). Greenhouse cultivation showed a great specialization in green pepper production, which is, in fact, the mainstay of this farming system. The survey information revealed that intensification, materialized in fixed capital investments aimed at enlarging and/or modernizing the greenhouses, had been the predominant farm trajectory in the study period (followed by 81.5% farms). External capital, in the form of bank credits, enabled these strong investments. As for the farmland, the surface expansion was hampered by the high land value in this coastal zone – where agriculture competed with a strong development of residential tourism. Thus, 55.6% the survey farms remained operating in their little original surface, although a notable 39.2% managed to increase their territorial base, either by renting or purchasing land. The remaining 5.2% of the holdings opted by downsizing the farm surface (typically elderly farmers with no successor).

It is worth noting that the most aggressive strategies of growth (i.e., capital intensification combined with land expansion, undertaken by 35.6% farms) were carried out by farms jointly held by several kin households. These multifamily structures were the most appropriate to cope with the complexity of this farming system, which demands a great organizational effort. Thus, several members of the extended family can deal with the multiple technical and managerial aspects of the business – control of the hired labour, administrative paperwork, fertirrigation and other farm operations. Moreover, these holdings had a larger investment capacity, either by pooling their financial resources or by gaining easier access to bank credits. In contrast, in farms run by a nuclear family the growth is relatively constrained by its limited managerial and investment capacity.

Interestingly, the manifest advance towards intensification in this area has been fully compatible with the massive adoption of quality schemes, labels and protocols, most of the times induced by the strict requirements of the large European trade operators. The majority of the farms (81%) joined to the Integrated Production[10] scheme in the second half of the 1990s, and a significant part of them (16%) follow the more restrictive "organic farming" regulations. Other quality labels such a EUROPGAP or ISO 9001/2000 are also widespread in the zone.[11] Cooperatives and other farmers' organizations have played a major role in the diffusion of these labels, by providing farmers with technical advice on the changes in the management and farm practices required to comply with these schemes.

Remarkably, despite the improvement of the quality and health standards and the immense efforts made in improving the efficiency of water management, farmers remain obtaining exceptional yields under a great pressure for natural resources.[12] The demand of water for irrigation resulting from the disordered farm growth has led to the proliferation of a great number of irregular wells; the overexploitation of aquifers has brought about a severe salinization of soils and subsoil water, a critical environmental problem in this region (Sánchez-Picón et al., 2011). The generation of plastic and other residuals (Sevilla, Domene, Uceda, Buendía, & Racero, 2012), coupled with a dramatic landscape deterioration, are other negative externalities of the agricultural development in this region. Arguably, the apparent 'turn' towards more environment or health-sensitive farm practices can be considered, rather than a real progress of a different agricultural regime or mind-set, a convenient strategy to accommodate the productivist rationale of this farming system to new economic stimuli.

SPATIAL AND FARM-LEVEL PATTERNS OF CHANGE IN VINEYARD FARMING SYSTEMS

As discussed above, vine grape production is the agricultural orientation registering the most striking relative increase in the number of "hard core" farms. A better understanding of the complex dynamics underlying this piece of data will be reached by giving a closer look to this farming system, focusing on particular territories.

Cots-Folch et al. (2009) carried out an interesting analysis of the agricultural change in the mountainous county of *Priorat* – a less favoured area of NE Spain with an important forest surface, were wine grape vines were the most important crop (20% of the land surface), followed by nuts and olive trees. A Protected Designation of Origin[13] labels the long tradition of elaboration of high-quality wines in *Priorat*. Elaborating on a land-use cover database of 1986 and 2005, these authors reported a trend of intensification in this area based on mechanized terrace cultivation, which have been replacing the traditional, low-intensity farm practices.

Farm-level information was collected by means of a survey to a representative sample of 50 farms; the trajectories of change being captured and analyzed by means of multivariate techniques. The results show that medium intensification and specialization was chosen by 40% farms, whereas 48% remained diversified and close to traditional management practices. Outstandingly, whereas only 12% farms opted by strong intensification, this group was the most important one in terms of surface (accounting for 42% of the total agricultural land and 61% of the new vine plantations). These farmers, who were fully specialized in vine production and also run their own winery, were identified as newcomers, since they had no land in 1986. As the authors report, a worldwide revaluation of the wines of this county could have attracted these new investors. The mechanization of farm practices required the construction, with retro-excavators and bulldozers, of unstable bench terraces unprotected by stonewalls. The profound alteration of the terrain morphology that these movements entailed had negative effects in the original soils, the landscapes and the natural drainage. Moreover, it is remarkable that this small group gathered 68% the subsidies perceived by all farms. Considering the crescent emphasis placed by the CAP on environmental goals and production contention from the 1980s, it is something of a contradiction that restructuring and conversion subsidies apparently have encouraged all these changes.

Some other interesting workings on the transformations in wine grape vineyards have been carried out by Spanish geographers for other regions. For example, Alonso, Aparicio, and Sánchez-Hernández (2003) and more recently Molinero (2011) focus on different counties with a long tradition of wine grape production within the vast inland region of Castilla y León. The traditional vineyards portrayed by these authors were oriented to the bulk elaboration of wines for mass consumption in the local markets.

Such systems experienced a long crisis that can be roughly dated between mid-1950s and early 1980s, related, first, with the strong rural exodus that took place in the 1960s (following a national trend), and second with a fall in the domestic demand of substandard wines. In this context, extensive areas of traditional vineyards were gradually substituted by arable crops, more profitable and suitable for mechanization. This decline was generalized throughout the region, with the exception of some experiences aimed at increasing the quality of the wines, mainly localized in the banks of the river Duero. These initiatives, which involved the cooperation between farmers and wine producers, laid the foundations of the first quality labels in these territories.

The turning point in this progress took place in the 1980s, when the national and foreign demand of high-quality wines (coinciding with the Accession of Spain to the EEC), grew in parallel to the spread of geographical indications – notably Protected Designations of Origin (PDO). The earliest PDOs in Castilla y León appeared in the counties of *Rueda* and *Ribera del Duero* at the beginning of the decade, and managed to take advantage of the favorable market conditions. Following their example, these schemes spread across the region, to the point that a great part of its vineyard surface was covered by some territory-based regulation by mid-2000s. Taken as a whole, the territories with PDO have developed differently from those that remained out of any initiative of this kind, although they have had a different degree of success.

Indeed, the turn towards differentiated, high-quality wines brought about dramatic changes in the spatial patterns of vineyard production in Castilla y León. On the one hand, the territories constituting a PDO which gained international prestige throughout the 1990s started to expand their vine surface, colonizing fertile meadows formerly given over to arable crops or acquiring planting rights in marginal nearby areas (Alonso et al., 2003). This dynamics was led by the two above-mentioned PDOs located close to the axis of the river Duero, which enjoy certain agroclimatic advantages, and to a lesser extent in others arisen in nearby counties. Some distant territories managed to join this dynamics to a certain degree and, at least, to limit the vineyard decline – for example, the PDO *El Bierzo*, where the limitations imposed by the sloped relief and the fragmented farm structure have been partly alleviated by collective action. Conversely, other zones of the region were unable to emulate this behavior. This is the dominant trend of the most remote and steep territories, where any PDO prospered[14] and other entrepreneurial initiatives were probably hampered by the awkward natural conditions. These zones deepened in their inexorable regression;

furthermore, the reduction of vine surface has been catalyzed by the CAP grubbing-up premiums.

Interestingly, Molinero (2011) states that the most flourishing areas are the ones which have distanced the most from the traditional vineyard systems, not only in the type of wine produced but also in terms of farm structure and practices – as Cots-Folch et al. (2009) found in *Priorat*. In contrast to the fragmented and relatively diverse agricultural landscape prevailing in the past, the counties located in the Duero banks are now pervaded by large plots with ordered lines of vines trained in espalier, ready for mechanization.[15] The planting density has been remarkably increased (reaching more than 3,000 vines/hectare nowadays), and trickle irrigation has been also introduced. Specialization is another common feature of these farms; most of them are given over to vine monoculture. In the zones of red wine such as *Ribera del Duero*, the autochthonous *Tempranillo* variety has massively replaced other varieties which had certain presence in the past; in *Rueda*, success has followed the specialization in the white wine obtained from the traditional variety, *Verdejo*. In these territories, a more expert management has enabled substantial gains in yields without jeopardizing the quality of the grapes.

Obviously, alongside the territorial differentiation we have outlined, farm-level disparities are also observed. When trying to capture this diversity by means of an all-reaching farm typology embracing the farms of the whole region, two broad groups of farms can be firstly distinguished: the large-scaled productive units that integrate both the vine grape growing and the winery, on the one hand, and the holdings only concerned with agricultural production, on the other. Molinero portrays two different profiles of agro-business belonging to the former group. The first one is the so-called *hacienda vitícola*, a medium-sized to large farm (up to 300–400 has) where all the above-mentioned transformations towards vineyard modernization and specialization are fully present. First-class wines are elaborated in their modern wineries – many of which have gained international prestige. Several hired workers settled in nearby villages are permanently occupied in these farms, and the labour force is greatly enlarged with eventual workers from other zones during the grape-harvesting season. Moreover, these holdings may house multipurpose buildings, sometimes with high historical value or an eye-catching architecture, where complementary tourist-oriented activities are developed. Although some of these *haciendas* may have resulted from a sound expansionary strategy of the major territorially rooted wine producers, a great part of them were newly set up after the 1980s by external investors (again, as happened in *El Priorat*). These exogeneous actors[16] frequently acquired their own vine surface in order to control all the productive process – occasionally buying old vineyards to smallowners at high price.

Despite their strong emphasis on quality, the position of the *haciendas* toward PDOs is ambivalent. Many of them are firmly committed to their own brand strategy and keep themselves out of PDOs. Notwithstanding the autonomous behavior of the haciendas, these complex holdings play a leader role as drivers of agricultural change, landscape transformation, technological innovation, and rural dynamism in their areas of influence. As one may expect, they have a stronger presence in the core zone of vine expansion of the region, where they first started to develop, although they have gradually disseminated to other areas. However, the most inaccessible territories of Castilla y León with little vine surface remained largely marginalized from these entrepreneurial initiatives.

Alongside the haciendas, there are numerous "family-based wineries" widespread across the region, mostly run by autochthonous wine producers who consolidated in the last decades. In many cases, these are enterprising farmers who withdrew a cooperative and embarked in their own winery activity – either based on their own vines or buying grapes from other farmers. These holdings typically have 20–40 has of vines (sometimes newly planted in brought land), and family labour plays a more important role in them than in the case of the *haciendas*. However, the most thriving ones can eventually jump to such category.

The former two types of holding are of a great relevance in terms of the vine surface they manage. As Alonso et al. (2003) observe, their strategy of acquiring their own vineyards has greatly contributed to the land consolidation in their areas of influence. Meanwhile, a myriad of farmers that do not transform the grapes in the farm remain agglomerated around wine cooperatives. These organizations play an important role in maintaining great number of smallholdings in some areas with natural and structural limitations, but some of them are also found in the "booming" territories which have successfully joined quality strategies and experienced an intense growth. In these cases, they provide their members with sound support to restructure or enlarge their vineyards. Finally, there is a heterogeneous array of farms which do not belong to any cooperative and try to sell their grapes to private wineries. Although disseminated throughout the region, they are more present in the most inaccessible and marginal (sloped, sandy, graveled) areas. They are typically fragmented in very small plots which, precisely, occupy those lands that in the past were only possibly given over to vine growing. Mainly in charge of elder farmers with scarce perspectives of continuity, they are branded by Molinero as "loser vineyards."

SMALL RUMINANT LIVESTOCK SYSTEMS: BETWEEN INTENSIFICATION AND MARGINALIZATION

Spain occupies a prominent role in small ruminant farming in Europe, as the second country in sheep flock number after United Kingdom and second in goats flock number after Greece (Ministerio de Medio Ambiente y Medio Rural y Marino, MARM, 2009). When analyzing these farming systems in Spain, as happens in other countries, we struggle with the great diversity of productive models existing in the field. Sheep and goats can appear separate or together in the same farm – often mixed with other livestock (e.g., beef cattle) and agricultural activities (pasture, crop growing). Similarly, milk orientation coexists with meat or meat/milk farms, and a spectrum of management practices can be found between the most extensive systems to the most intensive ones (De Rancourt, Fois, Lavín, Tchakérian, & Vallerand, 2006). Within this complex framework, our purpose here is to outline some characteristics and general trends in this farming system in Spain, as well as to discuss some meaningful trajectories observed in specific areas recently investigated in the country.

Broadly speaking, the sheep flock in Spain has traditionally been more meat-oriented, whereas milk orientation prevails for goat flock. Dairy production is more concentrated in some territories – in the South (Andalusia) and the Canary Islands in the case of goats; in the inland plateaus, more in the North, in the case of dairy sheep. In turn, the meat or meat/milk systems are more disseminated all over the country (De Rancourt et al., 2006). Table 4 summarizes the evolution of the aggregate figures of small ruminant farming in Spain. The remarkable increase in the overall goat number until mid-2000s has been related with a rise in the milk demand (Gaspar, Escribano, Mesías, Escribano, & Pulido 2011). However, this flock descended at the end of the series, and this downward evolution was even more noticeable in the case of sheep. According to the literature, this trend continued in subsequent years owing to the 2007–2008 crisis in the input prices (MARM, 2009) and the effect of the decoupling of CAP subsidies[17] (Castel, Mena, Ruiz, Camúñez-Ruiz, & Sánchez-Rodríguez, 2011; Pardos, Maza, Fantova, & Sepúlveda, 2008; Riedel, Casasús, & Bernués, 2007). In addition, the need to comply with strict animal health and food safety

Table 4. Basic Data on Small Ruminant Livestock Farms in Spain (1997–2007).

	1997	2003	2005	2007	Var. 1997–2007 (%)
Sheep					
No. holdings	102,832	91,946	85,247	79,137	−23.0
No. sheep heads (000)	19,669.4	19,970.4	19,660.1	18,758.6	−4.6
Sheep heads/holding	191.3	217.2	230.6	237.0	23.9
Goats					
No. holdings	51,453	40,631	38,648	35,976	−30.1
No. goat heads (000)	2,062.1	2,392.6	2,527.3	2,475.7	20.1
Goat heads/holding	40.1	58.9	65.4	68.8	71.7

Source: Own elaboration from the *Surveys on Farm Structure* (INE).

regulations may also played a role in this decline (Estrada, Nadal, & Iglesias, 2010; De Antonio, 2005). The regression is particularly intense for meat-sheep heads, due to the low price of the lambs, which has been related with a drop in the consumption (Martín-Cerdeño, 2012). As for farm structure, the national statistics register a strong adjustment of goats and sheep farms, which is observable both in terms of a sharp drop in the number of farms and a substantial increase in the average flock size.

Coupled with the process of farm amalgamation, the recent literature reports a decline in the traditional small-ruminant systems in favour the most intensive productive models in Spain. A number of factors have been reported by commentators as drivers of this shift, within which the labour issues play an important role (Castel et al., 2011; Riedel et al., 2007). Indeed, the harsh working conditions in grazing livestock systems result in a lack of generational succession – even in a shortage of shepherds in certain areas. By means of intensification, significant improvements are achieved in terms of workload (e.g., feeding mechanization) and of work schedule along the year (by means of changes in reproduction management). Critically, labour-saving practices also make it possible to increase the flock size without increasing labour requirements and, ultimately, to raise the incomes per unit of labour.

The changes toward intensification in goats are commonly related with a greater specialization in milk production – following the increase in the milk prices from mid-1990s – and, therefore, with the requirements of the cheese industry.[18] Traditional goat farms were oriented to the production of kids for meat in winter, and the milk production was concentrated in spring. However, the manufacturing industry demanded to reduce the milk seasonality, which entailed an alteration in the reproductive practices (Castel et al., 2010; Oregui & Falagán, 2006). The lengthening of the milk production period has profound implications in the rest of the management practices of the farms. As Castel et al. (2011) summarize, *the more intensive the reproduction, the greater the decrease in grazing* (p. 89). Thus, the increase in the nutritional requirements of animals forces pasture-based systems to turn to more intensive ones, where goats are permanently confined and fed with concentrates – or, at very least, to "false grazing systems," where daily grazing is complemented with indoor feeding. The same rationale lies behind the production of off-season lambings in meat sheep systems, which also bring about an increase in housed pregnant and lacting ewes, animal confinement and shortening in the grazing periods (Bernués, Ruiz, Olaizola, Villalba, & Casasús, 2011; Riedel et al., 2007). The decrease in grazing that these changes entail leads, eventually, to pasture abandonment, with adverse environmental effects in the form of loss of biodiversity and increased risk of forest fire (Henle et al., 2008; MacDonald et al., 2000).

Besides, interesting cross-effects in intensification processes arise when these changes in small ruminant farms are related with the reduction of the available pastureland which, in turn, is caused by the pressure of agricultural crops (Castel et al., 2011; Pardos et al., 2008) or the intensification in other livestock farming systems. For example, Oregui and Falagán (2006) explain how sheep grazing in the Basque Country used to be associated to the rental of pastures that the dairy cattle could not utilize in winter (because the terrain was too humid for them). However, the intensification in dairy cattle production resulted in a decline in pastures and subsequently in sheep grazing. Reforestation or competition with other land uses are also reported as causes of the fall in the area of usable pasture (Estrada et al., 2010). Finally, sometimes intensification in small ruminants is aimed at giving over the natural pastures to the other livestock species present in the farm (Castel et al., 2011).

Further elements of analysis have been highlighted by research works focussed on specific areas of Spain. In a traditional goat production system of a mountainous county of western Spain (*Villuercas-Ibores*, province of Cáceres), Gaspar et al. (2011) assessed a farm typology and reported an important movement towards intensification (specialization in milk production, confinement of animals in long periods, use of nonautochthonous breeds) in a significant group of farms (28% of the sample). The modern equipments have been installed in small, owned plots with good road connections to the near urban areas. Interestingly, these farms were not eligible for the local Protected Designation of Origin (PDO) "Ibores Cheese" created in 2005, but the good productivity per goat that they obtained and the availability of milk throughout the year largely compensated the differential with the PDO milk price. These farms coexist with large, extensive farms (25%) following a strategy of livestock diversification (goats were combined with sheep and, occasionally, with other species) and increase in the heard size, in order to make the most efficient use of labour and maximize CAP subsidies. Finally, a large group of small, mixed goat-sheep farms (47%) remained very close to the traditional grazing system with seasonal milk and meat production. Paradoxically, even if most of them sold the milk to cheese industries covered by the PDO label, they are described by authors as the most prone to abandon the activity.

Moving to the Central Pyrenees, Riedel et al. (2007) examined the meat-sheep farms of *Sierra y Cañones de Guara* Natural Park. Here, the land from the farms which have disappeared is rented by the "surviving" ones, which have importantly increased their surface. However, as these authors emphasize, the availability of farmland does not reduce the need for intensification aimed at improving the work schedule and the animal productivity. Therefore, as happened in the previous case, the farms with less incorporation of technology are the ones with the lowest chances of family continuity (excepting a small group of very large ones). Interestingly, two patterns of land use intensification were found in this area: (i) farms with pasture surface concentrated the grazing in small areas of productive and accessible pastures in order to minimize shepherding effort, whereas large pasture areas were abandoned or less used and sheep were housed in longer periods[19]; (ii) small farms oriented to agriculture embarked in crop intensification in order to maximize the production of on-farm forage.

CONCLUSIONS

The principal conclusion we can draw from this analysis is that many of the chief patterns of change in Spanish agriculture over the last decades are still pervaded by a productivist rationale.

Thus, as agricultural statistics reveal, the farm structure is experiencing from the 1990s an intense process of adjustment. Strong rates of disappearance of farms have been registered in this period, whereas the remaining productive units increase their average size. More meaningfully, a reduced segment of "hard core" holdings in continuous expansion gathers increasing shares of labour, land and agricultural output – a phenomenon of concentration which is generally observed in developed countries (Martins & Tosstorff, 2011; Hoppe, Korb, & Banker, 2008). In addition, most of the farms which are swelling this "hard core" in the last years have an intensive productive profile, what implies an alteration in the classic model of farm adjustment prevailing in the 1990s.

The progress in farm intensification strategies has been tackled by focusing on some specific farming systems. In the horticultural production of SE Spain, the model of agriculture which has consolidated involves the alteration of the natural cycles, strong investments in sophisticated technology, massive use of immigrant hired labour and a high consumption of external inputs. As Sánchez-Picón et al. (2011) note, it shows more similarities to an industrial production than to a traditional farming system – in fact, the term "super-productivism" has been related with indoor horticultural production in recent studies (Calleja et al., 2012). The essential features of this model were already observable by early 1990s; however, a closer look to the most recent dynamics has shown some noteworthy nuances. On the one hand, the farming practices have been adapted to comply with the quality and health standards imposed by the European food chains. Although these changes bring about undeniable environmental benefits (e.g., less use of agrochemicals), one could hardly state that they have altered the fundamentals of this style of farming. Indeed, these changes have not sufficed to resolve other major environmental problems originated by agriculture in these territories.

Besides, unconventional versions of family farms arise in this context, better suited to face both the external pressures and the internal complexity of this farming system. Organizational forms made up by several kin households, capable to undertake aggressive strategies of growth (those combining both territorial expansion and intensification), have become widespread in the zone and challenge many of our old conceptions on family farming (Moreno-Pérez et al., 2011).

The traditional farming systems of Mediterranean permanent crops have also played a focal role in the recent transformations of Spanish agriculture.[20] Focusing on particular areas with vineyard orientation, evidence has been shown on the striking processes of differentiation, both at territorial and farm level, lying behind the overall picture. Indeed, changes in the spatial patterns of production have been observed, in terms of a dramatic regression of vineyards in some areas in parallel with their spectacular expansion, intensification and modernization in others. This differentiation has been related both with disparities in the agroclimatic conditions in different areas and with their uneven capability to respond to the market demands – for example, by developing territorially based quality labels for wine production. At the micro level, commentators have stressed the prominence of certain holdings (either family-based or set up by external investors) in leading the chief transformations of this farming system.[21]

Finally, intensification trends have been also observed in traditional grazing livestock systems. In small ruminant farms, intensification has functioned as a strategy to avoid farm marginalisation, since it reduces the farm labour requirements. Thus, several studies have identified the most extensive, pasture-based farms as the ones under a highest risk of abandonment, with the environmental repercussions that this could entail (Gaspar et al., 2011; Riedel et al., 2007). On occasions, the abandonment of marginal pastures runs parallel with the intensification in the use of the most productive ones, and these divergent trajectories are even observable within the same farm. Therefore, as found in vineyards, this empirical evidence backs the supposition of a sort of polarization in the intensity of the land use within the Mediterranean farming systems, which may entail undesirable environmental effects (Caraveli, 2000; Henle et al., 2008; MacDonald et al., 2000).

Two overall reflexions can be highlighted from the three farm systems analysed. First, the Accession to ECC in 1986 acted as a formidable driver of agricultural change in Spain, both amplifying the market signals (e.g., increase in the demand of vegetables or wine) and by way of the CAP implementation (e.g., vine restructuring subsidies and grabbing-up premiums; decoupled payments), which have accelerated the intensification in certain agricultural systems and, sometimes, catalyzed the abandonment of marginal farms. Second, a critical assessment of certain trajectories of change – namely the spectacular diffusion of quality labels – is necessary to avoid over-optimistic lectures pointing at the advance of a "postproductivist" model of agriculture in Spain. As Evans et al. (2002) emphasized, the quality has become a part of productivist food systems, rather than substituting them. Indeed, evidence shows that these schemes (either territorially based or *green* labels) are, on occasions, fully compatible with intensive styles of farming which provoke soil degradation, overexploitation of water resources and irreversible alterations of traditional farm landscapes. This would fit in a model of "neoproductivism" (Anderson, 2009; Burton & Wilson, 2012), in the sense that it capitalizes on the consumers' awareness as a marketable commodity.

NOTES

1. What scholars refer to as "cost-price squeeze". This is, in fact, a long-running issue, although it has been alleged to be more noticeable from late 1980s (Van der Ploeg & Roep, 2003).

2. The reversal of the processes with which Bowler (1985) had formerly characterized productivism: intensification, concentration, and specialization.

3. As Armesto (2005) relates, the "importation" of the concept resulted from direct connections between scholars of both countries, which culminated in the I Symposium of British and Spanish Rural Geographers held in 1996 in Leicester.

4. These statistics complement the most comprehensive information registered by the Agricultural Censuses performed every 10 years. Although the most recent structural data available are those of the last Census conducted in 2009, we do not utilize this source here because the universe of reference of this Census has been reduced with respect to that of 1999 and 1989 – hence the data contained in them are not directly comparable.

5. Certainly, some farms may have jumped to upper strata of economic dimension uniquely due to the effect of the inflation, without having experienced major changes in real terms. However, the swelling of the segments over the threshold of 40 ESU is so pronounced that it cannot be merely ascribed to this cause.

6. The 'new' farms given over to cattle for meat in 2007 could have come from a former specialization in milk cows – which has undergone an intense regression. As Sineiro, López, Lorenzana, and Valdés (2006) found in NW Spain, this change in the productive profile is frequently undertaken as a strategy of gradual withdrawal from farming undertaken by bovine-oriented holdings.

7. 42% the value of Spanish exportations of fruits and vegetables in 2010 came from these three provinces (Montaner-Salas, 2011)

8. Nowadays, this concrete county of only 27.000 hectares house more than a half of the indoor surface existing in all Spain – which is, in turn, the leader country in hectares of greenhouse farming in the Mediterranean basin. It is, in fact, the area with a greatest concentration of greenhouses in the world (Aznar-Sánchez et al., 2011).

9. The average farm size is 9.5 has, although a half of them do not reach 5 has. 80% farms have at least one greenhouse, the average indoor surface being 2.9 has.

10. To put it briefly, this scheme certifies the production under farming practices that reduce the consumption of natural resources and the environmental pollution, and also take advantage of the natural mechanisms of production (e.g. biological control of pests). It is considered an intermediate alternative to the organic production label, more restrictive in terms of pesticides and nutrients.

11. Similar results are found by Aznar-Sánchez (2011) in Almería – where more than 65% the greenhouse surface is covered by some quality label (GlobalGap, Naturane, etc.) and more than 75% the greenhouses have incorporated biological control of plagues.

12. The productivity of green pepper greenhouse growing in Campo de Cartagena rounds 100.000kg/ha per crop cycle, and two cycles are cultivated per year (Pérez-Morales & Gil-Meseguer, 2011).

13. In fact, it was distinguished in 2006 as "Qualified Designation of Origin". There are only two wine territories with this label in Spain, conceded under strict conditions – e.g. the PDO should have been operating for more than 10 years, the attached wineries must be located within the PDO territory, etc.

14. These zones are only sheltered by the Protected Geographical Indication "Vinos de Castilla y León".

15. Espaliers ease the mechanization of pruning and other farm works, making use of adapted technology (i.e. small tractors or specialized implements). As for grape harvest, manual work is often preferred in certain zones because it allows the grape selection and a more delicate handling.

16. Interestingly, Alonso et al. (2003) enumerate investors of a different kind that may be behind these businesses: non-agricultural private actors investing in wine, agro-food enterprises entering into wine production, national or international wine companies aiming at diversifying their catalogues (they are sometimes present in different PDOs), middle-size wineries from other Spanish vine areas that replicate their business model in this region, or even prestigious enologists that embarked in their own enterprise.

17. Following the CAP Mid-Term Review in 2003, ovine and caprine First-Pillar subsidies were 50% decoupled in Spain in 2006 and fully decoupled in 2010.

18. Castel, Ruiz, Mena, and Sánchez-Rodríguez (2010) mention the toughening of the health legal standards as an obstacle for the production of cheese at a farm level, so that farmers opted to channelize the whole raw milk production to the commercial cheese industry.

19. This "dualization" in the use of pastures within the same farm was also found by Estrada et al. (2010) in another Pyrenees area, the county of *Alta Ribagorça*.

20. Indeed, 42% the holdings that joined to the 'hard-core' of Spanish farming between 1997 and 2007 were specialized on viticulture, olive grove production or fruit trees and citrus fruits.

21. Some of these tendencies could be extensible to other permanent crops, notably olive trees in the South of Spain, where the CAP has triggered the productive intensification and colonization of land previously given over to arable crops (Gallardo & Ceña, 2006).

REFERENCES

Alonso, J. L., Aparicio, L. J., & Sánchez-Hernández, J. L. (2003). Los espacios vitivinícolas en Castilla y León: La evolución hacia un sistema productivo de calidad. *Boletín de la Asociación de Geógrafos Españoles*, *35*, 101–122.

Anderson, D. (2009). Productivism and ecologism: Changing discourses in TVET. *Work, Learning and Sustainable Development*, *8*(1), 35–57.

Armesto, X. A. (2005). Notas teóricas en torno al concepto de postproductivismo agrario. *Investigaciones Geográficas*, *36*, 137–156.

Arnalte, A. (Coord.) (2006a). *Políticas agrarias y ajuste estructural en la agricultura española*. Madrid: MAPA.

Arnalte, A. (2006b). Economía política del proceso de ajuste estructural en la agricultura de los países desarrollados. In E. Arnalte (Coord.) *Políticas agrarias y ajuste estructural en la agricultura española* (pp. 19–54). Madrid: MAPA.

Arnalte, E., Ortiz, D., & Moreno, O. (2008). Cambio estructural en la agricultura española. Un nuevo modelo de ajuste en el inicio del siglo XXI. *Papeles de Economía Española, 117*, 59–73.

Aznar-Sánchez, J. A. (2011). El cluster agroindustrial de la horticultura intensivade Almería: Surgimiento, dinámica y perspectivas. *Cuadernos de Estudios Agroalimentarios, 2*, 199–217.

Aznar-Sánchez, J. A., Galdeano-Gómez, E., & Pérez-Mesa, J. C. (2011). Intensive Horticulture in Almería (Spain): A counterpoint to current european rural policy strategies. *Journal of Agrarian Change, 11*(2), 241–261.

Bernués, A., Ruiz, R., Olaizola, A., Villalba, D., & Casasús, I. (2011). Sustainability of pasture-based livestock farming systems in the European Mediterranean context: Synergies and trade-offs. *Livestock Science, 139*, 44–57.

Blanc, M. (1987). Family and employment in agriculture; recent changes in France. *Journal of Agricultural Economics, 38*, 289–301.

Bowler, I. (1985). Some consequences of the industrialization of agriculture in the European Community. In M. Healey & B. Ilbery (Eds.), *The industrialization of the countryside* (pp. 75–98). Norwich: GeoBooks.

Burton, R., & Wilson, G. A. (2012). The rejuvenation of productivist agriculture: The case for "cooperative neo-productivism". In R. Almas & H. Campbell (Eds.), *Rethinking agricultural policy regimes: Food security, climate change and the future resilience of global agriculture. Research in Rural Sociology and Development* (Vol. 18, pp. 51–72). Bingley, UK: Emerald Group Publishing Limited.

Calleja, E. J., Ilbery, B., & Mills, P. R. (2012). Agricultural change and the rise of the British strawberry industry, 1920–2009. *Journal of Rural Studies, 28*(4), 603–611.

Caraveli, H. (2000). A comparative analysis on intensification and extensification in Mediterranean agriculture: Dilemmas for LFA's policy. *Journal of Rural Studies, 16*(2), 231–242.

Castel, J. M., Mena, Y., Ruiz, F. A., Camúñez-Ruiz, J., & Sánchez-Rodríguez, M. (2011). Changes ocurring in dairy goat production systems in less favoured areas of Spain. *Small Ruminant Research, 96*, 83–92.

Castel, J. M., Ruiz, F. A., Mena, Y., & Sánchez-Rodríguez, M. (2010). Present situation and future perspectives for goat production systems in Spain. *Small Ruminant Research, 89*, 207–210.

Cots-Folch, R., Martínez-Casasnovas, J. A., & Ramos, M. C. (2009). Agricultural trajectories in a Mediterranean mountain region (Priorat, NE Spain) as a consequence of vineyard conversion plans. *Land Degradation & Development, 20*, 1–13.

De Antonio, F. (2005). Revisión Intermedia de la PAC: Repercusión para el futuro del sector ovino y caprino. *Pequeños Rumiantes, 6*(3), 36–41.

De Rancourt, M., Fois, N., Lavín, M. P., Tchakérian, E., & Vallerand, F. (2006). Mediterranean sheep and goats production: An uncertain future. *Small Ruminant Research, 62*, 167–179.

Estrada, F., Nadal, E., & Iglesias, J. R. (2010). Twenty-first century transhumants: Social and Economic Change in the Alta Ribagorça. In I. Vaccaro & O. Beltran (Eds.), *Social and ecological history of the Pyrenees: State, market, and landscape* (pp. 105–126). Walnut Creek: Left Coast Press.

Evans, N., Morris, C., & Winter, M. (2002). Conceptualizing agriculture: a critique of postproductivism as the new orthodoxy. *Progress in Human Geography, 26*(3), 313–332.

Gallardo, R. & Ceña, F. (2006). El proceso de ajuste estructural en el olivar de Andalucía. In Arnalte, A. (Coord.) *Políticas agrarias y ajuste estructural en la agricultura española* (pp. 159–197). Madrid: MAPA.

Gaspar, P., Escribano, A. J., Mesías, F. J., Escribano, M., & Pulido, A. F. (2011). Goat systems of Villuercas-Ibores area in SW Spain: Problems and perspectives of traditional farming systems. *Small Ruminant Research, 97*, 1–11.

Gómez-Espín, J. M., López-Fernández, J. A. & Montaner-Salas, M. E. (Coord.) (2011). Modernización de regadíos: Sostenibilidad social y económica. La singularidad de los regadíos en el Trasvase Tajo-Segura, Murcia (Spain): Universidad de Murcia.

Henle, K., Alardb, D., Clitherowc, J., Cobbd, P., Firbanke, L., Kullf, T., ... Youngl, J. (2008). Identifying and managing the conflicts between agriculture and biodiversity conservation in Europe–A review. *Agriculture, Ecosystems & Environment, 124*(1–2), 60–71.

Hill, B. (2006). Structural change in European agriculture. In D. Blandford & B. Hill (Eds.), *Policy reform and adjustment in the agricultural sectors of developed countries* (pp. 5–21). Wallingford: CAB International.

Hoppe, R., Korb, P., & Banker, D. (2008). Million-dollar farms in the new century. *Economic Research Service*, USDA. Bulletin Number 42.

Ilbery, B., & Bowler, I. (1998). From agricultural productivism to post-productivism. In B. Ilbery (Ed.), *The geography of rural change* (pp. 57–84). Harlow: Pearson.

Lowe, P., Murdoch, J., Masden, T. K., Munton, R., & Flynn, A. (1993). Regulating the new rural spaces: The uneven development of land. *Journal of Rural Studies, 9*, 205–222.

MacDonald, D., Crabtree, J. R., Wiesinger, G, Dax, T., Stamou, N., Fleury, P., ... Gibon, A. (2000). Agricultural abandonment in mountain areas of Europe: Environmental consequences and policy response. *Journal of Environmental Management, 59*, 47–69.

Marsden, T. K., & Symes, D. (1987). *Survival strategies in post-productionist agriculture*. United Kingdom Analytical Paper No. 2. The Socio-economic Consequences of Agricultural Development in Europe series. Vienna: European Coordination Centre for Research and Documentation in Social Sciences.

Martín-Cerdeño, V. J. (2012). El consumo de carne de ovino en España. *Distribución y consumo, March-April*, 89–93.

Martins, C., & Tosstorff, G. (2011). Large farms in Europe: Less than 1% of European farms occupy 20% of the Utilised Agricultural Area. EUROSTAT. Statistics in Focus, 18/2011.

Mather, A., Hill, G., & Nijnik, M. (2006). Post-productivism and rural land use: Cul de sac or challenge for theorization? *Journal of Rural Studies, 22*, 441–455.

Menor-Toribio, J. (2000). Reflexiones en torno a los modelos productivista y postproductivista en la Vega de Granada. *Cuadernos Geográficos, 30*, 415–427.

Ministerio de Medio Ambiente y Medio Rural y Marino, MARM. (2009). *Caracterización del sector ovino y caprino en España. Año 2009*. Madrid: MARM, Subdirección General de Productos Ganaderos.

Molinero, F. (2006). La evolución de la agricultura en España: Tradición, modernización y perspectivas. *Norba. Revista de Geografía, XI*, 85–106.

Molinero, F. (2011). Los paisajes del viñedo en Castilla y León: Tradición, renovación y consolidación. Polígonos. *Revista de Geografía*, *21*, 85–117.

Moreno, O., & Ortiz, D. (2008). Understanding structural adjustment in Spanish arable crop farms: Policies, technology and multifunctionality. *Spanish Journal of Agricultural Research*, *6*(2), 153–165.

Moreno-Pérez, O., Arnalte-Alegre, E., & Ortiz-Miranda, D. (2011). Breaking down the growth of family farms: A case study of an intensive Mediterranean Agriculture. *Agricultural Systems*, *104*(6), 500–511.

Oregui, L. M., & Falagán, A. (2006). Spécificité et diversité des systèmes de production ovine et caprine dans le Bassin Méditerranéen. In Y. Mena, J. Castel & P. Morand-Fehr (Eds.), *Analyse technico-économique des systèmes de production ovine et caprine: Méthodologie et valorisation pour le développement et la prospective. Options Méditerranèennes (Série A, Séminaires Méditerranéens)*, (Vol. 70, pp. 15–21).

Ortiz, D., & Moreno, O. (2007). La política agraria en España en el contexto del debate europeo productivismo/post-productivismo. Communication presented at the VI Spanish Congress of Agricultural Economics held in Albacete (Spain).

Pardos, L., Maza, M. T., Fantova, E., & Sepúlveda, W. (2008). The diversity of sheep production systems in Aragón (Spain): characterisation and typification of meat sheep farms. *Spanish Journal of Agricultural Research*, *6*(4), 497–507.

Pedreño, A. (1999). *Del jornalero agrícola al obrero de las factorías vegetales*. Madrid: Ministerio de Agricultura, Pesca y Alimentación, Serie Estudios.

Pérez-Morales, A., & Gil-Meseguer, E. (2011). Estado de la modernización de regadíos en la zona regable del Campo de Cartagena. In J. M. Gómez-Espín, J. A. López-Fernández & M. E. Montaner-Salas (Coord) Modernización de regadíos: Sostenibilidad social y económica. La singularidad de los regadíos en el Trasvase Tajo-Segura. Murcia, Spain: Universidad de Murcia.

Riedel, J. L., Casasús, I., & Bernués, A. (2007). Sheep farming intensification and utilization of natural resources in a Mediterranean pastoral agro-ecosystem. *Livestock Science*, *111*, 153–163.

Sánchez-Picón, A., Aznar-Sánchez, J. A., & García-Latorre, J. (2011). Economic cycles and environmental crisis in arid southeastern Spain. A historical perspective. *Journal of Arid Environments*, *75*, 1360–1367.

Sevilla, A., Domene, M. A., Uceda, M., Buendía, D., & Racero, J. L. (2012). ¿Residuos? Productos con otras calidades y distintas aplicaciones. *Cuadernos de Estudios Agroalimentarios*, Julio, *2012*, 161–180.

Shucksmith, M. (1993). Farm household behaviour and the transition to post-productivism. *Journal of Agricultural Economics*, *44*(3), 466–478.

Sineiro, F., López, E., Lorenzana, R., & Valdés, B. (2004). La tipología de las explotaciones en función de su viabilidad económica y demográfica; aplicación a las explotaciones de bovino en Galicia. *Economía Agraria y Recursos Naturales*, *4*(8), 63–85.

Sineiro, F., López, E., Lorenzana, R., & Valdés, B. (2006). El proceso de ajuste en la ganadería bovina de la Cornisa Cantábrica. In Arnalte, A. (Coord.) *Políticas agrarias y ajuste estructural en la agricultura española* (pp. 261–289). Madrid: MAPA.

Sumpsi, J. M. (Coord.) (1994). *Modernización y cambio estructural en la agricultura española*. Madrid: MAPA.

Van der Ploeg, J. D., & Roep, D. (2003). Multifunctionality and rural development: The actual situation in Europe. In G. Van Huylenbroeck & G. Durand (Eds.), *Multifunctional agriculture. A new paradigm for European agriculture and rural development* (pp. 37–53). Aldershot: Ashgate.

Walford, N. (2003). Productivism is allegedly dead, long live productivism. Evidence of continued productivist attitudes and decision-making in South-East England. *Journal of Rural Studies*, *19*, 491–502.

Ward, N. (1993). The agricultural treadmill and the rural environment in the post-productivist era. *Sociologia Ruralis*, *33*, 348–364.

Ward, N., Jackson, P., Russell, P., & Wilkinson, K. (2008). Productivism, post-productivism and European agricultural reform: The case of sugar. *Sociologia Ruralis*, *48*(2), 118–132.

Wilson, G. A. (2007). *Multifunctional agriculture. A transition theory perspective*. Wallingford: CAB International.

CHAPTER 7

THE AMBIGUITIES OF FRENCH MEDITERRANEAN AGRICULTURE: IMAGES OF THE MULTIFUNCTIONAL AGRICULTURE TO MASK SOCIAL DUMPING?

Catherine Laurent

ABSTRACT

Which is the main characteristic of the French Mediterranean Agriculture (FMA)? The recognition of the multifunctionality of agriculture, supported by the institutions of the territorial development? Or the development of social dumping considered as a necessity by many institutions of the sector? To answer this question the analysis is based on three main sources of data: agricultural statistics, monographs and administrative reports. The results show that the structural diversity is still important in the FMA. A significant proportion of the farms have based their economic strategy on making the most of the multifunctionality of agriculture. Some have built real success stories. But this development path cannot

Agriculture in Mediterranean Europe: Between Old and New Paradigms
Research in Rural Sociology and Development, Volume 19, 149–171

ISSN: 1057-1922/doi:10.1108/S1057-1922(2013)0000019009

guarantee the viability of a large range of holdings: the number of farm holdings in FMA has decreased by 27% since 2000 and 57% since 1988. Due to the specificities of the Mediterranean productions, the cost of labour is still considered as a major adjustment variable to secure farm income in the region. Many situations are reported where the situation of casual labour is concerning, in particular for migrant workers. However, the working conditions of temporary migrant workers remain invisible and the image of the multifunctional agriculture is put forward as a marketing asset by all types of actors. This image is misleading. It makes invisible, issues that are essential for the future. Thus, it generates knowledge gaps and leads to the depoliticization of debates on the development models of agriculture in masking the contradictions and the conflicts of interest that they generate.

Keywords: Agriculture; France; farm workers; multifunctionality; policy; knowledge

INTRODUCTION

Impressionist landscapes, fields of flowers on the Riviera, quality wines, olive oil and the health benefits of a Mediterranean diet, small peri-urban farms, the development of short supply chains around towns and cities, pluriactivity and agro-tourism ... This image of French Mediterranean Agriculture (FMA) seems to correspond to the wishes of the Council of Europe (2003) which has recommended the promotion of sustainable and multifunctional agriculture for the Mediterranean, contributing to local employment, poverty alleviation and a supply of quality products.

This multifunctionality of agriculture may be defined (Laurent, 2001a, 2001b) as the full range of contributions of agriculture to economic and social development considered as a whole. The official recognition of multifunctionality reflects the intention that these different contributions be sustainably and coherently associated according to modalities deemed satisfactory by citizens. These contributions may be described in a positive list. In France such a list may include, in particular: commodity production, safety (product quality, consumers' and farm workers' health), environmental conservation, services (amenities for urban populations, landscape management, etc.), contribution to social and economic cohesion in rural areas (through job creation and diversification of activities), and political functions (occupation of land, food security) Fig. 1.

This conception of multifunctionality that was promoted by the French government for several years attests to a political strategy to reshuffle relations between the economic, societal and environmental issues. Reappropriated by the actors of territorial development (local authorities, NGOs, etc.), it has also been the basis for designing new development strategies for production units that are not competitive on international markets, but contribute to rural development and fulfil functions of public interest that society regards as worth maintaining.

There are success stories showing that in France, in the Mediterranean area, these various functions of agriculture may be combined in a win-win game. A plethoric literature praising the development potential of these economic strategies proposes monographs showing how commodity production can be associated with high added-value services such as agro-tourism, and generate complex systems of activity that generate new employment (Laurent, 2006).

Yet the fairy-tale appeal of the win-win games of multifunctionality should not mask the extent to which situations exist in which contradictions between the various objectives assigned to agriculture are, on the contrary, exacerbated.

In particular, a set of recent studies has shown how the working conditions of a significant number of farm labourers in this area have deteriorated (e.g. special issue of *Etudes Rurales*, 2008). The Bouches du Rhône, one of the French *départements* that employs the most occasional workers, is considered to be 'a real laboratory for increasing employment

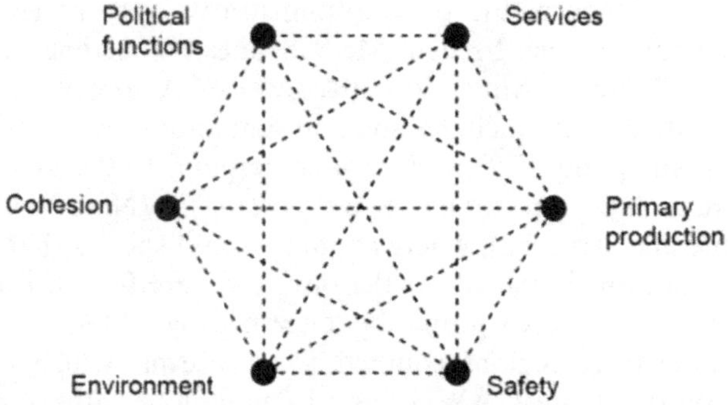

Fig. 1. The Multifunctionality of Agriculture.

insecurity in the agricultural sector' (Mesini, 2008; Clary & Van Haecke, 2001).

What exactly is French Mediterranean agriculture? Is the region emblematic of a successful development strategy based on recognition of the multifunctionality of agriculture, or is it a special experimental device to test new recipes for social dumping?

To answer this question the analysis will be based on three main sources of data: agricultural statistics, monographs and administrative reports. I first present the main structural characteristics of French Mediterranean agriculture (first section). I then show that it always combines very different types of farm holding which occupy unequal places in the logic of territorial and sectoral development (second section). Throughout the area, farms have taken advantage of the region's resources to multiply the number of quality labels and to improve the added value of their production (third section). But the profitability of a significant number of the farms relies on the adjustment of wages and on very low wages for occasional workers (fourth section). The working conditions of the latter remain largely invisible (fifth section). This invisibility stems partially from a convergence of the objective interests of the actors of territorial development and sectoral development, at the expense of those workers with the most precarious jobs (sixth section).

THE MAJOR STRUCTURAL FEATURES OF FRENCH MEDITERRANEAN AGRICULTURE

By convention, in this chapter I consider that French Mediterranean agriculture (FMA) covers the three administrative regions (Nuts 2) that include a portion of the French Mediterranean coastline: Languedoc-Roussillon, the Provence-Alpes-Côte d'Azur (PACA) region, and Corsica. This administrative entity includes some mountainous areas which, from a strictly bioclimatic point of view, do not correspond to the Mediterranean. It furthermore excludes a portion of two *départements* (Nuts 3) of the Rhône Valley (Drôme and Ardèche), where certain vegetal species that are typical of the Mediterranean (in particular the olive tree) are found. But overall it corresponds well to the area primarily concerned by FMA.

In 2010, these three regions counted 55,623 farms, which used 85,404 Agricultural Working Units (AWU)[1] and 1.6 million hectares of agricultural area (AC (Agricultural Census), 2010).

FMA belongs to the half of French agriculture in which the value produced by AWU is lowest (Fig. 2). In 2010 FMA accounted for 11% of the farms and the working time (AWU) registered in France, but for only about 8% of the value of the total national production (excluding subsidies) (AC, 2010).

In addition, the French Mediterranean area has always had a concentration of productions which receive little support from the CAP (Lauret 1992; Maillard, Daniel, & Colson, 2000): in 2010, 68% of the farms belonged to the types of farming that are characterized by vineyards, fruits, olive, permanent crops and horticulture (Fig. 3).

The number of farms and the amount of labour used on these farms has declined significantly over the past two decades, as in the rest of France.

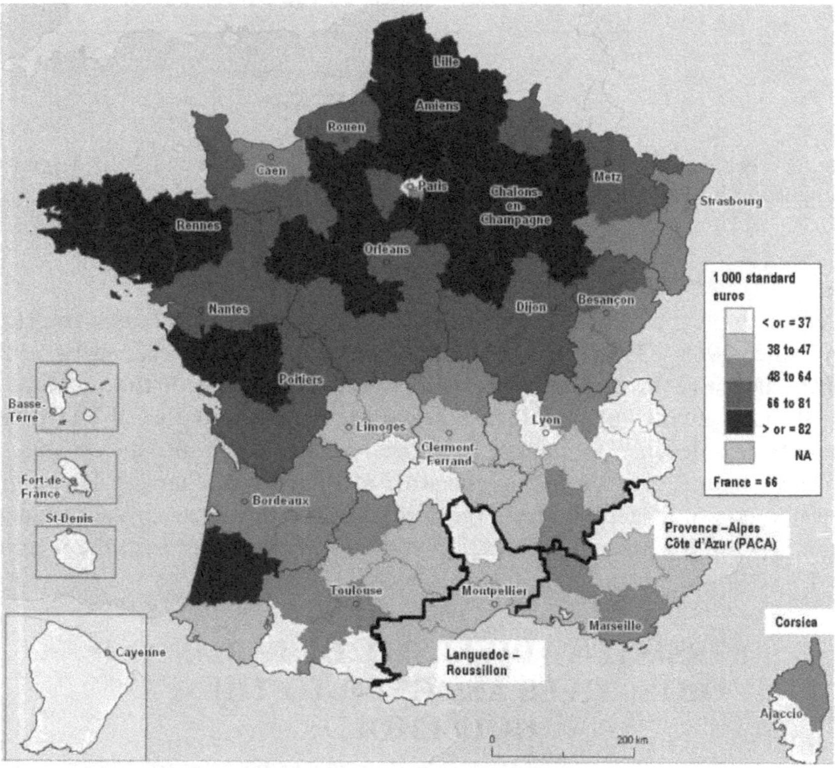

Fig. 2. Standard Output per Agricultural Working Unit. *Source*: Agricultural Census 2010, Agreste.

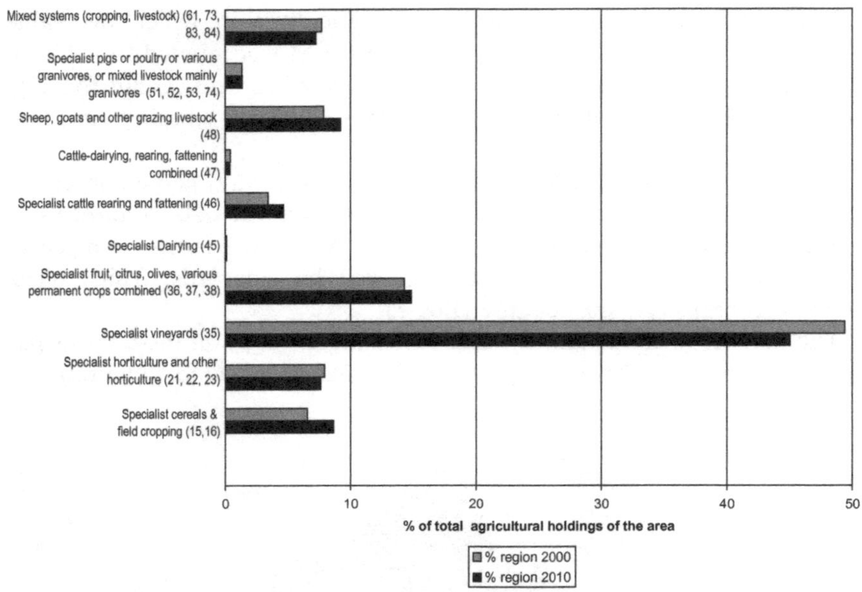

Fig. 3. Percentage of Holdings per Type of Farming in 2000 and 2010. Regions: Languedoc-Roussillon, PACA and Corsica. *Source*: Agricultural Census 2000 and 2010, Agreste.

Since the Agricultural Census in 2000, 27% of farms have ceased to exist, and since 1988, 53% (Fig. 4). The number of AWU has declined by 25% since 2000 and 46% since 1988. Although a percentage of the remaining farms have greatly enlarged their acreage, the agricultural area of FMA has nevertheless declined by almost 10% (against 3% for France as a whole).

These figures give an idea of the scale of the current changes. However, one should not forget the very large degree of structural heterogeneity which remains, and the plurality of development models that coexist in the region.

PERSISTENT DIVERSITY OF FARMS AND OBJECTIVES ASSIGNED TO THEM BY HOUSEHOLDS

The farms of FMA are still widely diverse in terms of their size. Although France is the Mediterranean country with the lowest proportion of small

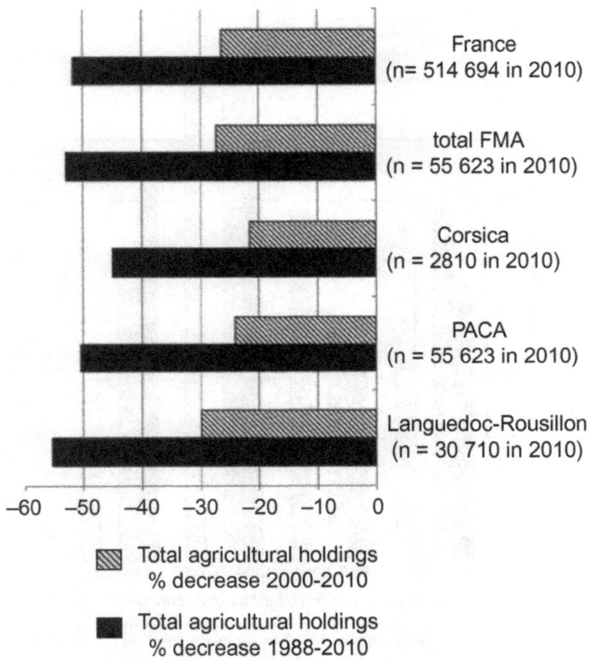

Fig. 4. Decrease of the Total Number of Agricultural Holdings over the Last Two Decades. *Source*: Agricultural Census, Agreste (1988, 2000, 2010).

farms, the figures nevertheless remain significant: 47.3% of the holdings have a small economic size (less than 25,000 Euros per year of standard output),[2] against 38.5% for the country as a whole (Fig. 5).

Farms in the FMA are also widely diverse in terms of objectives assigned to farming by households regarding their systems of activities and income.

Indeed, agricultural activity may have very different functions for the households concerned, especially for small-scale farms (Laurent et al., 1998; Laurent & Rémy, 1998; Laurent & Mouriaux, 2008). Schematically, the farms of the region can be classified into four main groups:

A first group aims at generating a commercial income from specialized agricultural activity; it includes most of the large-scale farms and some of the medium-scale ones (Agreste, 2011).

A second group includes mainly medium- and small-scale farms; it is composed of households for which specialized farm income is considered only as one part of a complex system of activities and income. Such systems

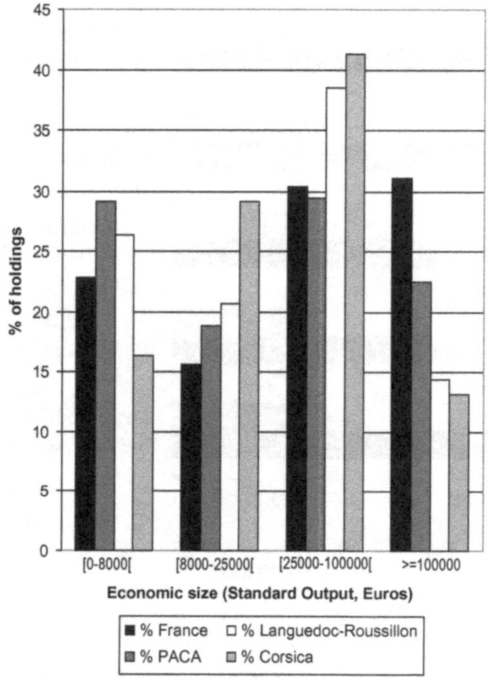

Fig. 5. Distribution of Agricultural Holdings in France and FMA According to Their Economic Size, in 2010. *Source*: Agricultural Census 2010, Eurostat.

may combine widely diverse sources of income, derived from other occupational activity or through the diversification of the farm activities (agro-tourism, food processing, commercial activities such as direct sales, etc.) (Agreste, 2011).

A third group of small farms is managed by low-income households which maintain a farm activity for subsistence and bartering, and may derive some extra income seasonally from the sale of their products. These farms are associated to various types of households (retired and unemployed people with low pensions, households with low occupational income).

A last group is composed of hobby farms; these are usually (but not exclusively) small farms. They are found in various productive orientations (vineyard, fruit, olives, grazing livestock, horses, etc.) (Jarrige, 2004). Expenses can be disproportionately high compared to the output of the farm. In some cases the farm may generate a small income, but this is not essential for the survival of the household.

When households meet problems in starting or maintaining an agricultural activity (to get access to land, for example), they refer to different institutions and different system of norms to solve the conflicts; some of them designate the sector-based institutions while others feel that their interests will be better taken care of by local authorities. Indeed the evolution of Mediterranean agriculture is thus situated at the interface of two institutional spheres which are concerned with different issues and challenges, and support different development logics (Laurent & du Tertre, 2008):

- sectoral development, which is based on the possibility of capital accumulation and the reproduction of activities of the sector, from income derived from agricultural activities. Sectoral regulation pertains above all to the way in which the conditions of the reproduction of capital are broadened by the institutional dynamics of the sector;
- territorial development, which implies the ability of a localized and institutionalized social system to guarantee processes of accumulation on its territory, and to maintain economic activities, promote and protect its natural resources and technological assets, and guarantee social peace.

Whereas sectoral regulation revolves around the conditions of accumulation of capital, territorial regulation deals with the necessity to guarantee the possibilities of reproduction of that capital accumulation in a given territory, and to regulate the social tensions implicit therein (living conditions of the unemployed population, management of the negative impacts on the environment, on health, etc.). In terms of agricultural development, it is therefore in the interests of the actors of territorial development to support those farms that allow for the best combinations of the various functions of agriculture.

THE MULTIFUNCTIONALITY OF AGRICULTURE AS AN ECONOMIC STRATEGY

We thus find a large number of situations in which local authorities work in partnership with different types of stakeholder to support initiatives that take advantage of and foster all the existing ties between agricultural activities and the territory (Laurent, 2006). These projects concern all farm holdings and not only those in Groups 2 and 3, even though the latter two groups are seen as the most emblematic of territorial development.

It would be a tedious task to draw up an inventory of all these projects. Most importantly, they concern widely diverse dimensions of territorial development, notably:

- the local anchorage of jobs, by facilitating the practice of pluriactivity (e.g. Tallon, 2012), and the diversification of activities based on agriculture (agro-tourism, transformation of farm products, etc.) (Volle, 2006);
- the recognition of the role of livestock breeding in mountainous and highland areas for the maintenance of landscapes, scrub invasion control, etc. (e.g. Paoli & Santucci, 2010);
- strengthening of ties to the region through specific marks of quality (AOC, quality labels, 'Mediterranean food') (Agreste, 2011; Lamine, 2005);
- the revival of town-countryside ties through concerted actions on the landscape (Perrin, 2009) or the development of short supply chains that enhance the value of proximity to urban centres (Montpellier, Marseilles, Nice, etc.) and to the tourist concentration of the coast in summer (Jarrige, 2004; Soulard & Thareau, 2009). Thus, in 2010 a significant proportion of farms in the region reported that they sold all or part of their production via short supply chains, that is, forms of commercialization involving one intermediary at the most (e.g. a cooperative where employees sell the products directly, or a supermarket): 20% in Languedoc Roussillon, 33.6% in PACA and 64.7% en Corsica (against 21% of all agricultural holdings in France in 2010) (Agreste, 2011).

When different types of actors join forces to promote one or other aspect of territorial development, the projects are often associated with a vision of local development that highlights the positive functions of agriculture but fails to raise the question of possible contradictions between these diverse functions. The spotlight is put on the supposed positive effects on the environment, social cohesion and the maintenance of local jobs, as if their virtuous articulation stood to reason.

For example, during the visit of S. Royal (a politician in the Socialist Party) to Corsica in 2011, her campaign website encouraged the development of short supply chains in the following terms: *'Consumers want products that don't damage their health, that add value to the notion of territory, tradition, authenticity and the social link. They want their consumption to be respectful of the environment, with local products that generate few carbon dioxide emissions (lower cost of transport) and less packaging (excessive packaging). They are also calling for greater traceability of products and better quality regarding taste and nutrition. Producers, on their side, get a stable income and recognition as social players through a direct link with consumers, enabling them to share and to promote as effectively as possible a specific mode of production that is a real alternative to mass retailing'* (Defending peasant agriculture and short supply chains. *Désir d'Avenir*. 17 August 2011. Visit of Segolène Royal to Corsica).[3]

Short supply chains do indeed appear to be a possible strategy in the struggle against the growing fragility of farm holdings, by increasing their profit margins. Monographs on the region have moreover gathered a body of testimonies showing the role that new forms of producer–consumer relations can play in creating networks of solidarity within low-income populations, or even in reconfiguring the modalities of food aid (Celavar, 2011).

Yet, without denying the existence of such experiences, the fact that short supply chains are always associated with better social and environmental performance has to be demonstrated. Schlich and Fleissner (2005) and Schlich et al. (2006) have shown that the most popular claims for regional food production and distribution, instead of global process chains, are not generally valid. Small farmers and short supply chains basically need much more energy to produce and distribute their products, compared with bigger units. These authors show how, for three products, the consumption of fuel per kilogram of produce delivered is much higher in short supply chains (e.g. ten times more for lamb produced locally than that imported from the southern hemisphere). Aubry and Chiffoleau (2009), who analyse this issue in several regions, especially in Mediterranean areas, also find that until now no data support the argument that short supply chains are more ecological, from the point of view of cultural practices implemented by the producers or of energy expenditures related to transport.

This example shows that it is not enough to juxtapose objectives for the virtuous circle of the win-win game to be set in motion. But it is possible to conceive of arrangements to make the contradictions acceptable, at least temporarily. A whole system of project support engineering has been set up, which contribute to that. It concerns both individual and collective projects. For example, for the Languedoc-Rousillon region alone, 250 organizations of various kinds (NGOs, local authorities, public advisory services, etc.) have been recorded (Tallon, 2012), which propose advisory services on pluriactivity to people who have to do two or more part-time jobs to earn an adequate income.

This type of support can help to regulate certain minor contradictions (e.g. in the framework of multi-activity, the adjustment of insertion in various regimes of social protection, or issues of work organization). But the durability of such arrangements can be problematic.

For instance, C. Lamine (2005) has shown how the notion of Mediterranean nutrition is a social construction, and she has described the many coordination efforts that were necessary to build a brand such as *'Nutrition méditerranéenne en Provence'* based on a qualification process encompassing three items: geographic origin, preservation of the environment, and preventive nutrition. The production of such a brand results from compromises and the organization of the compatibility of the interests of the various stakeholders in the supply-chain, the advisory services, and the scientists. However, as Lamine points out, such a system is fragile. At any moment new contradictions may resurface, brought for instance by the medical community which could easily deny the claim that one isolated product of this brand may have positive effects on human health – the importance of the balance of a person's whole diet, rather than the quality of one of its components, usually being a matter of consensus.

These territorial development projects are sometimes accompanied by real success stories. But the considerable decline in the number of farm holdings in recent years shows that they have not (not yet?) led to a development model that guarantees the economic viability of small and medium-sized farms. Studies have also revealed the fragility of mechanisms designed to temporarily regulate certain contradictions (e.g. between the demands of competitiveness and those of quality) but do not necessarily contribute to reducing them.

Moreover, the ubiquity of the image of the multifunctional agriculture may hide these contradictions. This phenomenon is quite striking when we consider the transformation of farm labour situation on many holdings. While the cost of labour is a significant adjustment variable to guarantee the farm's income, the actors of the territorial development give little attention to the growing invisibility of these farm workers' actual employment conditions.

THE IMPORTANCE OF THE FLEXIBILITY OF LABOUR

It has been known for a long time (e.g. Berlan, 1983) that analyses of French Mediterranean agriculture must consider the specificities of the intensive crops that predominate in the area (fruits, vegetables and vineyards, Fig. 3), and their implications for the organization of work. These three productive orientations demand a large amount of work per hectare and this demand is distributed very unevenly over the year. Labour needs therefore vary widely over time and their occurrence is partly random: the time of harvesting and that of various key interventions depends on climatic conditions which vary from one year to the next.

For small farm, the volume of work is adjusted by playing on the flexibility of family labour (including the use of the extended family and sometimes even neighbours). But over a certain threshold, it becomes necessary to hire labour. In all Mediterranean Europe, the massive and flexible availability of foreign labour force has been a key factor of competitiveness and consolidation of intensive agriculture (Arnalte-Alegre, Ortiz Miranda, this book, Chapter 3).

Ideally, an employer needs to have all the necessary manpower available as required (for treating crops, harvesting, etc.), but without having to pay a day more than is necessary; in other words, cost should be strictly proportional to the task at hand. The cost per hour of work should also be as low as possible. Employers therefore resort to a whole range of arrangements to tend towards this ideal. These arrangements concern the type of person employed as much as the nature of the contractual relations entered into. In certain cases, the farmer remains the employer, either directly or via intermediaries such as employer groupings (*groupement d'employeur*).[4] In other cases the employer is a third party, a temporary employment agency or a service provider.

Even though it is essential for efficiently running a farm, little is known about this type of labour. There are several reasons for the fact that this part of agricultural labour has always been underestimated and poorly described:

- The complexity of the situations concerned is difficult to grasp, even for in-depth monographic approaches. Periods in which this occasional manpower is inactive vary widely, from one day to several months, in continuous or fragmented intervals. The tasks to perform are also widely diverse, as are the skills required, and the range of people concerned is broad (occasional employment of people living in the region; trainees; specialized workers moving from farm to farm; migrant workers with differing statuses; etc.).
- Administrative data (e.g. registration with the social protection) are sorely lacking. This sector is known for the high rate of 'under-registration' (see, e.g. the reports of the commission to control illegal employment: Commission de lutte contre le travail illegal, 2007, 2012.

– Agricultural statistics such as agricultural censuses dissociate the labourer from the work when it is occasional. They consider not the individuals but the number of days that they work, measured in AWU. This measurement gives no indication of the individuals concerned, their numbers, their demographic characteristics, and their status. The 17,250 AWU of occasional labour recorded for FMA thus correspond to several tens of thousands of individuals whose exact numbers are unknown. Moreover, various comparisons suggest that statistical data are also under-informed (number of individuals, of days worked, of hours worked per day) and that the actual number of AWU worked is greater than that which is declared.

In other words, the contribution of occasional workers is always underestimated, whether they are employed directly by the farm or by other organizations (sub-contractors, temporary employment agencies, etc.). But even if this bias is taken into account, this type of work remains substantial for a large number of farm holdings in the Mediterranean region. Close to 20% of the total number of AWU used in FMA is provided by occasional workers and individuals who work on the farm but also have other employers (service providers, temporary employment agencies) (Fig. 6).

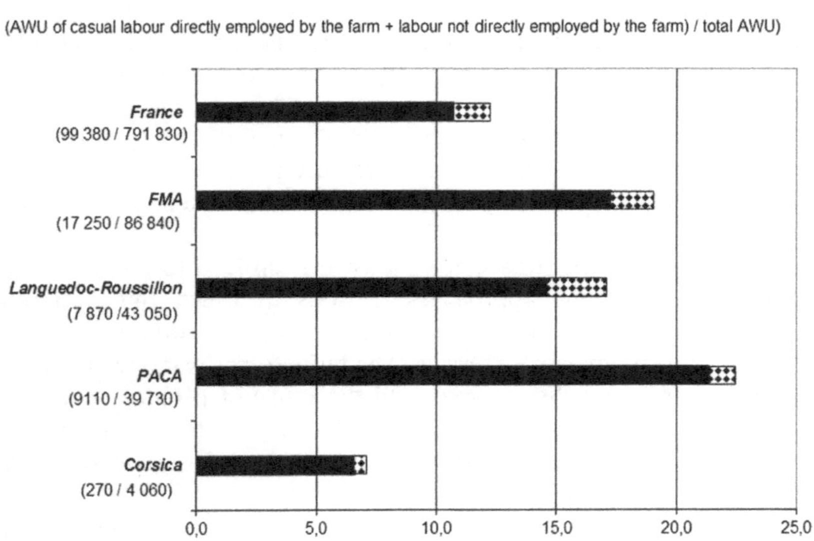

Fig. 6. Percentage of the Total Agricultural Working Units Used on Farms That Are Provided by Casual Labour and Workers Not Employed by the Farm (Subcontractors, etc.) in 2010. *Source*: Agricultural Census 2010, Eurostat Eurofarm.

This situation is not new, but a convergent set of data from monographs and the findings of the devolved services of the ministry in charge of labour strongly suggest that this population of agricultural workers is undergoing profound change (Agreste, 2011; Breem, 2012, DGT (Direction générale du travail), 2012; Mesini, 2008).

THE GROWING INVISIBILITY OF MIGRANT FARM LABOUR

The quest for maximal flexibility in the use of temporary labour has always caused employers to turn to foreign workers from countries where wages are low and working conditions less favourable. These differences afford them access to a more malleable labour supply than the domestic one. This logic has been described in numerous countries, on all the continents (e.g. Bauder, 2006).

In France the population of foreign workers in Mediterranean agriculture corresponds to three main situations:

1. workers who have arrived within the framework of institutional agreements between France and other countries (outside the EU), on a nominative contract. These agreements enable an official agency (currently the Office Français de l'Immigration et de l'Intégration, OFII)[5] to set up recruitment offices abroad. These workers may remain in France for a limited period only (six months per year maximum, unless they have an exemption) and the same person may return every year[6];
2. foreign workers employed directly by the farmer, who have organized their stay in France independently (mainly family networks (e.g. Michalon & Potot, 2008)) and who may or may not have a residence and work permit;
3. foreign workers recruited by temporary employment agencies or service providers (posted workers). This possibility was opened by the General Agreement on Trade in Services that was signed by the EU in 1994. The European Directive (96/71/EC) which regulates this possibility implies that workers who are employed by a foreign agency but work in France for delivering a service in France, have to benefit from the guarantees associated with French labour laws.

Since EU enlargement in 2007 the relative proportion of the latter category of workers has grown considerably. Since 1 July 2008, after transitional arrangements, France has opened its labour market to all workers of the

EU member states. In particular, people who formerly came from Eastern European countries in the framework of agreements set up by the agency in charge of immigration now benefit from the right to circulate freely. The relative proportion of workers from outside the EU in the framework of OFII agreements (especially Moroccan) has thus declined, while Polish workers who were formerly numerous have disappeared from the statistics of this organization since they may now circulate freely within the EU (Breem, 2012).

This trend could be considered a step towards employment situations more respectful of workers' rights. A set of monographic findings in France and in other European countries has indeed shown the limits of contracts obtained under institutional agreements organized by national immigration agencies to guarantee the rights of the workers concerned (Décosse, 2011; Mesini, 2008; Morice & Michalon, 2008; Rogaly, 2006). In many situations these workers have to be docile enough during their stay if they are to hope that they or their family can come back the following year. This docility may imply agreeing to particularly heavy workloads (up to 300 hours per month), inadequate accommodation, and a low profile when it comes to safety in the workplace. For example, Décosse (2008, 2011) has shown the difficulties encountered by these workers to protect themselves from very real chemical hazards related to the use of pesticides in the course of their work. He has also shown the impossibility of having health problems recognized and treated when they appear after the period of temporary work, once the worker has 'gone home'.

Thus, these contracts allow approaching the ideal of economic optimization of Mediterranean intensive agriculture: 'having the work without the worker', benefiting from labour as a source of profit under agreements regulated sectorally, but passing the responsibility to outside territories for workers who are sick, ageing and responsible for a family.

Will the new agreements on free circulation, and sometimes the substitution of workers from third countries by workers from the EU, radically change this situation? There is no evidence of that. As discussed for the Greek case by Kasimis, Papadopoulos, and Pappas (2010), the regulation potential of migrant labour is effective as long as it allows securing the profitability of the farms. As a matter of fact, new forms of triangulation of labour contracts are developing, with ambivalent employment statuses that are difficult to link up to precise social rights.

Various modalities are explored by temporary employment agencies and service providers to make foreign workers available for French farmers. There is a growing disconnection between the nationalities of the agency or service provider and those of the posted workers. For example, a large number of Spanish agencies or service providers make workers from Eastern European countries (Romania, Bulgaria, Poland, etc.) or beyond (Ecuador, etc.) available to farmers in the south of France (DGT, 2012). In the best of cases these firms simply play on the differences in the cost of social protection between countries, to offer services at a lower cost than French firms could. In other cases the reduced cost of services is also the result of fraud qualified by the ministry of labour as either 'simple' (failure to pay wages, exceeding the number of hours of work legally allowed, failure to pay overtime, etc.) or 'complex' (illegal work, failure to report labour accidents, even modern forms of modern slavery). In addition, *'The gravity of this fraud is accentuated by the lack of hygiene and safety, medical attention, and prevention of occupational hazards (work accidents, occupational diseases, etc.)'* (DGT, 2012, p. 53[7]).

How many individuals are concerned? It is totally impossible to answer that question. The firms that post these workers are supposed to officially inform the ministry of labour, but this is not always done, and the services concerned seem to be unable to enforce the law. The last report by the labour administration on this subject (DGT, 2012) is filled with conclusions of the following kind: *'Languedoc-Roussillon. 2nd semester 2011. Most of the agricultural firms operating on the Hauts-Cantons systematically fail to inform the labour inspectorate of posted workers. These firms never answer their mail'* (p. 10). *'Languedoc-Roussillon. 2nd semester 2011. The policy of domiciliation in Spain of firms that are mostly Romanian (or Bulgarian) continues to prosper as it allows for the employment of Romanian and Bulgarian workers registered with the police in Spain, who are then posted to French firms under service provision contracts equivalent to the lending of illicit manpower or even as "marchandages"'*[8] (p. 30). The powerlessness of the ministry in charge of labour to do anything about this situation is attested in its 2011 report where it renounced to estimate the number of cases of service provision not reported and the extent of the fraud (DGT, 2012, p. 22).

We are thus witnessing a growing segmentation of the status of individuals working on farm holdings (Rau & Mesini, 2007), along with a huge difficulty in obtaining precise data on this form of employment. The relative invisibility of these changes is striking. From the sectoral point of view, the question of the cost of work is fundamental for the future. But it is not necessarily in the profession's interest to attract attention to the growing numbers of occasional wage labourers when their employment conditions are often bordering on the illegal. From the point of view of territorial development, poor working conditions on farms can end up having perverse effects in the long run, which the community will have to deal with. This is, for example, the case of the diseases resulting from exposure to dangerous products such as pesticides, described by Decosse (2011). The long-term cost for local authorities can end up being high. One might therefore imagine that strong resistance would emerge from among the actors of territorial development at regional level (territorial authorities) or national level (health insurance system). But the use of migrant labour leads to this responsibility being dumped on the countries from which these workers have come. It also lessens this type of contradiction since long-term health problems, for example, have to be dealt with elsewhere. Here again, there does not seem to be an advantage to drawing attention to this issue.

CONCLUSION: UNEXPECTED CONVERGENCE OF INTERESTS TO PUT FORWARD A NEW TYPE OF VIRTUAL FARMER?

The institutional space of the territorial development provides attractive pictures of FMA which are valuable for all the stakeholders of the region. These pictures stress the positive role played by agriculture in defending a regulatory style based upon the old notion of 'public interest' or the new liberal motto of 'contribution to public goods'. Indeed, there are farms that contribute to maintaining magnificent landscapes, versatile rural employment and lively local markets, where they offer quality products. The multifunctionality of these farms is the showcase of regional agriculture.

But the development of these strategies has been unable to curb the considerable reduction in the number of farm holdings in FMA. In parallel, the transformation of the working conditions of a part of agricultural manpower attests to the difficulty that farms have in maintaining the profitability of their activities in the current conditions of international competition.

But this does not result in the denial of multifunctional agriculture as it was described in the UK by Marsden and Sonino (2008). Basically, both institutional spaces of the territorial and the sectoral development have been unable to design farm development models that might allow a wide range of farms to compete in internationalized markets and to fully meet the costs of the labour they need. As a consequence, and without it being the result of any deliberate strategy, there is an objective convergence of the interests of the two institutional spaces of regulation to ensure that the working conditions of temporary migrant workers remain invisible, and to put forward an image of multifunctional agriculture.

For the actors of territorial development, the use of this migrant labour ensures that part of the costs, which for French employees would be carried here (care for children, social housing, treatment of work-related chronic diseases, etc.) are transferred to other territories. On the other hand, it is important for the economy of the region, for tourism and for the sale of agro-food products, to preserve a positive brand image of Mediterranean agriculture.

For the dominant actors of sectoral development, it is necessary above all to preserve the possibility of maintaining labour at the lowest possible cost and of keeping attention away from actual working conditions. The image of multifunctional agriculture, harmoniously integrated into territorial development, is a marketing asset.

Thus, a new model of the 'virtual farmer' is born. This notion of 'virtual farmer' was proposed by J.D. van der Ploeg (2003) who showed that *'a shared image of agriculture and the countryside (…) may be based on a theoretical model of farming, where the characteristics of the real farmers, their work, their environment, their point of view, are further and further removed from this image'* (p. ix). The last two decades of the twentieth century were marked by deceptive representations of agriculture, dominated by the model of a 'virtual farmer' that excluded from the analysis and from public decisions a portion of small- and medium-sized farms contributing to maintaining a rural economic and social fabric (van der Ploeg, 2003; van der Ploeg, Laurent, Blondeau, & Bonnafous, 2009). Today's situation is tending the other way. The desirable image is that of a new 'virtual farmer' who, in a win-win game, combines all the functions expected from multifunctional agriculture while simultaneously meeting the profitability demands of intensive agriculture in productive orientations which receive little support from the CAP.

This image has several kinds of impact. It could well function partially as a self-fulfilling prophecy and trigger virtuous feedback loops. For example, in short supply chains some farmers in direct contact with consumers are encouraged to adjust their production methods to match their customers' representations.

But this image can also be misleading as it makes invisible, issues that are essential for the future. Thus, it generates knowledge gaps; the difficulties to get reliable data on the ecological print of short supply chains or on the real working situations of the casual workers are prominent examples. It also leads to the depoliticization of debates on the development models of agriculture in masking the contradictions and the conflicts of interest that they generate. Yet, an issue such as the new patterns of social dumping undermines not only the relationship between workers and employers, but also between regions within the EU itself.

NOTES

1. In the European statistics, one annual work unit (AWU), corresponds to the work performed by one person who is occupied on an agricultural holding on a full-time basis. Full-time means the minimum hours required by the relevant national provisions governing contracts of employment. If the national provisions do not indicate the number of hours, then 1,800 hours are taken to be the minimum annual working hours: equivalent to 225 working days of eight hours each. As the volume of agricultural labour is calculated on the basis of fulltime equivalent jobs, nobody can represent more than one AWU, even if someone works on agricultural activities for more than the maximum number of hours defining full-time work in that Member State. (Regulation/1200/2009/EC)

2. The standard output corresponds to the standard value of the gross production of the holding for one year. It is calculated at farm-gate price. It excludes direct payments. It is expressed in Euros. Its calculation is based on average regional values calculated over a reference period of five years. (Regulation 1242/2008/EC)

3. http://portail.123.fr/bis/DA91/2011/08/defense-de-lagriculture-paysanne-et-des-circuits-courts/

4. In this case the employee works for several farmers who belong to the employer grouping.

5. Formerly the *Office des migrations internationales* (OMI) which then became the *Agence nationale de l'accueil des étrangers et des migrants* (ANAEM), and finally the OFFI.

6. There are additionally a number of variants depending on bilateral contracts signed by France, or arrangements between federations of entrepreneurs and territorial administrations.

7. This report concerns the whole of France and all sectors of employment.

8. 'Marchandage'. The lending of manpower is strictly controlled by French labour law which distinguishes between the prohibition on the lending of manpower for profit (L.125-1) and the 'offence of merchandise' (L.125-3). There is an 'offence of merchandise' when the operation of providing manpower entails damages for the employee, for example the payment of wages inferior to the current minimum wage in France (Mouriaux, 2006).

REFERENCES

AC (Agricultural Census). (2010). On line data. Agreste.

Agreste (2011). Recensement agricole (2010). Premières tendances. Fiches départementales et régionales. *Agreste données*. Passim.

Arnalte-Alegre, E., & Otiz-Miranda, D. (2013). The southern model of European agriculture revisited: Continuities and dynamics. In *Agriculture in Mediterranean Europe. Between old and new paradigms* (pp. 37–74). Bingley, UK: Emerald.

Aubry, C., & Chiffoleau, Y. (2009). Le développement des circuits courts et l'agriculture péri-urbaine: Histoire, évolution en cours et questions actuelles. *Innovations agronomiques*, *5*, 53–67.

Bauder, H. (2006). *Labor movement: How migration regulates labor market*. New Oxford University Press.

Berlan, J. P. (1983). L'agriculture méditerranéenne de la France: Dynamiques et contradictions (I. Théorie du marché du travail). *Economie rurale*, *153*, 42–49.

Breem, Y. (2012). *Immigration et présence étrangère en France en 2011*. Rapport du SOPEMI pour la France. p. 76.

Celavar. (2011). *Circuits courts & Cohésion sociale*. Capitalisation et analyse d'expériences conduites sur les territoires ruraux et périurbains. Synthèse et annexes, 40pp.

Clary, G., & Van Haecke, Y. (2001). *Enquête sur l'emploi des saisonniers agricoles étrangers dans les Bouches-du-Rhône. Inspection générale de l'Agriculture*. Rapport n° 2001-118. p. 26.

Commission Nationale de Lutte contre le Travail Illégal. (2007). *Bilan 2006 du Plan National de lutte contre le travail illégal*. p. 11.

Commission Nationale de Lutte contre le Travail Illégal. (J.M. Ayrault Président). (2012). *Plan national de lutte contre le travail illégal*. p. 34.

Council of Europe. (2003). *The challenges for Mediterranean agriculture*. Parliamentary Assembly. Resolution 1331, p. 3.

Decosse, F. (2008). La santé des travailleurs agricoles migrants: Un objet politique ? *Etudes rurales*, *182*, 103–120.

Decosse, F. (2011). *Migrations sous contrôle. Agriculture intensive et saisonniers marocains sous contrat OMI*. Thèse de sociologie soutenue à l'EHESS le 1er décembre 2011 sous la direction d'Annie Thébaud-Mony.

DGT (Direction générale du travail). (2012). *Analyse des déclarations de détachement des entreprises prestataires de service en France en 2011*. p. 65.

Jarrige, F. (2004). Les mutations d'une agriculture méditerranéenne face à la croissance urbaine: Dynamiques et enjeux autour de Montpellier. *Cahiers Agricultures*, *13*(1), 64–74.

Kasimis, C., Papadopoulos, A., & Pappas, C. (2010). Gaining from rural migrants: Migrant employment strategies and socioeconomic implications for rural labour markets. *Sociologia ruralis*, *50*(3), 258–276.

Lamine, C. (2005). Santé et qualification des produits. Coopération de scientifiques et d'agriculteurs autour de la nutrition méditerranéenne. *Nature, Sciences et Société*, *13*, 403–412.

Laurent C., Cartier C., Fabre C., Mundler P., Ponchelet D., & Rémy, J. (1998). L'activité agricole des ménages ruraux et la cohésion économique et sociale. *Economie Rurale n*, *244*, 12–21.

Laurent, C. (2001a). La multifonctionnalité de l'agriculture. In M.-F. Durand, P. Giordano & A. Valladao (Eds.), *vers un accord entre l'Europe et le Mercosur* (pp. 409–425). Presses de Sciences Po.

Laurent, C. (2001b). Multifunctionnality of agriculture. In P. Giordano, A. Valladao & M.-F. Durand (Dir.), *Towards an agreement between EU and the Mercosur* (pp. 409–425). Paris: Presses de Science Po.

Laurent, C. (2006). *Multifunctionality of activities, plurality of identities and new institutional arrangements. State of the art for France*. Multagri report. p. 30.

Laurent, C., & du Tertre, C. (2008). Secteurs et territoires dans les régulations émergentes. *Paris, L'harmattan*, 236.

Laurent, C., & Mouriaux, M.-F. (2008). *Secteurs, territoires, rapport social d'activités*. In *Laurent, du Tertre dir. Secteurs et territoires dans les régulations émergentes*. L'Harmattan, 21–42.

Laurent, C., & Rémy, J. (1998). Agricultural holdings: Hindsight and foresight. *Etudes et Recherches sur les Systèmes Agraires et le Développement*. Numéro spécial 'Nouveaux regards sur l'exploitation agricole'. J. Brossier et B. Dent ed. n 31, pp. 415–430.

Lauret, F. (1992). L'agriculture méditerranéenne française et l'évolution de la politique agricole commune. *Economie Rurale*, *211*, 53–60.

Maillard, L., Daniel, K., & Colson, F. (2000). Géographie de l'agriculture communautaire: Les productions soutenues par la PAC ne se concentrent pas. *Agreste Cahiers*, *4*, 37–44.

Marsden, T., & Sonino, R. (2008). Rural development and the regional state: Denying multifunctional agriculture in the UK. *Journal of Rural Studies*, *24*, 422–431.

Mesini, B. (2008). Flexi-insécurité dans un secteur en tension: Processus de segmentation statutaire et éthique du marché des saisonniers étrangers dans l'agriculture. *Revue Asylon*, *4*, 20p. http://www.resear-terra.eu/article.html

Michalon, B., & Potot, S. (2008). Quand la France recrute directement en Pologne. Réseaux transnationaux et main d'oeuvre agricole. *Etudes rurales*, *182*, 87–102.

Morice, A., & Michalon, B. (2008). Introduction Special issue Les migrants dans l'agriculture: Vers une crise de main d'oeuvre ? *Etudes rurale*, *182*, 9–28.

Mouriaux, M.-F. (2006). *Du fait au droit, diverses figures du temps partagé*. Working paper. Centre d'Etude de l'Emploi. p. 23.

Paoli, J.-C., & Santucci, P.-M. (2010, June). *Le problème des parcours méditerranéen au regard du dispositif de prévention des incendies. Le cas de la Corse*. In Innovation et Développement Durable dans l'agriculture et l'agroalimentaire. ISDA, Montpellier, p. 10.

Perrin, C. (2009). *Construire les campagnes méditerranéennes. Usages, aménagement et valorisations du foncier agricole périurbain en Provence et en Toscane* (1950–2010). PhD in Geography. University of Provence - Aix-Marseille I. 663 p. + annexes.

Rau, V., & Mesini, B. (2007, December 6–8). *Segmentation statutaire et ethnique du marché de l'emploi en agriculture: Le cas des saisonniers migrants dans la production de fruits et légumes en Méditerranée*. Proceedings of the International conference « New migration dynamics: Regular and irregular activities on the European labour market». University of Nice-Sophia Antipolis. pp. 14–37.

Rogaly, B. (2006). *Intensification of work-place regimes in British agriculture. The role of migrant workers*. Sussex migration working paper n°. 36. p. 18.

Schlich, E., Biegler, L., Hardtert, B., Luz, M., Schröder, S., Schoeber, J., & Winnebeck S. (2006). La consommation d'énergie finale des différents produits alimentaires, un essai de comparaison. *Courrier de l'environnement*. INRA. n° 53, 111–120.

Schlich, E., & Fleissner, U. (2005). The ecology of scale: Assessment of regional energy turnover and comparison with global food. *Int J LCA*, *10*(3), 219–223.

Soulard, C., & Thareau, B. (2009). Les exploitations agricoles périurbaines: Diversité et logiques de développement. *Innovations Agronomiques*, *5*, 27–40.

Tallon, H. (2012). *L'accompagnement à la pluriactivité en Languedoc-Roussillon*, Projet PSDR INTERSAMA, Région Languedoc-Roussillon, Série Les focus PSDR3.

van der Ploeg, J.-D. (2003). *The virtual farmer past, present and future of the dutch peasantry*. Uitgeverij Van Gorcum.

van der Ploeg, J. D., Laurent, C., Blondeau, F., & Bonnafous, P. (2009). Farm diversity, classification schemes and multifunctionality. *Journal of environmental management*, *90*(Suppl. 2), S124–S131.

Volle, A. (2006). Agritourisme et productions biologiques dans les Alpes-de-Haute-Provence: Indicateurs de la redynamisation des campagnes ? *Méditerranée* (107), 67–71.

CHAPTER 8

REGIONAL PATTERNS OF STRUCTURAL CHANGE IN ITALIAN AGRICULTURE

Biagia De Devitiis and Ornella Wanda Maietta

ABSTRACT

This chapter highlights some features of the current structure of Italian agriculture by focusing on the regional patterns of agrarian change. These patterns are followed mainly by comparing the data of the 5th and 6th Census of Agriculture and the data of holdings registered to the Chambers of Commerce. The analysis confirms the Northern-Southern dichotomy of Italian agriculture as the physical and economic dimensions of Northern regions' holdings are appreciably higher than those in the South. Other traits of farms, not usually included in most traditional analyses, help explain that Northern-Southern dichotomy: the farmers' educational level and the ICT availability on farms. The agriculture of Southern regions has been affected less by the structural adjustment and has maintained some traits of more 'traditional' farming. However, important innovations, such as organic farming and direct selling to 'consumers in house', have been adopted more readily by Southern farms. The marked regional duality of Italian agriculture corresponds to the several ways in which

Agriculture in Mediterranean Europe: Between Old and New Paradigms
Research in Rural Sociology and Development, Volume 19, 173–205

ISSN: 1057-1922/doi:10.1108/S1057-1922(2013)0000019010

farmers and their activities interconnect with territorial development models that have shown a deep regional differentiation.

Keywords: Structural adjustment; regional patterns; Italian agriculture; innovation

INTRODUCTION

The objective of this chapter is to highlight some features of the current structure of the Italian agriculture by focusing on the regional patterns of agrarian change.

The chapter consists of four parts. The first part provides a description of the traditional indicators of the current physical structure of Italian agriculture, mainly by making use of the data of the 6th Census of Agriculture, and of its evolution by comparing these data with data from the 5th Census of Agriculture. The aim is to analyse the dynamics of some aspects of farm structures in order to assess their capacity of attracting resources into the sector.

Next, a more detailed description of the immaterial resources used in the Italian agriculture is provided by focusing on farmers' human capital and on the use of ICT, which are inputs of the innovative capacity of farms. Following this section, a detailed analysis of the different types of innovations is presented, comprising official innovation definitions to include as well the diffusion of geographically based certifications, the expansion of organic farming and agritourism.

The fourth part tackles the relationships of farms with agro-industry and also with final consumers. The aim of this section is to assess the presence and the size of new types of chain integration at a regional level. The set of features compiled throughout the chapter are considered particularly relevant indicators of the competitiveness of Italian farms in light of the future lowering of direct public support to agriculture. Finally, remarkable conclusions on agrarian regional differentiation in Italy are highlighted.

ITALIAN FARMS: SIZE, LAND PROPERTY, LEGAL STATUS AND MIX ORIENTATION

According to the figures of the Census of Agriculture in Italy, 1,620,844 farms (Table 1) were active in 2010. These farms use 12.9 million hectares

Table 1. Farms, Utilised Agricultural Area and Total Area by Geographical Area[a] in 2000 and 2010.

Area	Farms		Variation	UAA		Variation	Total Area		Variation
	2000	2010		2000	2010		2000	2010	
	Number		%	Ha		%	Ha		%
North-West	22,0145	14,5243	−34.0	2,243,193	209,6985	−6.5	3,127,737	2,745,985	−12.2
North-East	367,052	251,859	−31.4	2,632,288	2,471,852	−6.1	4,003,085	3,538,563	−11.6
Centre	423,085	252,012	−40.4	2,435,200	2,191,651	−10.0	3,898,892	3,349,801	−14.1
South	929,514	691,281	−25.6	3,571,517	3,554,349	−0.5	4,683,196	4,426,634	−5.5
Islands	456,478	280,489	−38.6	2,299,662	2,541,211	10.5	3,053,985	3,020,116	−1.1
Italy	2,396,274	1,620,884	−32.4	13,181,859	12,856,048	−2.5	18,766,895	17,081,099	−9.0

Source: Istat, 6th and 5th Census of Agriculture.

[a]The correspondence between regions and geographical areas is the following: North-West: Pimeonte, Valle d'Aosta, Liguria and Lombardia; North-East: Trentino Alto Adige, Friuli Venezia Giulia, Veneto and Emila Romaggna; Centre: Toscana, Marche, Umbria and Lazio; South: Abruzzo, Molise, Campania, Puglia, Basilicata and Calabria; Islands: Sicilia and Sardegna. See also the map in Fig. 1.

of land (Utilised Agricultural Area, UAA) which is only the 43% of the national territory, while the total farm area amounts to 17.1 million hectares.

The number of farms has substantially decreased (−32.4%) compared to the number registered in the previous census (2000). However, the reduction in UAA and in total farm area – due to the non-agricultural use of land or to the land abandonment – has been modest; 2.5% in the former and 9% in the latter.

As a result, the average size of Italian farms has increased and is equal to 7.9 hectares of UAA. This value is closer to the average farm size in the European Union (EU), which in 2007, was equal to 12.6 (Greco & Di Cristofaro, 2011).

The observed decrease in the number of farms has been weaker in the South of Italy (25.6% compared to a national average of 32.4%), and consequently has resulted in a higher concentration of farms in this geographical area, compared to the past. In fact, in 2000, 39% of Italian farms were located in the South of Italy while in 2010, this percentage was 43%.

The main structural dynamics emerging from the Census data show a process of land concentration. In terms of farm size, the decrease in the number of farms has been strong among farms of small and medium size with less than 30 hectares of UAA (Table 2), while those with more than 30 hectares have increased their presence both in absolute and in relative terms. The number of farms with more than 10 hectares of UAA was 252,419 in 2010, while the number of farms with more than 30 hectares of UAA was 85,617 in the same year.

Table 2. Farms and UAA by UAA Class in 2000 and 2010.

UAA Classes (Ha)	Farms				UAA			
	2000	2010	2000	2010	2000	2010	2000	2010
	Number		%		Number		%	
Without UAA	3,113	5,294	0.1	0.3				
<2	1,466,484	819,358	61.2	50.6	1,138,212	726,994	8.6	5.7
2–10	677,996	543,813	28.3	33.6	2,955,282	2,415,142	22.4	18.8
10–30	175,453	166,802	7.3	10.3	2,910,315	2,792,463	22.1	21.7
30–50	36,688	40,915	1.5	2.5	1,395,351	1,556,922	10.6	12.1
50–100	23,944	29,214	1.0	1.8	1,634,060	1,994,065	12.4	15.5
>100	12,596	15,488	0.5	1.0	3,148,639	3,370,461	23.9	26.2
Total	2,396274	1,620,884	100	100	13,181,859	12,856,048	100	100

Source: Istat, 6th and 5th Census of Agriculture.

The average size of the farms has grown in all regions. The largest increases are recorded in the Islands (+79.8%) and in the Centre of Italy (+51.1%), where the average size reaches respectively 9.1 and 8.7 hectares of UAA by farm. Nevertheless, farms in the North continue to have the largest average size (14.4 hectares of UAA by farm in the North-West and 9.8 in the North-East). The lowest value is reported in the South (5.1 hectares by farm), where the lowest relative increase over the past decade also is registered (+33.8%). Therefore, the size of the farms in the Centre of Italy and in the Islands is closer to that of farms in the North, while the gap in relation to the size of the Southern farms is increasing.

Fig. 1 reports the map of Italian regions.

Fig. 1. The Italian Regions.

The process of land concentration affects all regions, although to a different degree. In Sardegna, Sicilia, Puglia and Abruzzo (Table 3), a decrease of the number of production units corresponds to a positive change in the UAA. The remaining regions, however, are characterised by negative changes in both the number of farms and the UAA; the process of land concentration is mainly due to a change in the number of farms, which is more negative than that in the UAA.

Table 3 shows the average farm size by region: the lowest farm sizes are in Southern regions as a result of different co-determinants, such as better pedo-climatic conditions, a weaker economic system and a less urbanised and more land-owner population (see Table 4), whose link with small plots is still strong.

Table 3. Percentual Variation in the Farms' Average UAA, Number of Farms and UAA by Italian Region (2010/2000).

Regions	Farms' Average UAA			Farms	UAA
	2000	2010	Var.	Var	Var
	Ha		%	%	%
Piemonte	10.1	15.1	49.6	−36.8	−5.4
Valle d'Aosta	12.0	15.6	30.3	−40.0	−21.8
Liguria	1.7	2.2	24.0	−45.4	−31.4
Lombardia	14.6	18.2	25.6	−23.5	−5.1
Trentino Alto Adige	8.1	10.3	27.3	−28.3	−8.8
Veneto	4.8	6.8	41.1	−32.4	−4.6
Friuli Venezia Giulia	7.2	9.8	36.1	−32.5	−8.2
Emilia Romagna	10.6	14.5	36.1	−30.8	−5.8
Toscana	7.1	10.4	47.0	−40.0	−11.8
Umbria	7.1	9.0	27.3	−29.9	−10.8
Marche	8.1	10.5	29.6	−26.1	−4.2
Lazio	3.8	6.5	71.0	−48.2	−11.4
Abruzzo	5.6	6.8	20.7	−12.8	5.2
Molise	6.8	7.5	10.5	−16.7	−8.0
Campania	2.5	4.0	60.6	−41.6	−6.2
Puglia	3.7	4.7	27.6	−19.3	3.0
Basilicata	7.1	10.0	41.7	−31.8	−3.4
Calabria	3.2	4.0	25.3	−21.0	−1.0
Sicilia	3.7	6.3	72.3	−37.1	8.4
Sardegna	9.5	19.0	99.8	−43.4	13.1

Source: Istat, 5th and 6th Census of Agriculture.

Table 4. The Percentage of UAA by Type of Land Property Deed in Italian Regions (2000 and 2010).

Regions	Only Ownership		Only Rent		Only Free Use		All Combinations	
	2000	2010	2000	2010	2000	2010	2000	2010
Piemonte	28	17	3	6	1	1	68	76
Valle d'Aosta	17	5	7	19	1	1	76	75
Liguria	40	28	3	6	4	2	53	65
Lombardia	23	17	8	12	1	1	67	70
Trentino Alto Adige	43	39	0	1	2	3	54	58
Veneto	36	26	4	5	1	6	59	63
Friuli Venezia Giulia	29	23	3	6	1	1	67	70
Emilia Romagna	30	24	7	9	0	0	62	66
Toscana	37	29	5	6	2	6	57	59
Umbria	38	30	4	10	1	0	57	60
Marche	35	25	5	8	1	4	60	62
Lazio	43	33	2	7	1	1	54	59
Abruzzo	42	33	2	5	1	1	55	62
Molise	36	29	2	3	2	3	60	64
Campania	40	30	2	7	2	2	56	62
Puglia	43	35	2	4	1	3	54	58
Basilicata	41	32	2	5	2	2	56	61
Calabria	44	37	2	4	3	2	52	57
Sicilia	43	31	2	7	1	3	54	59
Sardegna	34	26	4	8	2	2	60	64
North-West	26	17	6	9	1	1	68	73
North-East	34	27	5	6	1	3	60	64
Centre	38	29	4	8	1	3	56	60
South	42	33	2	5	2	2	55	60
Islands	39	29	3	7	2	2	56	61
Italy	37	28	4	7	1	2	58	63

Source: Istat, 5th and 6th Census of Agriculture.

Small economic farm sizes generally correspond to small physical farm sizes, but average economic results are very different at a regional level. Information on economic sizes are available from the EU FADN sample which reports the number of Italian farms by region and by standard production.[1] The universe is represented by the Italian farms which have a standard production higher than 4,000 €: there were 833,552 such farms in 2010.

Among these, the number of small farms, defined as the class of 4,000–25,000 th.s € of standard production, is quite high: 70% of the EU universe of Italian farms in 2010. Even if small farms, as defined above, are more than two-thirds of total farms, they realise only one-third of the total value added. The highest percentage of small farms is observed in Calabria and the lowest in Piemonte.

Furthermore, small-sized farms have different farm family income[2]; on average, farm family income of the Italian small farms is 9,560 € in 2010, but it is more than twice this value for Valle d'Aosta (23,137 €) and Liguria (21,312 €). The lowest family farm income is observed in small farms in Emilia Romagna (5,004 €).

On the other hand, the higher regional percentage of large farms (>500 th.s € of standard production) is observed in Lombardia and Emilia Romagna; nevertheless, in many regions of Italy large farms are not even present (Valle d'Aosta, Liguria, Trentino Alto Adige and all Southern regions).

The correlation between the regional average farm size, reported in Table 3, and the regional average farm family income, derived from the FADN data, is quite high (around 0.60 in 2010).

Large-sized farms gather distinct farm resources through more modern and flexible contracts. This is the case of land because the percentage of sole property, as land property deed, has decreased in the last decade (Table 4), particularly in the North. All the other forms of land property deed have increased, and particularly sole rent has increased as a consequence of rent agreements which have been more flexible in length and rental fees (Greco & Di Cristofaro, 2011). 'Only ownership' is not diffused in Valle d'Aosta and Lombardia where only around 40% of farms follow this regime, while it is widespread in Calabria and Puglia, where more than 80% of holdings are only ownership.

A similar kind of reasoning can be conducted for farm capital as corporate management also is more widespread than in the past (Table 5): among the 859,808 Italian farms, registered to the provincial Chambers of Commerce in 2010,[3] sole proprietorship is still the prevalent form, but corporations and cooperatives (or other partnership associations) are more widespread than in the past. The regions where corporations and cooperatives have grown more in terms of UAA are Valle d'Aosta (12%), Lombardia (11%), Emilia Romagna (11%) and Friuli Venezia Giulia (10%).

The regions where 'sole proprietorships' are less diffused among the holdings are Lombardia and Emilia Romagna.

Mix orientation has not substantially changed compared to the past (Greco & Di Cristofaro, 2011), confirming the process of geographical productive specialisation at a regional level, which is summarised in

Table 5. The Percentage of Farms by Legal Status in Italian Regions (2000 and 2010).

Regions	Corporations		Partnership Associations		Sole Proprietorships		Other Legal Forms	
	2000	2010	2000	2010	2000	2010	2000	2010
Piemonte	0.4	0.7	5.2	7.1	94	91	0.7	0.8
Valle d'Aosta	0.2	0.4	1.3	6.0	97	92	1.2	1.5
Liguria	0.5	0.7	2.4	4.6	96	93	1.3	2.0
Lombardia	1.9	2.8	17.6	19.1	79	77	1.4	1.2
Trentino Alto Adige	0.1	0.4	1.6	2.4	97	96	1.2	0.9
Veneto	0.5	1.3	7.2	11.2	92	87	0.7	1.0
Friuli Venezia Giulia	0.6	1.2	5.1	9.3	93	88	1.0	1.3
Emilia Romagna	0.8	1.4	9.8	13.9	88	84	1.2	1.2
Toscana	1.7	3.1	8.1	10.1	89	85	1.3	1.4
Umbria	1.1	1.9	7.1	10.4	91	86	1.1	1.3
Marche	0.4	0.9	6.2	8.9	93	89	0.9	1.0
Lazio	2.4	3.8	3.2	5.1	92	89	2.1	2.4
Abruzzo	0.5	0.9	1.7	2.9	97	95	1.0	1.1
Molise	0.2	0.5	0.9	1.3	98	97	0.8	0.9
Campania	0.7	1.7	1.0	1.5	96	95	2.0	2.1
Puglia	0.4	1.2	0.9	2.3	98	95	1.1	1.9
Basilicata	0.5	1.0	1.1	1.6	97	96	1.8	1.7
Calabria	0.7	1.2	2.3	2.2	94	93	2.7	3.5
Sicilia	0.5	1.1	1.5	2.8	95	92	2.6	3.9
Sardegna	0.4	0.9	3.7	6.1	95	91	1.3	1.9
North-West	1.0	1.6	9.6	11.7	88	86	1.0	1.1
North-East	0.6	1.2	7.2	10.6	91	87	1.0	1.1
Centre	1.5	2.7	5.9	8.2	91	88	1.4	1.6
South	0.5	1.3	1.2	2.0	97	95	1.5	2.0
Islands	0.5	1.0	2.0	3.7	95	92	2.3	3.4
Italy	0.8	1.5	4.8	6.7	93	90	1.4	1.8

Source: Unioncamere – Movimprese (www.infocamere.it/movimprese).

Table 6. Geographical productive specialisation is very marked and even sub-regional: for example, more than the 50% of national cows' milk production is realised in the the provinces of Lodi, Cremona, Mantova (Lombardia) and Parma, Reggio Emilia and Modena (Emilia Romagna). It is mainly devoted to the production of the PDO (Protected Designation of Origin) cheeses Grana Padano and Parmigiano-Reggiano. In the same provinces, pork production is concentrated, while poultry production is

Table 6. Regional Productive Specialisation.

Production	Regions
Common wheat	Piemonte, Emilia Romagna
Durum wheat	Puglia, Sicilia, Baslicata
Tomato	Emilia Romagna, Campania, Puglia
Flowers	Liguria, Toscana, Campamia
Apple	Trentino Alto Adige
Apricot	Campania
Peach	Emilia Romagna
Olive groves	Southern regions
Citrus trees	Southern regions
Table grapes	Puglia
Wine grapes	Sicilia, Puglia, Piemonte, Friuli Venezia Giulia, Veneto
Cow's milk	Lombardia, Emilia Romagna
Buffalo farming	Lazio, Campania
Pig meat	Lombardia, Emilia Romagna
Poultry	Veneto, Emilia Romagna
Ewe farming	Sardegna, Lazio
Goat farming	Sardegna, Calabria, Basilicata

Source: Fanfani (2009), Messori and Ferretti (2010).

concentrated around two processing poles: Verona (Veneto) and Forlì-Cesena (Emilia Romagna). Durum wheat is produced in the provinces of Foggia (Puglia), Caltanisetta (Sicilia) and Matera (Basilicata), while common wheat mainly is produced in the provinces of Alessandria (Piemonte) and Bologna (Emilia Romagna). Tomatoes mainly are produced in the provinces of Parma, Piacenza (Emilia Romagna), and Salerno (Campania) (Fanfani, 2009; Messori & Ferretti, 2010). At any rate, some new territorial specialisations are emerging from the Census data as energy crops are concentrated in the North.

This specialisation in process has been accompanied by the productive intensification in plains and an even more marked territorial concentration in the food and drink industry, giving birth to agro-food districts whose relevance, in some cases, is not limited to the local economy. In 2010, exports of four agro-food districts overcome 500 million €, three of them (wines from Langhe, Roero & Monferrato, Alba & Cuneo Confectionary, and Verona Wines) are located in the North of Italy and one (Processed Crops of Nocera Inferiore) in the South of Italy (Intesa San Poalo, 2011).

HUMAN CAPITAL AND ICT INVESTMENTS IN ITALIAN AGRICULTURE

Human Capital

Economic literature recognises human capital as a key factor of productivity growth. The existence of an efficient agricultural education system is crucial for an internationally competitive agriculture. In a knowledge-based and globalised economy, countries have to be fast in the adoption of new technologies, in order to gain competitiveness and be able to exploit national comparative advantages. In the food and agricultural sector, new technologies cannot be imported fully because country-specific natural conditions require country-specific education and research to adapt new technologies (Csaki, 1999).

The level of human capital in EU agriculture can be tackled by using as a proxy variable the percentages of farms' managers with full agricultural training, sourced from the Eurostat Farm Structure Survey conducted in 2005.

Table 7 reports the percentage of farmers with basic or full agricultural training for the EU-27 countries. The Italian rank is very low: it is the 21st country for the level of the variable.

Table 7. The Percentage of Farmers with Basic or Full Agricultural Training in 2005 in EU Countries.

Countries	% Farmers	Countries	% Farmers
Belgium	47.7	Luxembourg	55.9
Bulgaria	5.3	Hungary	13.4
Czech Republic	44.7	Malta	0.4
Denmark	44.5	Netherlands	71.5
Germany	68.5	Austria	48.1
Estonia	32.9	Poland	38.5
Ireland	30.7	Portugal	11.8
Greece	5.4	Romania	7.4
Spain	10.5	Slovenia	28.0
France	54.3	Slovakia	14.6
Italy	11.2	Finland	40.6
Cyprus	6.4	Sweden	33.6
Latvia	34.1	United Kingdom	23.2
Lithuania	30.9	EU-27	20.0

Source: Eurostat, Farm Structure Survey 2005.

A first explanation of this sharp difference is the different demand for higher skills in the national labour market and the way in which the degree and qualification structures are organised within countries. Some countries have not introduced the differentiation into two kinds of tertiary qualifications (OECD, 2003): qualifications at the tertiary-type B level (ISCED 5B) and qualifications at the tertiary-type A level (ISCED 5A). Tertiary-type A programmes are largely theoretically based and designed to provide qualifications for entry to advanced research programmes and professions with high skill requirements. Countries differ in the length and in the way in which tertiary type A studies are organised. This means that the percentage of farmers with tertiary education is not completely comparable between countries because the shares of the two kinds of tertiary qualification are different in the EU nations,[4] and a low level of the tertiary graduates could mainly reflect a higher presence of the tertiary-type A level, to which correspond a higher number of education years. Countries like Germany, for example have a practical and agricultural-specific school programme for those who intend to become a farm manager. A second explanation is the effect of different farm structures and farm size distributions: big farms are more likely to be endowed by a high level of farmer's human capital but also, in a strongly polarised farm size distribution, any increase in the level of human capital in the few professional farms is hidden by the absence of increase or by the low increase in the level of human capital in the numerous marginal farms.

Table 8 reports the percentages of famers by education levels and by regions in Italy in 2000 and 2010. The first observation is that the level of human capital has improved in 10 years. In 2010, 92,693 farmers have a bachelor qualification, of whom 11,528 are agricultural-specific and 81,165 not specific, and the percentage of farmers with a bachelor qualification has increased from 3.5% in 2000 to 6% in 2010.

However, by and large, the level of human capital is still low: as generally for Italy, the tertiary qualifications, both agricultural and non-agricultural bachelors, are not so diffused. The percentages of farmers with an agricultural-specific tertiary qualification in 2010 are higher in Lombardia, Emilia Romagna and Umbria. In Campania, this percentage has even decreased from 0.7% in 2000 to 0.4% in 2010.

A second observation is that Italian agricultural education system is not efficient as agricultural schools, at any level, are not attractive for farmers, and the percentages of farmers with a non-agricultural qualification are much higher than the percentage of farmers with an agricultural-specific qualification of the same level. The reasons of this low efficiency are

Table 8. The Percentages of Farms' Managers by Region and by Education Level in Italian Regions.

Regions	2000					2010				
	Primary	Agr. secondary	Secondary	Agr. bachelor	Bachelor	Primary	Agr. secondary	Secondary	Agr. bachelor	Bachelor
Piemonte	82.3	2.4	13.0	0.4	1.9	71.3	5.2	19.4	0.9	3.2
Valle d'Aosta	83.1	1.7	13.9	0.2	1.1	71.7	4.6	21.0	0.6	2.1
Liguria	79.0	2.0	16.2	0.3	2.5	63.7	2.7	28.0	0.5	5.1
Lombardia	77.7	4.4	14.5	0.9	2.5	65.1	7.9	21.1	1.5	4.5
Trentino Alto Adige	85.5	3.0	9.1	0.4	2.0	56.2	15.1	25.1	0.7	2.8
Veneto	80.7	3.5	13.6	0.4	1.9	77.8	4.4	14.6	0.6	2.5
Friuli Venezia Giulia	84.5	2.8	11.0	0.4	1.3	70.9	5.9	19.6	0.8	2.8
Emilia Romagna	77.0	5.0	13.8	1.1	3.1	68.6	7.2	18.3	1.4	4.5
Toscana	77.1	2.0	15.4	0.6	4.8	68.0	2.8	20.8	0.9	7.5
Umbria	73.6	1.7	18.8	0.7	5.2	62.4	2.8	25.8	1.2	7.8
Marche	79.9	2.0	14.1	0.5	3.5	76.0	2.7	15.5	0.7	5.0
Lazio	79.1	1.7	16.0	0.4	2.7	70.3	1.7	22.5	0.4	5.0
Abruzzo	83.4	2.2	11.6	0.4	2.4	71.7	2.4	20.6	0.5	4.8
Molise	83.0	1.4	12.8	0.3	2.5	73.7	2.6	17.9	0.6	5.3
Campania	83.8	1.8	11.3	0.7	2.4	73.7	1.7	19.1	0.4	5.1
Puglia	80.2	1.9	14.5	0.4	3.1	75.3	2.0	16.4	0.6	5.7
Basilicata	83.4	2.1	11.9	0.4	2.2	72.6	2.9	19.0	0.7	4.8
Calabria	79.9	1.7	15.1	0.4	2.9	73.5	2.6	17.8	0.6	5.5
Sicilia	79.9	1.7	13.7	0.5	4.1	70.4	2.3	18.7	1.0	7.7
Sardegna	85.6	1.8	10.2	0.5	2.0	78.8	3.2	13.5	0.9	3.6
Italy	80.6	2.2	13.7	0.5	3.0	72.1	3.3	18.6	0.7	5.3

Source: Own elaboration on Istat – 5th and 6th Census of Agriculture data.

several: substantially, the programmes of the agricultural-specific secondary schools have not been targeted to the training of a modern farmer and the competences supplied have not been perceived as sufficient and adequate (Maietta, 2004; Se.Ri.Fo., 2010). More recently, agricultural schools also have been affected by the general cutting of public expenditure for education.

A further relevant aspect of human capital relates to the age and the gender of farm managers and, for corporations or cooperatives, to the composition, in terms of age and of gender, of the corporate governance bodies (Table 9).

The percentage of women is still low but it has increased in all positions relevant for firm management in the Italian agriculture; on the contrary, the percentage of people 50 years old or more has not decreased in the decade.

More precisely, according to the Census data, in the intercensus period some improvement has been observed in the age class distribution of farm

Table 9. Age and Gender of Italian Farm Managers and Similar Positions in 2000 and 2010.

Position	2000	2010
Male		
Main owners	78.1	73.4
Co-op members/Other owners	6.1	7.8
Directors	10.6	15.0
Others	5.3	3.8
Total	100.0	100.0
Female		
Main owners	29.1	31.0
Co-op members/Other owners	30.5	33.1
Directors	15.1	19.0
Others	15.6	17.2
Less than 30 years old		
Main owners	4.2	3.7
Co-op members/Other owners	8.1	5.4
Directors	6.1	4.7
Others	6.4	1.3
50 years old or more		
Main owners	63.7	63.7
Co-op members/Other owners	46.4	51.6
Directors	46.6	51.7
Others	41.4	51.7

Source: Unioncamere – Movimprese.

managers as the percentage of farm managers in the age class 'less than 30 years old' is stable (2.1%) and the percentages of farm managers in the age classes '30 to 34' and '35 to 39' have increased (respectively, from 2.9% to 3.2% and from 4.9% to 5.5%). The reason for this improvement can be found in the effect of the public aids for young farmers, defined as less than 40 years old. The percentage of young farmers increased from 9.9% in 2000 to 10.9% in 2010 as a consequence of two effects: an increase in the entry of farms held by young farmers, and a lower exit of farms with the same characteristics.[5]

The presence of young farmers is higher in the Southern regions: Pierri (2013) analyses the percentage of youth-run farms from Movimprese source in 2012. More particularly, Movimprese defines a business as 'youth-run' when the percentage of persons 35 years old or less in the corporate governance bodies is more than 50%. The percentage of youth-run farms in the Italian agriculture was 6.7% in 2012 but this percentage was around 14% to 15%, in decreasing order, in Calabria, Campania, Sicilia and Puglia due to a younger population and to the weaker economic situation.

Information and Communication Technology (ICT) Investments in Italian Agriculture

ICT investment in the Italian agriculture, even if increased compared with the last census (Cioffi & Gorgitano, 1998), is still very low: only 60,945 farms own ICT in 2010. They represent the 3.8% of total farms.

Generally, the ICT use is multi-purpose (see Table 10), in any case the three motivations are, in decreasing order: administrative duties, crop management and livestock management; 26,254 Italian farms use ICT for electronical sales (selling and purchasing).

The number of ICT-owning farms is higher in the North-West (10.9%) and very low in the South (1.3%). Twenty-five per cent of Italian holdings with Information and Communication Technology (ICT) are in only two regions: Lombardia and Emilia Romagna. In particular, Lombardia is the region with the highest number of farms owning ICT for administrative duties, livestock management and e-commerce, and Emilia Romagna is the region with the highest number of ICT-owning farms for crop management.

Toscana is the region with the highest number of farms using ICT for a farm homepage (due to the huge presence of agritourism farms), for electronical selling and simply for Internet use.

Table 10. The Number of Italian ICT-Owning and ICT-Using Farms by Region in 2010.

	With ICT	With ICT for			Using Internet	With Homepage	Electronic Selling	Electronic Purchasing
		Administrative duties	Crop management	Livestock management				
Piemonte	5994	4638	2031	1475	1599	2572	808	1313
Valle d'Aosta	259	222	55	58	137	92	51	126
Liguria	1205	1038	423	57	527	648	253	425
Lombardia	8322	5833	2521	3401	1747	2568	806	1472
Trentino Alto Adige	4889	3098	2214	721	1356	2325	793	795
Veneto	6493	4797	2452	1707	1288	2807	704	1037
Friuli Venezia Giulia	1946	1517	826	465	527	802	261	443
Emilia Romagna	7084	4994	3130	1521	1670	2252	833	1368
Toscana	4292	3650	1490	450	2141	4559	1515	1433
Umbria	1384	1160	567	273	675	1071	472	481
Marche	1472	1153	563	262	520	1097	334	388
Lazio	3005	2203	1347	580	1387	1283	728	1200
Abruzzo	898	726	349	183	446	577	260	377
Molise	385	289	144	95	168	264	63	142
Campania	2585	2140	1058	512	880	1201	519	740
Puglia	2798	2317	1153	277	997	1535	596	773
Basilicata	856	688	374	219	365	340	178	327
Calabria	1391	1219	598	226	585	600	348	520
Sicilia	3709	3132	1447	531	1401	1581	881	1128
Sardegna	1978	1502	720	551	1041	869	462	901
North-West	15780	11731	5030	4991	4010	5880	1918	3336
North-East	20412	14406	8622	4414	4841	8186	2591	3643
Centre	10153	8166	3967	1565	4723	8010	3049	3502
South	8913	7379	3676	1512	3441	4517	1964	2879
Islands	5687	4634	2167	1082	2442	2450	1343	2029
Italy	60945	46316	23462	13564	19457	29043	10865	15389

Source: Istat, 6th Census of Agriculture.

It is clear that Italian farmers do not use ICT as a tool for communication, as more traditional tools are prevalent. A possible and partial explanation is the presence of bad infrastructure in rural areas (Se.Ri.Fo., 2010).

INNOVATIVE CAPACITY OF ITALIAN FARMS

Innovation in Italian Agriculture

Information on the number of Italian innovative farms is scarce. The 'Osservatorio sull'innovazione nelle imprese agricole' of Agrisole2000 in Bologna collects information on this subject through a sample of farms defined as professional. The criterion is that the farmer is full-time (2,220 hours per year) and earns, after social security pro-rata payments and without public subsidies, at least the gross compensation of a manager in the food and drink industry (which is at minimum 35,000 €). In 2010, the farms defined as such have been estimated by 'Osservatorio sull'innovazione nelle imprese agricole' to be around 120,000 and represent 14% of Italian farms registered with the local Chambers of Commerce. Farms able to become professional are estimated to be one-third of Italian farms, registered with the local Chambers of Commerce, while the remaining ones are considered marginal by the 'Osservatorio sull'innovazione nelle imprese agricole'; they are one-half of the farms registered to the local Chambers of Commerce.

The survey in 2010 of a sample of 1,200 farms, revealed that 29% of farms introduced innovation in that year. This percentage was 61% in 2007, 38% in 2008 and 35% in 2009.

Fig. 2 reports the kind of innovation introduced by the Italian professional farms in 2010. Most of them still innovate through the introduction of new machinery, equipment and plants. Agro-energy, particularly photovoltaic energy, is the second important source of innovation for Italian professional farms.

The source of finance for innovation is represented by banks. In 2009, 83% of innovative farms have been financed by banks, 72% of them have a debt with a bank, the amount of which is higher than half of the investment. Among the other finance sources used for the rest of the investment, 40% of farms self-finance it, 30% use public aids, particularly from the Rural Development Plans, 26% use both self-financing and public aids, and the remaining 4% are financed totally by banks (Agrisole2000, 2009a, 2009b).

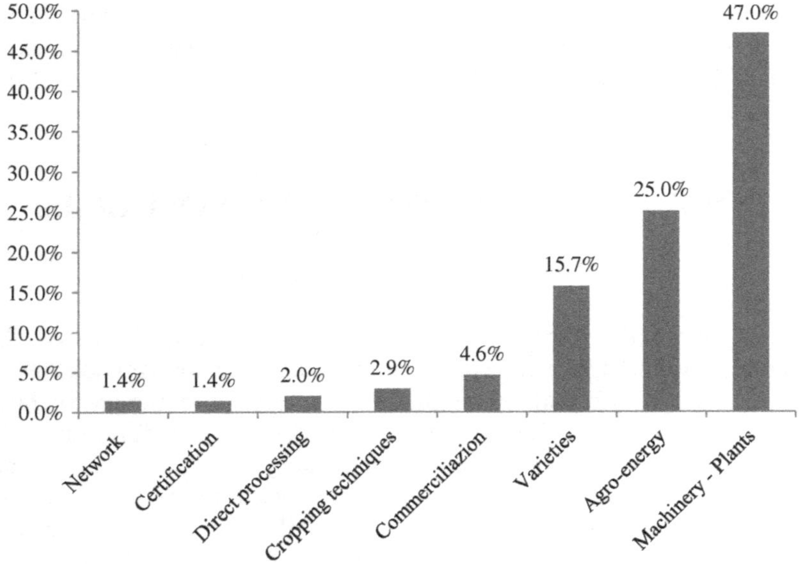

Fig. 2. The Innovation Adopted by Italian Professional Farms in 2010. *Source*: Gardini and Lazzarin (2011).

Farms which have not introduced innovation declare that it was their deliberate choice not to innovate (45%) or they encountered problems for credit rationing (28%), market instability (18%) or bureaucracy burden (6%).

The networks useful for innovation are still traditional (De Devitiis, Lopolito, Maietta, & Sisto, 2009): only 28% of farms state that they have stable relationships with other firms, universities or public research labs. Even the ties with cooperatives are weak: among the farm members of cooperatives, only 15% know whether their own cooperative has introduced initiatives to increase its own and its members' profitability.

In synthesis, the percentage of innovative farms in 2010 was equal to 4% of the total number of Italian farms, registered to the Chambers of Commerce, and to 2% of the total number of Italian farms, reported by the Census of Agriculture.

A low percentage of innovative farms of the total number generally is justified by the consideration that agricultural firms are less dynamic than those of the other sectors. For example, Becattini (2000) states that the benefits stemming from innovation are less appropriable in agriculture because production occurs outdoors and can be witnessed publicly, and the dynamics of the decomposition of productive process into very specialised and separate phases, which can be quickly converted towards the market-driven direction, is slowed in agriculture by technical and socio-cultural peculiarities. The socio-cultural determinants of lower dynamism in agriculture are: the habit to a less-quickly changing demand, the sense of belonging to an area, the major role of hierarchic relationships with respect to market relationships, the constraint of land property which blocks social and professional mobility within an area and prevents land consolidation. Spatio-temporal analysis confirms the presence of neighbourhood effects on the adoption of innovation in agriculture (Nyblom, Borgatti, Roslakka, & Salo, 2003). Pilati (2006) emphasises that the immobility of farm fixed assets implies a very high firm-specific investment in human capital which prevents plant delocalisation and transfer of the firm with its knowledge stock. Workers' skills are firm-specific because of the spatial dispersion of agricultural activities and of the seasonal diversification of tasks during the year.

All these factors, which can be used to explain a low presence of over-technologised farms, also could represent the determinants of the choice of other innovative patterns, linked to the local specificity of the new rural economy and accessible to small-sized farms (Brunori, Rossi, Cerruti, & Guidi, 2009).

The Italian Products of Designated and Guaranteed Origin

By and large, Italian foodstuffs have been associated with high-quality products worldwide. As Parrott, Wilson, and Murdoch (2002) identified, Italy is a clear example of South European food culture and agricultural production, which strongly associates quality with place, artisanal elaboration and tradition. Italy is the EU country with the greatest number of products of certified origin (Protected Designation of Origin (PDO) and Protected Geographical Indication (PGI)). Generally speaking, it can be said that the local culture of food is very strong and still represents the basis of regional identification as a consequence of the historical fragmentation of the Italian territory, where the cities have always been important markets for the peri-urban farm production (Dickie, 2007). This strong food culture is reflected in the birth of important citizen movements, like

Slow Food, the Solidarity Purchasing Groups (in Italian, Gruppi di Acquisto Solidale or GAS) or Libera, a non-profit association which contributes to the diffusion of a culture of legality in agricultural areas, fallen prey to organised crime, through the reuse with social aims of land belonging to criminal organisations after its confiscation (Ascione & Scornaienghi, 2009).

In this regard, many academic works have tackled the qualification process of Italian agrarian production from different perspectives. From the production side, many studies have tackled how farmers embed their products in specific territorial attributes and build up direct relationships with consumers through short food supply chains (Brunori & Rossi, 2000; Ventura & Milone, 2000). These analyses have mostly contributed to expand and underpin formulations on the new rural development paradigm (Kanemasu & Sonnino, 2009; Van der Ploeg, 2000, 2008). More recent works have tackled the consumption side, being particularly focused on the innovative Solidarity Purchasing Groups (Brunori, Rossi, & Guidi, 2012; Fonte, Eboli, Maietta, Pinto, & Salvioni, 2011). In a more comprehensive study, the importance of food in Italian identity has been shown through the analysis of the evolution of food discourse in Italy, showing the progressive integration of long-established food quality meanings with the food security framework (Brunori, Favilli, & Rossi, 2013).

The products of designated and guaranteed origin are not equally geographically distributed in the EU because the countries bordering the Mediterranean Sea have a clear advantage in terms of product number and variety. In fact, five countries (Italy, France, Spain, Portugal and Greece) account for nearly 80% of European PDO and PGI. In particular, Italy, with a total of 246 PDO and PGI, holds the record. Most are fruit, vegetables and cereals, cheese, oils and fats and prepared meats (Table 11).

Italian geographical indications are concentrated in few Italian regions: 47% of PDO and PGI are in five regions (Veneto, Emilia Romagna, Lombardia, Sicilia, and Toscana) with a different product specialisation. Cheeses are found in mostly Lombardia, Piemonte and Veneto, prepared meats in Emilia Romagna and Lombardia, oils and fats in Sicilia, Toscana, Campania and Puglia while fruit, vegetables and cereals are particularly present in Veneto, Sicilia, Campania and Emilia Romagna.

From an economic point of view, concentration is even stronger as 62% of turnover, evaluated at production prices, realised for PDO and PGI products in 2010, is due to two PDO cheeses (Grana Padano and

Table 11. Number of PDO and PGI Products, DOCG[a], DOC[a] and IGT[a] Wines by Italian Region.

Regions	PDO – PGI[b]						DOCG[d]	DOC[d]	IGT[d]
	Fruit vegetables and cereals	Oils and fats	Cheese	Prepared meats	Other products[c]	Total	Wines		
Piemonte	5		9	4	1	19	16	40	
Valle d'Aosta	0		2	2		4		1	
Liguria	1	1			1	3		8	4
Lombardia	2	2	12	10		26	5	21	15
Trentino Alto Adige	3	1	5	2	0	11		10	5
Veneto	17	2	8	7	1	35	14	26	8
Friuli Venezia Giulia	1	1	1	3		6	4	10	3
Emilia Romagna	11	2	5	12	5	35	2	17	9
Toscana	8	5	2	4	5	24	11	36	5
Umbria	2	1	1	2	1	7	2	12	6
Marche	2	1	2	4	1	10	5	15	1
Lazio	7	4	3	4	5	23	3	26	6
Abruzzo	2	3		1	2	8	1	6	7
Molise	0	1	2	1	2	6		4	2
Campania	11	5	3		2	21	4	15	9
Puglia	6	5	3		2	16	4	22	5
Basilicata	4	1	3		1	9	1	3	1
Calabria	5	3	1	4	2	15		9	9
Sicilia	14	6	4	1	2	27	1	20	7
Sardegna	1	1	3		2	7	1	15	15
Italy[b]	98	43	44	36	25	246	73	330	118

Source: MIPAAF, 15 December 2012.
[a]Situation updated to July 2012.
[b]Some products belong to more than one region, therefore, the sum of the DOP/IGP by regions does not match the total for Italy.
[c]Bread, honey, cheese, spices, vinegars, meats, fish, and non-food products.
[d]The national totals for DOC and IGT wines are lower than the sum of the regional totals because some wines are inter-regional.

Parmigiano Reggiano) and to two PDO hams (Prosciutto di Parma and Prosciutto San Daniele) (Vieri, 2012).

Italy is second in the EU, after France, for the number of PDO wines (Italian DOCG and DOC) and PGI (Italian IGT) and it is third, after Spain and France, for area cultivated with PDO and PGI vineyards.

The area under PDO in Italy was 272,433 ha in 2010, nearly 40% of the total area planted with vineyards. By adding the area cultivated with PDO and PGI vineyards, the share of quality grape is growing by more than 70%, still far from Spain and France, whose PDO and PGI vineyards' area reaches and exceeds 90% of total vineyards' surface area (Inea, 2012).

Italy is the second leading wine producer in the world, after France. The production of wine labelled with a PDO amounted to 15.743 million hectolitres, which represents almost 34% of the total wine produced in Italy in 2011 (Inea, 2012). Wine is the most important item of the Italian agrifood trade balance; in the years 2009 and 2010, Italian wine exports were respectively 3.6 and 4 billion € (Vieri, 2012).

PDO wines (especially red) represented more than half of the wine exports' value, with a total value of over 2 billion € realised in 2011 (Inea, 2012). In terms of hectolitres, the first PDO wine producers in 2010 in Italy have been, in decreasing order, Veneto, Piemonte and Toscana. The first PGI wine producers have been, in decreasing order, Veneto, Emilia Romagna and Sicilia (Vieri, 2012).

Both for wines as for other designated and guaranteed origin items, the productive concentration in the Northern regions is strong.

Other typical Italian food and beverage products are the ones labelled under the *Prodotto agroalimentare tradizionale* (*PAT*) scheme, which is only applicable within Italy and varies from one region to another, as summarised in Table 12.

These products are not protected through a trademark, as the previous ones, but simply are listed by the Ministry of Agriculture, through the information supplied by the regional administrations. The first classification, made in 2000, registered 2000 PAT. The aim of the classification is to preserve the knowledge of the local production techniques. In decreasing order, the most represented PAT are the bread and bakery products, the vegetable products, the meat products and the cheese. Some PATs are produced in the Slow Food Presidia (Fonte, 2006; Petrini, 2009), which are small producers' associations whose aim is to preserve traditional production techniques, biodiversity, local culture and identity, which are at risk of disappearing. In Italy, they are 223 of a total of 400 all over the world in 2012 (Slow Food, 2012).

Table 12. Number of Traditional Food and Beverage Products (PAT) by Italian Region in 2012.

Regions	
Piemonte	363
Valle d'Aosta	32
Liguria	295
Lombardia	242
Trentino Alto Adige	201
Veneto	370
Friuli Venezia Giulia	150
Emilia Romagna	285
Toscana	465
Umbria	69
Marche	150
Lazio	383
Abruzzo	147
Molise	159
Campania	370
Puglia	231
Basilicata	77
Calabria	269
Sicilia	234
Sardegna	178
Italy	4606

Source: MIPAAF, National List of Traditional Agro-food Products, 12th revision 17 June 2012.

Organic Farming in Italy

While production of designated and guaranteed origin products is concentrated in the North of Italy, the organic production is concentrated in the Southern regions.

In 2010, 37 million hectares of land were devoted to organic farming in the world. Italy covers 3.4% of this land, representing one of the top 10 producers in the world and is the leader in Europe, after Spain.

According to the 6th Census of Agriculture (Table 13), organic farmland has assumed a growing share in the Italian agriculture in the last decade, from 7.9% of UAA in 2000 to 9.7% in 2010. The number of organic farms accounted for 2.8% of total farms in 2010, as compared to 2.1% 10 years earlier.

Sicilia, with 229,204 organically cultivated hectares and Puglia, with 158,366 hectares, continue to be the regions most involved in this method

Table 13. Organic Farming as Share of Total Farms and Total UAA by Region, 2010.

Regions	Farms	UAA
Piemonte	3.0	4.4
Valle d'Aosta	2.1	3.3
Lombardia	1.9	8.9
Liguria	1.7	2.5
Trentino Alto Adige	2.9	2.3
Veneto	0.9	2.8
Friuli Venezia Giulia	1.3	2.5
Emilia Romagna	3.7	9.5
Toscana	3.4	10.3
Umbria	3.5	11.8
Marche	4.2	11.1
Lazio	2.8	12.5
Abruzzo	2.2	7.9
Molise	0.7	2.5
Campania	1.3	3.5
Puglia	1.9	12.3
Basilicata	6.2	20.3
Calabria	4.9	24.0
Sicily	3.6	16.5
Sardegna	2.6	9.1
Italy	2.8	9.7

Source: Istat, 6th Census of Agriculture.

of production. However, in Calabria, Basilicata, Lazio and Sardegna, the share of land involved in organic farming of total regional UAA is more important in relative terms. The territorial specialisation of the Southern regions in organic farming production can be explained by the regional public support given to this kind of production (Zanoli, 2007), by independent private initiatives (Milone, 2004) and by better climatic conditions.

Agritourism in Italy

Agritourism in Italy has experienced rapid growth and is different from the rural tourism widespread in other European countries because the link with the agricultural activities is prevalent: actually, agricultural activities must be predominant in respect to the tourism activities in order to receive public, national and regional, aids for this initiative.

The first Italian initiatives in the field of agritourism date back to the 1960s: the first and the largest national agritourism organisation in Italy, Agritourist, was created in 1965, but it is only during the 1980s that these initiatives became widespread, particularly in Trentino Alto Adige and Toscana. In these regions, which display the highest number of agritourism farms (36% of Italian agritourism holdings in 2010), their presence is capillary and historically rooted: in 1973 the Autonomous Province of Trento passed the first law in favour of agritourism farms.

From the supply side (Table 14), distinctive local diversity in farm and rural cultures and the entrepreneurship of Italian farmers have been cited as the determinants of agritourism development in Italy (Ohe & Ciani, 2012). From the demand side, the crisis of the mass-tourism demand, the choice of alternative locations and the emergence of new markets segments (i.e. educational services provided to school pupils) have worked towards the development of the agritourism market.

Table 14. Number of Agrotourism Farms in Italian Regions by Year.

Regions	1998	2001	2003	2004	2005	2006	2007	2010
Piemonte	180	515	515	619	761	772	882	1068
Valle d'Aosta	49	53	53	54	56	58	57	59
Liguria	288	279	279	347	339	361	368	488
Lombardia	907	914	914	1021	990	1050	1064	1412
Trentino Alto Adige	3944	3639	3639	3797	3816	3259	3071	3517
Veneto	766	914	914	962	1044	1148	1198	1120
Friuli Venezia Giulia	245	365	365	391	423	450	443	516
Emilia Romagna	927	880	880	932	1069	761	809	1005
Toscana	1512	3111	3111	3383	3527	4009	3977	3487
Umbria	440	672	672	656	890	952	1026	1229
Marche	312	431	431	516	537	681	747	683
Lazio	231	877	877	890	487	666	552	747
Campania	132	608	608	666	729	753	750	849
Abruzzo	311	458	458	526	459	535	600	540
Molise	21	68	68	74	78	82	82	97
Puglia	302	203	203	203	207	265	257	390
Basilicata	238	270	270	259	249	240	236	257
Calabria	113	266	266	299	313	330	461	488
Sicilia	155	323	323	332	342	396	422	679
Sardegna	385	567	567	593	393	656	718	673
Italy	11458	15413	15413	16520	16709	17424	17720	19304

Source: Istat (2008, 2009) and 6th Census of Agriculture.

Agritourism farms are mostly located in hills and in mountains, and only 15% of the total are in plains.

More than one-third of agritourism farmers are women, and are particularly concentrated in Toscana, where 25% of Italian female agritourism managers can be found.

Farms that offer rooms or apartments are concentrated in the North and in the Centre of Italy, while farms with restaurants are mainly located in the North and in the South of Italy (ISTAT, 2012).

THE SUPPLY CHAIN OF ITALIAN AGRICULTURAL PRODUCTS

According to the 6th Census data, 1,037,211 Italian farms sell their production in 2010, while 583,673 farms (36% of total farms) do not sell but self-consume their production.

The prevalent channels of product sale for Italian farms is still represented by wholesalers and farmers' associations or cooperatives, used respectively by 33% and 24% of Italian farms (see Table 15). Associations or co-ops are the prevailing channel in the North-East, where co-ops are historically more widespread.

New channels are emerging: 22% of Italian farms sell directly to consumers, in-house or in farmers' markets, particularly in the Centre and in the South of Italy. The regions where direct selling to 'consumers in house' are higher are Calabria, Campania, Liguria, Abruzzo and Toscana.

Recent short supply chains are those realised through the 'citizens-consumers' organised in Solidarity Purchasing Groups, which buy food products directly from farms. The number of these groups registered to the 'retegas' network[6] were only two in 1994, while their current number is 909 (Fig. 3).

According to the figures of Biobank (2012), the regions, with a higher number of GAS, are Lombardia (273), Toscana (110), Emilia Romagna (87), Veneto (82) and Lazio (76), which represent two-thirds of the total number of GAS, registered in 2011.

The scarce presence of GAS in the Southern regions can be explained by their lower urbanisation and their higher presence of consumers, buying directly and individually from farms.

Taking into account that the average annual purchase per group for the Solidarity Purchasing Groups interviewed in Rome was 31,000 € in 2010

Table 15. The Percentage of Farms by Channel of Products Sales in Italian Regions in 2010.

Regions	Other Farms	Associations Co-ops	Consumers in Markets	Consumers in House	Manufactures	Wholesalers
Piemonte	16	17	7	12	10	37
Valle d'Aosta	36	28	3	12	5	16
Liguria	7	13	9	26	7	37
Lombardia	22	16	6	16	13	27
Trentino Alto Adige	6	64	3	8	7	13
Veneto	14	34	2	6	7	36
Friuli Venezia Giulia	12	40	3	10	3	32
Emilia Romagna	19	38	3	9	9	22
Toscana	12	31	7	24	4	21
Umbria	12	25	7	22	6	28
Marche	11	22	5	16	7	39
Lazio	14	24	9	21	9	22
Abruzzo	14	22	7	25	4	29
Molise	15	13	6	11	12	44
Campania	15	9	6	27	8	34
Puglia	7	28	5	6	18	36
Basilicata	10	21	7	13	5	44
Calabria	5	6	12	45	9	23
Sicilia	9	19	10	10	7	45
Sardegna	9	21	8	16	19	27
North-West	18	17	7	15	11	33
North-East	14	40	3	8	7	28
Centre	12	26	7	21	7	27
South	10	18	7	20	12	33
Islands	9	19	9	11	10	41
Italy	12	24	7	15	10	33

Source: Istat, 6th Census of Agriculture.

(Fonte et al., 2011), the economic size of these groups can be considered to be of some relevance.

Also, farmers' markets have become successful in Italy only in the last few years, mostly by the initiative of Coldiretti, the main association of Italian farmers. Similar initiatives are organised by organic farmers: the 'bio' markets (213 in 2011, according to Biobank), particularly present in the North and in the Centre of Italy.[7]

Figs. 4 and 5 report the number of Coldiretti's farmers markets by regions and by category of products sold in 2012. The farmers' markets are

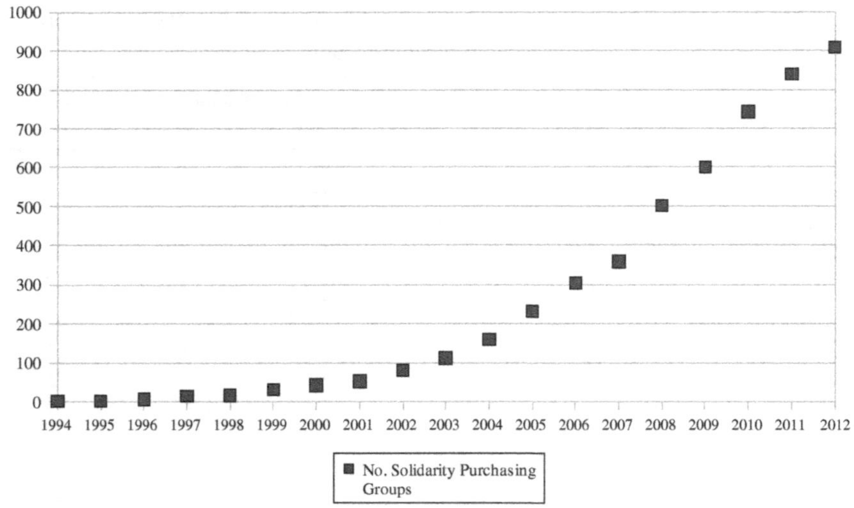

Fig. 3. The Evolution of Solidarity Purchasing Groups in Italy.

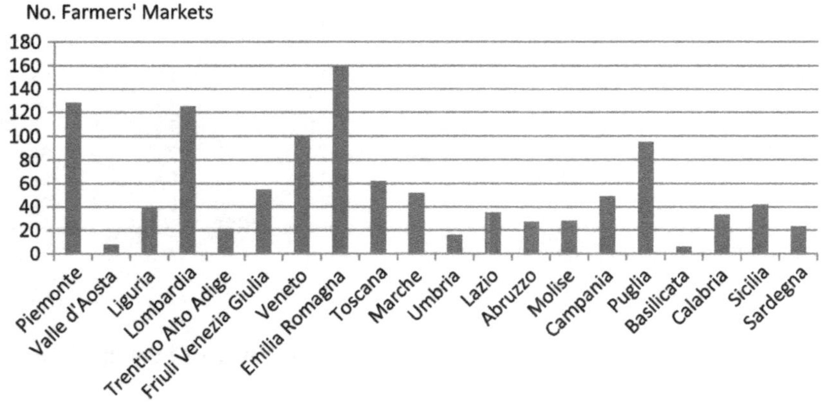

Fig. 4. The Number of Coldiretti Farmers' Markets by Italian Region in 2012. *Source*: Coldiretti.

widespread, particularly in the big cities and in the North, and the categories of most sold products are fruit and vegetables, cheese, wine and vinegar. The main consumer motivations of the purchase at the farmers markets are freshness, quality and price (Vecchio, 2009, 2010).

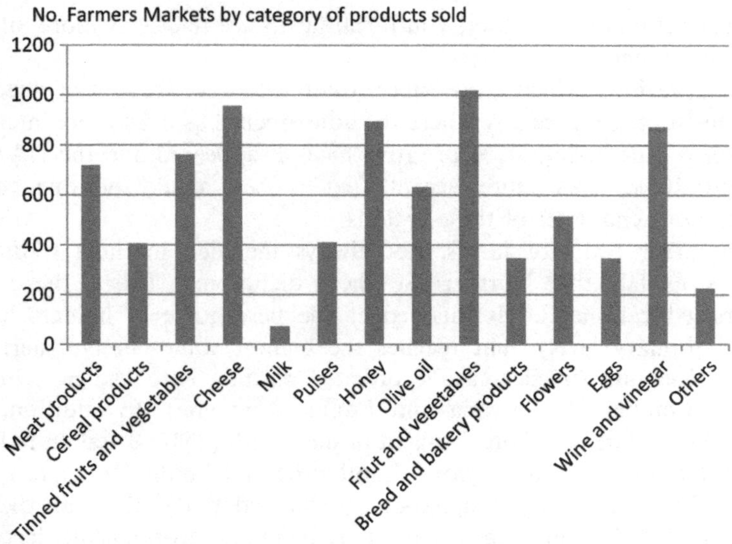

Fig. 5. Number of Coldiretti Farmers' Markets by Category of Products Sold in 2012. *Source*: Coldiretti.

CONCLUDING REMARKS

The 6th Census of Agriculture of 2010 confirms the Northern-Southern dichotomy of Italian agriculture. In addition, the differences become more accentuated due to the considerable structural adjustment that has led to the disappearance of almost one-third of holdings in the period 2000 to 2010.

Northern regions' holdings continue to have physical and economic dimensions appreciably higher than those from the South (e.g. 18.2 hectares UAA in Lombardia and only 4.0 hectares in Calabria or Campania). They show more diffusion of land renting and of mixed tenancy regimes. 'Only ownership' particularly decreases in Valle d'Aosta and Lombardia, where only around 40% of farms are sole property. Other forms of legal status – corporations and associations (21.9% of Lombardia's holdings and 15.3% of Emilia Romagna's ones) also are more diffused.

Globally, the agriculture of Southern regions has been less affected by the adjustment (reduction of 25% of farms and UAA stable) and maintains some traits of more 'traditional' farming; that is to say, smaller farms, lower diffusion of land renting (e.g. in Calabria and Puglia more than 80% of holdings own land by only ownership) and of legal status forms other than sole proprietorship. However, youth-run farms are relatively more diffused in Southern regions.

Central regions, which show intermediate values for that set of structural characteristics, are precisely where the adjustment has been more intense in the inter-census period: 40% of farms have disappeared and the UAA has dropped 10%. This more accentuated process could be due to the mountainous character of these regions.

Also, other traits of farms, not always included in most traditional analyses, explain that Northern-Southern dichotomy. This is the case of farmers' educational levels (defined as the percentage of holders having beyond primary level), that reaches the highest values in Northern and Central Regions (in decreasing order: Trentino Alto Adige, Umbria, Liguria, Lombardia, Toscana and Emilia Romagna). In addition, ICT availability in farms is more frequent in the North (25% of Italian holdings with ICT are in only two regions: Lombardia and Emilia Romagna).

Regarding the quality designations (PDO and PGI), although they are found in all the regions, they are more frequent and economically larger (in terms of turnover) in the North and Centre of Italy, where they are located in important agrifood clusters. Another element of innovation and diversification of holdings, agritourism, is growing in the whole country, but again only two regions (Trentino Alto Adige and Toscana, where these initiatives stemmed earlier) concentrated 36% of agritourism holdings in 2010 (48% in 1998).

On the contrary, Southern regions lead in two examples of innovation in farming practices. On the one hand, two regions from the South (Calabria and Basilicata) lead in organic farming, with more than 20% of UAA enrolled in this productive model, while in absolute terms and in decreasing order, Sicilia, Puglia, Calabria, Sardegna and Basilicata have more hectares cultivated thusly in Italy. The other comparatively strong point of Southern regions is the development of short food supply chains, particularly in the case of direct selling to 'consumers in house' (45% of Calabria's farms used this channel, followed by Campania (27%) and Abruzzo (25%)). Other forms of direct selling (like the farmers' markets organised by Coldiretti) are more extended in the North (Emilia Romagna, Lombardia and Piemonte).

The analysis carried out in this paper has tried to show the interconnectedness among some structural characteristics of Italian agrarian holdings, the penetration of several forms of innovation and some elements related to commercialisation and product differentiation. This interconnectedness has been shown to be complex and requires further research to identify clearly the nature of the relationships among those and other variables. Furthermore, the marked regional duality of Italian agriculture responds to several ways in which farmers and their activities interconnect with territorial development models that also show a clear regional differentiation.

NOTES

1. It is used to measure economic size since 2010.
2. Farm family income is defined as the value of farm net value added less the total external costs for wages, rent and interest paid.
3. The registration with the Chambers of Commerce is necessary in order to be exempted from agricultural fuel taxes. Generally, small farms are not registered: in 2010, the Italian farms registered to the provincial Chambers of Commerce were 53% of the farms registered by the Census of Agriculture.
4. For example, the share of the tertiary-type A level graduates on the total tertiary graduates in Agricultural Sciences was equal to 36% in Germany and to 82% in Spain (OECD, 2003).
5. The correlation between the number of farms, registered to the Chambers of Commerce during the 2000–2010 period, with the dummy 'farm manager less than 40 years old' is 0.42 while the correlation between the number of farms, cancelled from the Chambers of Commerce during the same period, with the dummy 'farm manager less than 40 years old' is −0.37.
6. http://www.retegas.org
7. 36 in Emilia Romagna, 35 in Lombardia, 30 in Veneto and 24 in Toscana.

ACKNOWLEDGEMENTS

We wish to express our heartfelt thanks to Eladio Arnalte-Alegre, Ana Moragues-Faus and Dionisio Ortiz-Miranda for their helpful suggestions, to Maria C. Fonte for generously providing us with research material and to Andrea Saroldi for the data relative to the evolution of the Solidarity Purchasing Groups.

REFERENCES

Agrisole2000 – Osservatorio sull'innovazione nelle imprese agricole. (2009a). Sintesi del 3° Rapporto sull'innovazione nelle imprese agricole. Bologna.

Agrisole2000 – Osservatorio sull'innovazione nelle imprese agricole. (2009b). Atti dell'incontro di presentazione del 3° Rapporto sull'innovazione nelle imprese agricole. Seminar at the Agriculture Faculty, Univerisità di Bologna, 27/11.

Ascione, E., & Scornaienghi, M. (2009). L'agricoltura legale: i terreni agricoli confiscati alla criminalità organizzata. *QA – Rivista dell'Associazione Rossi-Doria*, *3*, 153–175.
Becattini, G. (2000). *Dal distretto industriale allo sviluppo locale. Svolgimento e difesa di un'idea*. Torino: Bollati Boringhieri.
Biobank. (2012). Tutto bio. Annuario del biologico. Retrieved from www.biobank.it
Brunori, G., Favilli, E., & Rossi, A. (2013). Il ruolo emergente dei civic food networks nell'innovazione attorno al cibo. *Agriregionieuropa*, *9*(32), 1–8.
Brunori, G., & Rossi, A. (2000). Synergy and coherence through collective action: some insights from wine routes in Tuscany. *Sociologia Ruralis*, *4*(40), 409–423.
Brunori, G., Rossi, A., Cerruti, R., & Guidi, R. (2009). Nicchie produttive e innovazione di sistema: un'analisi delle transizioni tecnologiche attraverso il caso dei farmers' markets in Toscana. *Economia agro-alimentare*, *13*(3), 143–170.
Brunori, G., Rossi, A., & Guidi, F. (2012). On the new social relations around and beyond Food. Analysing Consumers' Role and Action in Gruppi di Acquisto Solidale (Solidarity Purchasing Groups). *Sociologia Ruralis*, *52*(1), 1–30.
Cioffi, A., & Gorgitano, M. T. (1998). Local innovative capability and innovative adoption in agriculture: Computers on Italian farms. *European Review of Agricultural Economics*, *25*, 210–227.
Csaki, C. (1999). Agricultural higher education in transforming Central and Eastern Europe. *Agricultural Economics*, *21*, 109–120.
De Devitiis, B., Lopolito, A., Maietta, O. W., & Sisto, R. (2009). Innovazioni e capitale sociale: il ruolo delle relazioni nelle imprese di trasformazione dei prodotti di agricoltura biologica. *Rivista di Economia Agraria*, *64*(3-4), 291–318.
Dickie, J. (2007). Delizia! The epic history of the Italians and their food. Sceptre.
Fanfani, R. (Ed.). (2009). *Il sistema agroalimentare in Italia*. Milano: Edagricole.
Fonte, M. C. (2006). Slow food's Presidia: What do small producers do with big retailers? In T. K. Marsden & J. Murdoch (Eds.), *Between the local and the global: Confronting complexity in the contemporary agri-food sector* (Vol. 12). Research in Rural Sociology and Development. Amsterdam: Elsevier.
Fonte, M. C., Eboli, M., Maietta, O. W., Pinto, B., & Salvioni, C. (2011). Il consumo sostenibile nella visione dei GAS di Roma. *AgriRegioniEuropa*, *7*(27), 80–83.
Gardini, C., & Lazzarin, C. (2011). Il rating competitivo di AGRI2000 e l'innovazione nelle aziende agricole italiane. *Agriregionieuropa*, *7*(24), 1–8.
Greco, M., & Di Cristofaro, E. (2011). Il settore agricolo attraverso i dati provvisoi del 6° Censimento generale dell'agricoltura. *Agriregionieuropa*, *7*(26), 81–83.
Inea. (2012). L'agricoltura italiana conta. Il Sole 24 ORE – Agrisole.
Intesa San Poalo. (2011). Monitor dei Distretti.
Istat. (2008). Le aziende agrituristiche in Italia al 31 dicembre 2007. Roma.
Istat. (2012). Le aziende agrituristiche in Italia. Anno 2010. Roma.
Kanemasu, Y., & Sonnino, R. (2009). Dynamics of power and cooperation in rural development: The case of Chianina beef production in Italy. *International Journal of Sociology of Agriculture and Food*, *16*(2), 36–53.
Maietta, O. W. (2004). Crescita della produttività e capitale umano nell'agricoltura italiana: un'analisi con dati provinciali dal 1951 al 1991. *La Questione Agraria*, *4*, 105–140.
Messori, F., & Ferretti, F. (2010). *Economia del mercato agroalimentare*. Bologna: Edagricole.
Milone, P. (2004). *Agricoltura in transizione: la forza dei piccolo passi, un analisi neo-istituzionale delle innovazioni contadine*. Ph.D. thesis, Wageningen University, Wageningen.

Nyblom, J., Borgatti, S., Roslakka, J., & Salo, M. A. (2003). Statistical analysis of network data – An application to diffusion of innovation. *Social Networks, 25*, 175–195.

Ohe, Y. & Ciani, A. (2012). Hedonic pricing evaluation of agritourism activity in Italy. Paper prepared for presentation at the International Association of Agricultural Economists (IAAE) Triennial Conference, Foz do Iguaçu, Brazil, 18–24 August.

OECD. (2003). Education at a glance. Paris: OECD.

Parrott, N., Wilson, N., & Murdoch, J. (2002). Spatializing quality: Regional protection and the alternative geography of food. *European Urban and Regional Studies, 9*(3), 241–261.

Petrini, C. (2009). *Terra madre*. Firenze: Giunti Editore.

Pierri, F. (2013). Le imprese giovanili italiane: un confronto tra settore agricolo e altri settori. INEA, mimeo.

Pilati, L. (2006). Politica della conoscenza e competitività dell'agricoltura italiana. *Rivista di Economia Agraria, 61*, 455–478.

Se.Ri.Fo. (2010). Analisi dei bisogni di apprendimento dei giovani imprenditori agricoli, Project "V-3DAS (Virtual 3D Agri-Stage for Experiential Learning)", n. 505690-LLP-1-2009-1-IT-KA3-KA3MP, EU Programme Lifelong Learning, KA3, ICT for Learning.

Slow Food. (2012). *Guida ai Presidi Slow Food.* Torino: Slow Food Editore.

Van der Ploeg, J. D. (2000). Revitalising agriculture: Farming economically as starting ground for rural development. *Sociologia Ruralis, 40*(4), 497–511.

Van der Ploeg, J. D. (2008). *The new peasantries: Struggles for autonomy and sustainability in an era of Empire and Globalization*. London: Earthscan.

Vecchio, R. (2009). European and United States farmers' markets: Similarities, differences and potential developments. Paper prepared for presentation at the 113th EAAE Seminar "A resilient European food industry and food chain in a challenging world", Chania, Crete, 3–6 September.

Vecchio, R. (2010). Local food at Italian farmers' markets: Three case studies. *International Journal of Sociology of Agriculture and Food, 17*, 122–139.

Ventura, F., & Milone, P. (2000). Theory and practice of multi-product farms: farm butcheries in Umbria. *Sociologia Ruralis, 40*(4), 452–465.

Vieri, S. (2012). *Agricoltura settore multifunzionale allo sviluppo*. Milano: Edagricole.

Zanoli, R. (Ed.). (2007). *Le politiche per l'agricoltura biologica in Italia*. Milano: Franco Angeli.

CHAPTER 9

THE EMERGENCE OF NEW DEVELOPMENT TRAJECTORIES IN ITALIAN FARMS

Cristina Salvioni, Roberto Henke and Elisa Ascione

ABSTRACT

The persistence of different farm types in Italian agriculture shows that productivism is not the only possible development path that farms can follow, and that farms can successfully adopt strategies based on diversification rather than standardization of production.

The aim of this work is to provide evidence about the diffusion of different diversification and differentiation strategies in Italian agriculture, and to compare the characteristics associated with the targeted groups of farms, as well as their structural and economic evolution over time.

The analysis is performed on a panel of data built on the basis of information collected by the Italian FADN over the 2003–2009 period. For the purpose of the analysis we divided the population of Italian commercial farms into a fivefold innovative farm typology based on the extent of diversification and differentiation strategies adopted by farms.

The findings show that conventional farms are still by far the largest category within the population of Italian commercial farms, while only

Agriculture in Mediterranean Europe: Between Old and New Paradigms
Research in Rural Sociology and Development, Volume 19, 207–232

ISSN: 1057-1922/doi:10.1108/S1057-1922(2013)0000019011

13% of total commercial farms is classified as differentiated and/or diversified. Conventional farms are also the best off in terms of economic results. As for the differentiated and/or diversified farms, their structures are still changing, their profitability is improving and they follow a more sustained income growth path than conventional ones.

The analysis highlights that diversification and differentiation are not necessarily a viable solution to the low-income problem faced by many farms. Future research is needed to better understand the relationship between diversification strategies and policies.

Keywords: Income diversification; farm profiles; farm policy; Italian agriculture

INTRODUCTION

After WWII the Italian agriculture was interested by a wave of productivism and modernization. In this period the farming system underwent a major structural change encompassing a rapid decrease in the number of farms associated to a large increase in average farm size, as well as an increase in the dependence on purchased inputs (Fabiani, 1986). These phenomena were particularly intense in the Northern plains, while farms in the Southern and mountainous areas mainly conserved many of the traits associated to a backward agriculture: small size of holdings, low level of technological equipment, and aged and pluriactive holders.

Any attempt of agricultural product differentiation, based for example on geographical origin, or of on- or off-farm diversification (pluriactivity, multifunctionality, etc.), was originally interpreted by the supporters of productivism solely as a strategy for survival, prerogative of marginal, small, or residential farms. However, during the last three decades differentiation and diversification strategies proved to be not solely a survival strategy but, rather, a way to turn some attributes traditionally interpreted as backwardness in features that many consumers perceive as important and that they are keen on rewarding with a premium price. In other words, differentiation and diversification strategies proved to be a source of competitive advantage and effective in improving farms' profitability. These two results explain why they have been increasingly adopted by farmers.

Farmers have constantly tried to diversify their production, either for reducing the agronomic and climatic risk on the output or for creating new and interweaved sources of income originating from on-farm activities. This tendency has been enhanced by the renovated role of agriculture and rural areas in society, as well as by the increased perception of farmers as the main natural resource managers and landscape stewards. The growing integration of agriculture in the rest of the rural economy has contributed to create more opportunities of diversification within farms, even in fields distant from farming. In this process, the food market segmentation and the variety of services offered by farms have also been key in explaining the interest in the diversification of farm activities.

Starting from the 1990s, alongside the acknowledgment of the environmental role of agriculture and the rethinking of the public support in agriculture that led to measures oriented toward the reduction of negative externalities in agriculture and the control of production surpluses, a new "virtuous circle" has been set up around the concepts of farming sustainability and multifunctionality of the primary sector. As a consequence, a process of diversification has been increasingly involving farms, which now produce not only food and fibers but also, and in a conjunct way, services and new products (often unfairly recognized as "secondary products"). At the same time, rural development policies, as the "second pillar of the CAP," were highly emphasized as key in widening the action field of multifunctional diversification to the production of externalities and pure public goods, and also to other economic activities whose social and environmental effects are of an indirect nature (European Commission, 1997; Organisation for the Economic Cooperation and Development (OECD), 2005). In reality, Common Agricultural Policy (CAP) policies have been increasingly addressed to support the production of public goods and services in agriculture and to improve the environmental function of the agricultural activities, through a process of "greening" of the CAP (Matthews, 2012). However, how successful and through this process has been in the recent years is a matter of further discussion (Henke, Macrì, & Storti, 2005; Mardsen & Sonnino, 2008; Zahrnt, 2009).

By and large, the last decades have seen a proliferation of differentiation and diversification initiatives. Differentiation, by moving the farm production toward agricultural goods with unconventional characteristics (organic products, quality products, typical products, etc.) or along the supply chain (e.g., direct sales), results in a deepening of farming activities (Van der Ploeg & Roep, 2003). Diversification replaces farming with other productions that may be connected to farming (e.g., agritourism and recreational, educational, or social services, etc.) or not (energy production, land and territory stewardship, etc.), in this way broadening the scope of the firm.

The aim of this work is to provide evidence about the diffusion of different diversification and differentiation strategies in Italian agriculture, and to compare the characteristics associated with the targeted groups of farms, as well as their structural and economic evolution over time. Previous applied literature on differentiation and diversification has focused the attention either on the diffusion of the diversification strategies among farmers and the characteristics of diversified farms in a limited period of time (Belletti, Brunori, Marescotti, & Rossi, 2003; Henke & Salvioni, 2008; Meert, Van Huylenbroeck, Vernimmen, Bourgeois, & van Hecke, 2005) and the role played by idiosyncratic characteristics of farmers on the adoption of diversified activities (Aguglia, Henke, & Salvioni, 2009; Esposti & Finocchio, 2008; Jongeneel, Polman, & Slangen, 2008; McNally, 2001). We add to this literature by analyzing the performance of conventional to diversified farms over time. The analysis is performed on the data collected by the Italian FADN survey over the period 2003–2009; this is the time period in which the survey provides information on a sample of farms that is representative of the population of commercial Italian farms.

The rest of the chapter is structured as follows: in the next section we review the new diversification trajectories emerged during the last decades in Italy; then, we describe the data and the farm typology used to segment the population of Italian commercial farms in mutually exclusive categories that take under account the extent of differentiation and diversification activities in farms. The results of the application of this typology are presented in the last two sections where we first compare the structural and economic characteristics of Italian commercial farms by farm type, and subsequently we follow their evolution over the 2003–2009 time period.

NEW DEVELOPMENT TRAJECTORIES IN ITALY

One of the traditional problems affecting the primary sector is low income caused by a biologically based production process, its structural weakness, and, ultimately, the technological changes that progressively worsen the price-cost squeeze. The remedy traditionally suggested by economic theory to overcome this problem and increase farm income has been based on growth and specialization.

In this vision, agriculture of European Mediterranean countries has always been considered as paradigmatically "difficult," including in this specification different meanings (marginal, non-competitive, structurally weak, territorially tough, and so on) but all underlying the idea of a "lagging behind" sector. This is mainly due to the fact that the features of Mediterranean agriculture do all amplify the acknowledged difficulties of the process of productivist modernization of the primary sector: the small size of farms, their often marginal location, the prevalent extensive production, the relevance of part-time farmers and the aging farmers were all considered as features of an inevitable process of decline and progressive exit from the sector.

In other words, agriculture in Southern Europe, and particularly in the mountainous and marginal areas, would be considered to be the first and most illustrious victim of the process of agricultural industrialization and production intensification that were considered the only possible response to the slow but inevitable agriculture decline. The specific characteristics of farms and farmers in the remote areas, the micro-structures often devoted to self-consumption and anyway far away from access to markets, together with the geographical harsh conditions were all causes of what has been very effectively defined as a "difficult modernization" (De Benedictis, 2002, 2005).

In the case of Italy, the co-existence of different types of agriculture is quite evident: the Northern agriculture of the main fields has rather different features from the mountain agriculture of many Southern regions: the size of farms, the crop specialization and the market relationships were all envisaged as elements leading to different models of agriculture and of the role of agriculture in the development process.

The persistence of different farm types shows that productivism is not the only possible development path that farms can follow, and that farms can successfully adopt logics based on diversification rather than standardization of production.

Diversification in agriculture can be inserted in a broader debate about methods and potential results of transition and reorganization processes underway in agriculture (Van der Ploeg, 2002; Van der Ploeg & Roep, 2003; Wilson, 2007). According to this literature, the move toward multifunctionality and diversification in Western agriculture can be interpreted as a reaction to the dominant socio-technological regime of the agri-food system, with large-scale production, gradual replacement of natural capital by financial and man-made capital, a tendency to mono-crop agriculture, abandonment of artisan processes and products, centralization and vertical integration by which costs are externalized (environmental, social, etc.), and concentration of control and profits. These aspects of industrialization carry negative effects for the quality of the natural environment (pollution, loss of biodiversity, etc.) and the food produced, as well as a loss of autonomy for those involved in agricultural production, and a growing subjugation to the logic of the agri-industrial system.

In the same vein, other practices, such as direct sales can be viewed as instruments for reducing dependence on the industrial and financial system, increasing employment and value added in agriculture, and regaining autonomy. This process of change finds strong support in the demand expressed by social movements, concerned for example with relocation of food products (e.g., solidarity purchasing groups), environmental issues (organic movements, zero kilometer, etc.), or social equity (e.g. fair trade).

After decades of success of the modernist paradigm that has pushed farms into process of production intensification and specialization, the lack of sustainability of this process, stressed by the economic crises and the long-standing process of agriculture economic decline, has stimulated the search of new activities leading to other and differentiated sources of income, sometime off-farm but more and more often on-farm and anyway outside the mere constraints of the agricultural activity (Wilson, 2007, 2008). The result of this process has been that of a co-existence of different models of agricultural development, which have been differently catalogued as post-modernist, neo-modernist and multifunctional, according the specific aspect underlined within this more generalized decomposition of the dominant model (Ravenscroft & Taylor, 2009).

Some features of the Mediterranean agriculture, especially those that had been always considered the main constraints to the industrial development of agriculture in remote, marginal, and mountainous areas, can be reconsidered under a different light thanks to the post-productivist logic and according to the concept of multifunctional diversification. In other words, the small size of farms, the reduced use of industrial inputs, the specialization in permanent crops that strongly characterize the rural landscape and the family-based labor are all factors that should be reconsidered as providing new opportunities in agriculture, because they offer new keys to access to the CAP and because they are closer to what consumers and citizens require to a renovated primary sector.

In Italy, from the 1970s the real remuneration of production factors in agriculture has slowly increased from 20% to 40% of that realized on average in the overall economic system (Henke & Salvioni, 2011). In spite of that, nowadays only a third of farms is able to remunerate factors at a level that is as much or slightly above the opportunity cost. Moreover, this level clearly decreases if one considers the remuneration levels net of the public support granted by the European Union (Istituto Nazionale di economia Agraria (INEA), 2010).

One possible trajectory of the farm income diversification process is *pluriactivity* (Arkleton Trust, 1983; Evans & Ilbery, 1993; Fuller, 1990; Gasson, 1988; Gasson & Errington, 1993; Hill, 2000; Marsden, 1990). Along this pattern, farm households decide to allocate part of their labor resources to off-farm employment with the goal of maximizing family incomes, referred to as the sum of on- and off-farm family income.[1] In fact, the unit of analysis of pluriactivity is the farm family, not the farm business. However, due to the persistent dominance of small family holdings in the total farm population in most countries, the two units of analysis often coincide (Hill, 1999).

Pluriactivity has been a very successful model in Italy, as shown by many studies on the topic developed in the 1980s and 1990s (De Benedictis, 1990, 1995; Fabiani, 1991; Saraceno, 1985). These studies focused on the persistence of pluriactivity in agriculture as a new mode of organizing production factors between agriculture and other off-farm activities that would ultimately ensure the survival of primary activities, especially in remote marginal areas that would have been destined to abandonment otherwise.

However, pluriactivity works successfully where economic and social conditions for off-farm work for the family members exist and are prosperous: as a consequence, the success of pluriactivity directly depends, in the end, on the non-agricultural context (Fabiani, 1991)

Another possible path farms can follow to counteract the decline in income is on-farm *diversification*. This is the case when a farm business undertakes a new activity targeted to increase returns through resources available on farm. The current literature investigates several forms of on-farm diversification (Organisation for the Economic Cooperation and Development (OECD), 2009).

A common form of diversification sees farmers developing a new business related to the original agricultural production, such as in the case of conversion to organic farming, high-quality or specialty food, on-farm processing, and direct marketing. These activities aim to increasing the value added per unit of the overall agricultural products and usually require a minimum level of agricultural business in order to make the conversion to more remunerative types of production (organic farming or quality products) or the expansion toward other segments of agriculture-related activities (direct sale, on-farm processing) convenient.

On-farm diversification can also take the direction of a new business that not necessarily, or not entirely, relates to the agricultural one, like, for example agri-tourism, energy production (photovoltaic, wind-powered, etc.), or natural resource management. In these cases diversification can be interpreted as a process that widens the range of production possibilities of the farm business, beyond agriculture but using the same production factors and structures.

In both cases, strictly defined agriculture compete with the new activities in terms of production factors (land, labor, capital). In Table 1, elaborating on the OECD work, we have classified a certain number of activities according to the main input they compete for with the main farm activity. This is a very important aspect to have in mind when diversification policies as well as agricultural policies are designed and implemented.

The optimal combination of agricultural and non-agricultural products and consequently the degree of farm diversification (specialization) depends on a large extent on output prices and technology.

Previous works have highlighted the extent and the relevance of diversification in Italy (Belletti et al., 2003; Esposti & Finocchio, 2008; Henke & Salvioni, 2008). In particular, the taxonomy of the activities has pointed out the relevance of the integration of post-production activities such as processing and direct sale, as well as of the quality certification. Still limited to a few but significant cases are the more innovative activities that move farms away from the primary activities, such as tourism, services recreational activities, and so on. However, one specific feature captured by the analyses is the co-existence of activities within the same farm: "complex" systems of activities and functions are arising, which put farms under new lights and imply a reorganization of factors' use, of entrepreneurial skills and of territorial relationships.

The latest Italian census (2010) defines diversification as the presence of other gainful activities alongside the main agricultural production. According to this restricted definition, only 5% of the total farms are diversified. However, even within this little sub-universe of farms there is a wide spectrum of diversification profiles, featuring farms with several gainful activities combining with each other and a complex allocation of production factors (Henke & Povellato, 2012).

Table 1. On-Farm Activities (Input Competition with Agriculture).

Land	Labor	Capital
Aquaculture	Agri-tourism	Processing
Energy	Educational activities	Food preparation
Recreational activities	Artcraft	Wood
Afforestation	Contract work services to livestock parks and gardens	Feed processing

The change in the CAP has also contributed to the enhancement of diversification in Italy thanks to the generous financial support granted to farms that choose to diversify alongside this path. In particular, the rural development policy, with the agro-environmental programs and the investment plans, has ensured the acknowledged status of "diversified and multifunctional farms" to units that have favored the production of public goods introducing "green measures" and sustainable investments in green technology, animal welfare, and food quality improvement (INEA, 2010; Sotte, 2009).

The diversification scenario considered so far becomes more complex when one looks at it in terms of entrepreneurial skills. To successfully produce and sell an organic product, or a local traditional product, or service, as in the case of social agriculture, farmers must be prepared to take the risk of making very specific investments (*asset specificity*) which are, in addition, difficult to reverse (*non-recoverable costs*). Moreover, they must know how to move in a complex, niche market, often distant from potential customers and from the original agriculture environment. All this implies certification and advertising costs, but also real *transaction costs* associated with the search for sales and communication channels in order to reach and gain potential customers. In other words, the diversification path involves a very complex cost-benefit analysis for farmers to be able to identify and measure the economic value of potential external and indirect benefits of diversification, as well as various kinds of costs that may arise from choosing one on another path.

These considerations show not only that diversification is a rational economic choice but that it can be seen as more than just a strategy for survival, exclusively a characteristic of marginal, small or residential farms. Indeed, when a farm identifies the economic value of the benefits of multi-functionality and internalizes these benefits in the production planning process, diversification becomes part of the set of strategies available to farms to increase their income and profits.

METHODOLOGY AND DATA

This study relies on data collected by the Italian Farm Accountancy Data Network (FADN) survey. The field of observation is the total of commercial farms, which are farms with an economic size greater than four European Size Units (ESU) that is around 4,800 euro. Starting from year 2003, the survey is conducted on a statistically representative sample. The sample is randomly drawn from the National Institute of Statistics (ISTAT) structural survey and it is stratified[2] according to the criteria of geographical region, ESU and type of farm (TF).[3] The sample size is fixed at 17,000 (commercial) farms by the EC regulation Reg. (EC) 60/1997. As the sample is random, it is possible to extend the results from sample to commercial farms population by applying the weights calculated by ISTAT.

The FADN survey provides information revealing the presence on farm of differentiation and diversification strategies. For example, it is possible to know whether the farm makes use of organic farming or low-impact techniques, if it produces landscape or biodiversity conservation services, as well as agritourism and commercial (direct selling) or social (educational, therapeutic, etc.) services. In addition to these indications, FADN provides information about the use of designation of origin (PDO) and protected geographical indication (PGI) or if it produces traditional products.

Starting from 2008 the survey also recorded information about the value of total production due to (a) PDO and organic and other products covered by quality certification products, (b) the production of agritourism, on-farm recreational activities, on-farm processing (wine, cheese, and so on), and other services. This information has been recently used by INEA to develop a farm typology taking account of both farming and non-farming activities present on farm (Ascione, Carillo, & Vagnozzi, 2011). The criteria and thresholds to discriminate the FADN farms in categories were defined by a panel of experts.[4]

With the aim to reduce the complexity of the typology, we further elaborated on the criteria defined by the panel of experts retaining the thresholds suggested by them to identify the micro farms, and those suggested to define whether a farm is conventional, differentiated, or diversified. On the other side we decided (a) in order to reduce the number of different categories and to increase the number of farms covered by each of them, not to segment each category in small and large farms, and (b) in order to obtain a set of mutually exclusive categories, to introduce a new category containing those farms that make use of both differentiation and diversified farms. The result is a new typology that discriminates farms in five mutually exclusive different categories.

Each of them (Table 2) is relatively homogeneous in terms of economic size, represented by gross production (GP) and of the degree of diversification and differentiation of production:

- *Micro farms*. This category includes very small farms that are commercial (hence part of the FADN sample), since they recorded an economic size

Table 2. Categories and Identification Criteria of the Used Farm Typology.

Farm Types / Criteria	Micro	Conventional	Diversified Farms	Differentiated Farms	Differentiated and Diversified Farms
GP	<€15,000		>€15,000		
Percentage of QP over GP		<30%	<30%	≥30%	≥30%
Percentage of NFGS over GP		<30%	≥30%	<30%	≥30%
Percentage of commercial farms	15	71	6	5	2

Source: Our elaboration on FADN data.

greater than four Economic Size Units (ESU) in the 2000 Census, but they recorded a GP less than 15,000 euros in year 2008.

- *Diversified farms.* This category covers farms with a GP larger than 15,000 euros and with at least 30% of GP originated by non-farming goods and services (NFGS). More in detail, these goods and services include both broadening activities, namely agritourism, on-farm recreational activities, educational and therapeutic services, and deepening activities, namely on-farm processing and marketing.
- *Conventional farms.* This category includes farms that make none or marginal (less than 30% of the total production) use of strategies of diversification or differentiation of farm production.
- *Differentiated farms.* This category covers farms with at least 30% of total production originated by the production of quality products (QP). In other words this category refers mainly to deepening activities, namely the adoption of PDOs, PGI, and organic and other kinds of quality certification.
- *Differentiated and diversified farms.* It includes farms that make use both of diversification and differentiation strategies.

This typology has been applied to the FADN data collected in 2008, which is the first year from which the survey started to gather the new information needed to apply the typology described above. The application of the typology allows us to provide evidences about the diffusion of the different strategies, and the structural and economic characteristics by groups of farms.

In addition to this, in order to compare the economic performance over time of the identified groups of farms, we applied the same typology to a panel of more than 3,000 farms observed over the 2003–2009 period. This panel has been obtained from the Italian FADN sample: after removing those observations with incomplete and missing data, we have built a seven-wave balanced panel of farms for which continuous records are available for the period from 2003 to 2009. Each farm in the panel has been assigned to one of the categories of the typology defined above on the basis of the characteristics observed in year 2008 – the first year in which the information needed to apply the typology is available. This expedient allows us to follow the structural and economic evolution of each farm group over the 7 years under analysis, even if we have information about their possible adoption of diversification and differentiation strategies only for the last biennium.

STRUCTURAL AND ECONOMIC CHARACTERISTICS BY FARM TYPES

In this section we present the descriptive statistics of the Italian commercial farms by the farm types described in the previous section (Table 3(a–c)). The data refer to year 2008; this is the first year in which the FADN survey started to collect the information needed to apply the typology that discriminates farms by the extent of differentiation and diversification activities adopted by farmers. Focusing on only one year allows us to exploit the information gathered on the entire sample[5] that in year 2008 covered 11,234 Italian commercial farms. The sample data have been extended to the population of commercial farms by using the system of weights provided by ISTAT.

Micro farms cover 15% of Italian commercial farms. As expected, their size in terms of both total and used agricultural area (respectively TAA and UAA) is the smallest in the population under study. More in detail, on average they record 7.5 hectares of total land and 6.3 hectares of utilized land, though it is interesting to note that median[6] values are even smaller, respectively 5 and 4.5 hectares. The small size also reflects in a very small number of annual working units (AWU). As for the kind of production, the micro farms are relatively more specialized in the production of permanent (grape, olive, fruit, etc.) and cereals, oil, and protein seeds (COP) crops in comparison to the rest of the commercial farms, while they are relatively

Table 3. Structural and Economic Characteristics of Italian Commercial Farms by Farm Type (Weighted) (Economic Data in Deflated Euros using the GDP Deflator, 2005 = 100).

(a) Structural Characteristics

Farm Type	TAA	UAA	AWU	TAA	UAA	AWU	Cop	Horticultural Crops	Permanent Crop	Livestock
	Median			Mean			Percentage by Group			
Micro	5.0	4.5	0.7	7.4	6.3	0.7	29.9	1.5	49.4	4.9
Conventional	11.0	9.8	1.0	24.7	20.4	1.5	29.6	7.3	28.7	16.8
Diversified	7.5	6.6	1.1	24.2	18.6	1.4	4.2	0.2	69.5	12.8
Differentiated	11.1	10.4	1.0	22.4	20.0	1.4	26.4	7.2	44.6	5.4
Diversified–differentiated	12.5	11.8	1.2	46.7	36.6	1.5	5.8	0.0	39.2	32.7
Total	7.4	6.5	0.9	18.6	15.4	1.2	27.8	4.7	39.3	12.1

(b) Economic Characteristics (Median)

Farm Type	FNVA/GP	FNVA	FNVA/AWU	FNVA/UAA	Pillar 1/FNVA	Pillar 2 Payments/FNVA	Grant Assett	Grant Asset/FNVA
Micro	48.9	4,231	6,622	824	15.2	0.0	0.0	0.0
Conventional	58.4	17,605	18,245	2,005	11.0	0.0	0.0	0.0
Diversified	69.2	17,610	17,832	2,960	9.2	0.0	0.0	0.0
Differentiated	56.7	18,029	16,597	1,738	10.5	0.0	0.0	0.0
Diversified–differentiated	56.7	16,283	19,493	2,132	14.1	0.0	0.0	0.0
Total	56.0	10,424	13,142	1,371	11.9	0.0	0.0	0.0

(c) Economic Characteristics (Mean)

Farm Type	FNVA/GP	FNVA	FNVA/AWU	FNVA/UAA	Pillar 1/FNVA	Pillar 2 Payments/FNVA	Grant Assett	Grant Asset/FNVA
Micro	4.5	3,913	8,651	1,362	30.7	1.2	0.0	0.0
Conventional	54.4	42,639	34,153	7,612	25.8	5.1	0.2	0.1
Diversified	63.7	28,242	22,066	4,031	14.5	3.5	0.1	0.0
Differentiated	54.2	37,885	46,469	6,531	36.7	6.9	0.4	0.1
Diversified–differentiated	47.9	57,508	103,477	5,472	16.8	7.4	1.2	−1.4
Total	36.7	27,673	25,283	5,059	27.1	3.6	0.2	0.0

Note: FNVA, farm net value added.
Source: Our elaboration on FADN data.

de-specialized in more industrialized productions such as horticultural crops and livestock.

Micro farms have a median net value added per unit of labor (FNVA/AWU), that is farm income, half of the value recorded in the whole sample, as a consequence they present the worst economic results in the population. These finding cannot be straightly interpreted as a symptom of poverty of the farm household though, rather this is probably due to the fact that they are hobby, residential, or retirement farms, that is production units where the farmer behavior is not oriented toward the farm income maximization, rather to the maximization of other non-economic variables such as social welfare (Perali, Salvioni, & Tommasi, 2005). Farm income is often used by these farm households only to complement other sources of off-farm income. In addition to this it is interesting to note that on average the financial aid under Pillar 1 is more than the 30% of farm net value added; this is above the average in the observed population. On the contrary, the intensity of Pillar II benefits over the farm net value added is well below the population average. Finally, only a marginal percentage of farms in this category receives a grant related to capital assets under Pillar II; this may be due to different causes. On one side these micro farms may have little access to this kind of financial aid due to the difficulties and the lack of competences in following the procedure to make request for this kind of support, on the other side it can be the result of a lower propensity to invest, again they may easily find in the farm household the financial sources needed to cover their little investments. The analysis needed to test these hypotheses is beyond the scope of this study.

71% of the Italian commercial farms are conventional; this means they are farms that make little or no use of diversification and differentiation strategies. In other words this is the group of farms mainly associated with the productivistic approach. Their size is large both in structural and financial terms. Their median allowance of total and used land (respectively 11 and 9.7 hectares) as well as the annual working units are above the median of the farm population. The same applies to their economic results. A further symptom of their association to the productivist approach is their relatively strong specialization in the production of more intensive and industrial products, such as horticultural crops and livestock, while they are relatively less specialized in the production of permanent crops compared to the whole population. As for public support, it is interesting to note that on average they receive financial aid under Pillar I of the CAP in line with the one received by other non-micro farms, while support from Pillar II is relatively little. This latter result was largely expected since the payments granted under the second pillar of the CAP (not including investments support) are mainly associated with the adoption of multifunctional and diversification activities. On the contrary they attract more grants related to assets; this result was also expected since these grants mainly take the form of investment subsidies and they are mainly associated with the modernization axis of Pillar II.

Diversified farms cover 6% of the total population of Italian commercial farms. Their average land size is not very different from the size of conventional farms, while the median size reveals that most of the farms belonging to this groups are fairly small with total and used land respectively equal to 7.4 and 6.6 hectares, much closer to the values observed among micro rather than non-micro farms. Their labor use is above the population mean. In terms of production they are strongly specialized in permanent crops, while they are relatively de-specialized in the production of COP crops. Their economic results are above the average, still lower than the results recorded in the conventional farms.

Differentiated farms, which are the group of farms where at least 30% of total production is originated by the production of quality products such as PDO, organic and other products covered by quality certification, cover 4.6% of the total population of Italian commercial farms. They have both average and median large allowances of total and utilized land. The use of land is in line with the one observed in the rest of the non-micro farms' population. They present a strong specialization in the production of horticultural and permanent crops, while they are relatively under-specialized in livestock production. The average income per unit of labor is very large, but the median value shows that most of the farms in this group produce an income per unit of labor lower than the rest of the non-micro farms.

Finally, the group of farms that adopted both differentiation and diversification (D&D from now on) strategies covers 2% of total commercial farms and contains those farms that have the largest allowance of both land and labor. They present the best economic results in the population.

It is interesting to note that we find overall differentiated and/or diversified farms cover 13% of Italian commercial farms; this is more than the double of the percentage of farms with other gainful activities found in the census. This difference is largely due to the fact that the definition used by the census covers only diversified farms, while it does not cover differentiated farms.

THE PERFORMANCE OF ITALIAN FARMS BY FARM TYPES OVER TIME

We now pass to study the performance over time of the previously identified group of farms. To this end, we calculated the percentage changes[7] over the 2003–2004 and 2008–2009 periods in the main structural and economic characteristics by farm type: land, labor, the value added to gross production ratio, value added per unit of labor and per hectare of utilized land and farm income per unit of labor. Differently from the previous section, the statistics presented in this section are calculated by using the information contained in the balanced panel of farms for which continuous records are available for the period from year 2003 to 2009. This is because we want to follow the evolution experienced by the same farms by group. More in detail, the inspection of the percentage change between the beginning and the end of the period under study is used to analyze what kind of growth path has been followed by the different identified farm groups.

As for the structural characteristics, the data reveal that all farm groups, except that referred to farms using both diversification and differentiation strategies, experienced a reduction in the utilized land. It is relevant to point out, though, that the decrease in land is very small in the group of conventional farms, while micro and diversified farms are relatively decreasing the land use. The changes in labor confirm that all the Italian commercial farms are experiencing a process of expulsion of labor force that reaches its peak in correspondence of the micro farms. The groups of non-conventional farms are intensively disinvesting in farm land. The phenomenon is particularly relevant in the cases of the micro farms; this evidence suggests a progressive marginalization of this group of farms probably leading to their exit from the commercial segment of the sector. The large negative changes in land and labor observed in correspondence of both differentiated and diversified farms suggest that these two groups of farms are still looking for an optimal resource allocation that allows them to increase their efficiency and profitability. This is partly due to the fact that products of farms adopting diversification strategies (multifunctional services, quality products, etc.) are unique, since they rely on unique attributes of the farm. As a consequence each diversified farm has to find its own optimum. Under this circumstance the search of the optimal structure requires time, since the solution is farm specific and there are not well-defined rules to refer to. Finally, D&D are the only group that is increasing their land allowance.

As for the economic results we start looking at farm net value added to gross production ratio (FNVA/GP). This indicator is a proxy for the farm profitability; an increase in this indicator means a larger quota of revenues is used to cover the remuneration of inputs (labor, land, and capital), while the quota of raw and intermediate costs decreases. All farm types, apart from that referred to the D&D farms, have experienced an increase in profitability. This common positive trend is partly due to the increase in agricultural prices experienced at the end of the decade. The increase is particularly large in the groups of non-conventional farms; this indicates these farms have been able to reduce their dependence on off-farm input and this is reflected in an increase of the value added kept on farm and not passed to rest of the supply chain.

The conventional and micro farms have recorded a loss in income per unit of land (FNVA/UAA). The reduction has been very drastic in conventional farms, where at the end of the period the indicator recorded half the level of the beginning of the period. On the contrary, in all the farms that adopted a diversification and/or differentiation strategy the income per unit of land increased; the largest increase is recorded in correspondence of the diversified farms. Though, a closer look to Table 4(a,b) reveals that they have an income per hectare much lower than conventional farms. This result was largely expected since conventional farms are usually engaged in cost leadership strategies which in turn require scale and land use intensification. On the contrary, diversified and differentiated farms often make use of extensive farming associated with lower income per hectare of land.

The farm income per unit of labor, which is the most important indicator of farm economic results, presents an increase in all groups. The highest percentage increase is recorded in the group of conventional farms. Recall this group of farms is relatively less specialized in permanent crops; hence, they might have been facilitated in the adaptation of their cultural mix toward those crops which benefited more of the price boom recorded at the end of the past decade. Large percentage increases in the income per unit of labor are recorded even in the groups of the D&D and differentiated farms. Further information about this indicator, central for the evaluation of the economic performance of farms, can be obtained by the inspection (Figs. 1–3).

Fig. 1 is the scatter plot of income per labor unit by farm type. Conventional farms present the more disperse values of income; this is probably due to the fact that they are not specialized in specific productions, rather they are active in wide range of crop and livestock productions, each

Table 4. Structural and Economic Characteristics by Farm Type – Balanced Panel Data.

Farm Type	UAA	AWU	FNVA/ Gp	Deflated FNVA	Deflated FNVA/AWU	Deflated FNVA/UAA
(a) 2003–2004 Averages						
Micro	7.7	1.0	36.8	7,487	8,125	2,220
Conventional	37.2	2.4	51.8	90,007	31,056	21,890
Diversified	35.0	2.5	57.8	64,458	23,818	3,830
Differentiated	49.1	2.0	52.7	60,378	28,786	3,934
Diversified–differentiated	54.4	2.4	59.7	107,986	37,888	2,726
Total	33.4	2.2	50.0	74,809	27,131	16.619
(b) 2008–2009 Averages						
Micro	7.3	0.8	40.8	4,896	8,757	2,003
Conventional	37.1	2.2	53.9	85,624	39,350	10,873
Diversified	32.1	2.3	63.0	65,760	26,162	4,893
Differentiated	47.3	1.8	57.9	58,796	33,907	4,423
Diversified–differentiated	60.2	2.1	52.4	133,976	46,104	2,848
Total	33.1	2.0	52.5	71,805	33,764	8,700
(c) 2003–4 to 2008–2009 Percentage Change						
Micro	−5.1	−19.4	10.8	−34.6	7.8	−9.8
Conventional	−0.4	−7.9	4.0	−4.9	26.7	−50.3
Diversified	−8.4	−8.1	9.0	2.0	9.8	27.7
Differentiated	−3.7	−8.2	9.9	−2.6	17.8	12.5
Diversified–differentiated	10.7	−14.2	−12.3	24.1	21.7	4.5
Total	−0.8	−8.8	5.0	−4.0	24.4	−47.6

Note: The financial variables have been deflated by using the GDP deflator 2005 = 100.
Source: Our elaboration on FADN data.

of them associated with different levels of incomes. As a consequence, the distribution of the income produced in this group of farms appears very scattered. On the contrary the income per unit of labor in diversified and/or differentiated farms is scarcely disperse, even though in all groups there are farms presenting results quite far from the average. It is also interesting to note that the dispersion areas of these groups of farms lay mostly in the positive region of the figure; this is not the case in conventional farms which

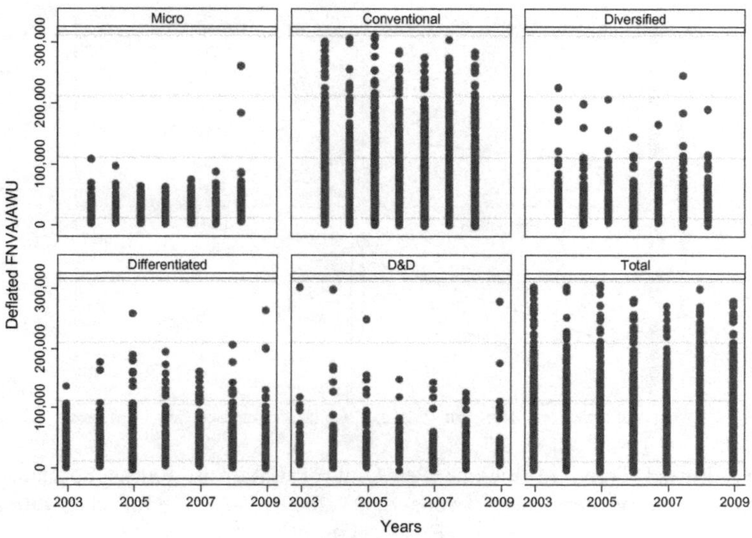

Fig. 1. Scatter Plot of Income Per Unit of Labor by Farm Types (2003–2009). *Source*: Our elaboration on FADN data.

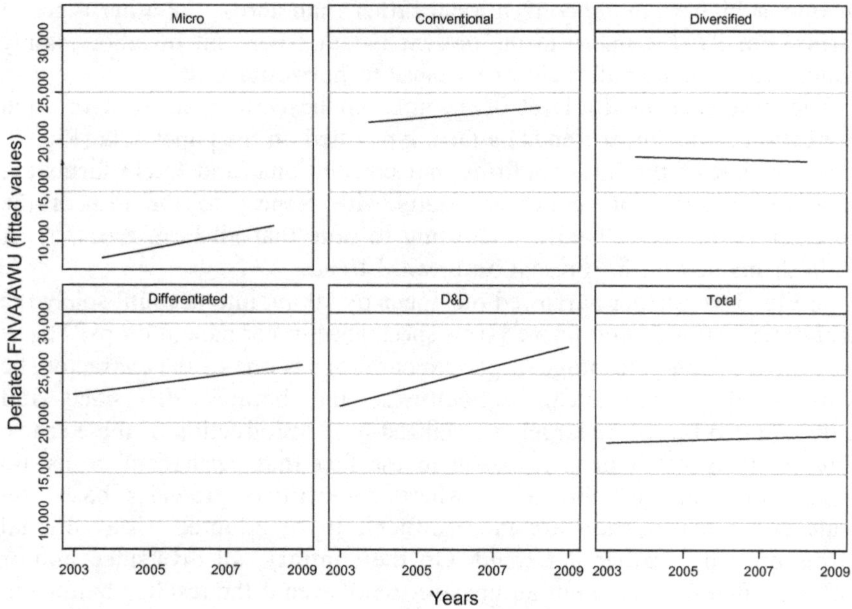

Fig. 2. Linear Trend of Income Per Unit of Labor by Farm Types (2003–2009). *Source*: Our elaboration on FADN data.

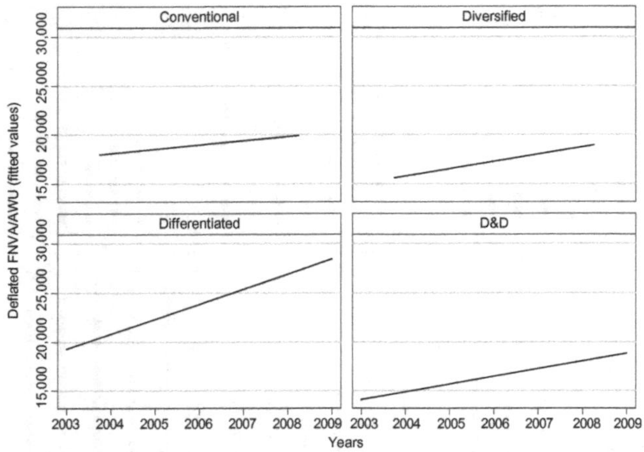

Fig. 3. Linear Trend of Income Per Unit of Labor in Farms Specialized in the Production of Permanent Crops (2003–2009). *Source*: Our elaboration on FADN data.

present many negative results. This means the probability to incur a loss of income is higher among conventional rather than across the other types of farms. Finally, the micro farms present incomes per unit of labor mainly concentrated in a small region very close to horizontal axis.

The difference in the level of income can be better perceived through inspection of the linear trends by farm type fitted on the panel data (Fig. 2). The position of the lines confirms that conventional and D&D farms are superior in terms of economic results with respect to the rest of the commercial farms. It is also interesting to note that all farm types, apart from diversified farms, present an upward trend.

In Fig. 3 we further narrowed our focus by taking into account solely the economic results of non-micro farms specialized in permanent crops. This is because the high percentage of permanent crops is one of the characteristic features of Mediterranean agricultures, and because diversified and differentiated farms are largely specialized in the production of these crops. This, in turn, often finds a reason in the fact that permanent crops are largely produced in hilly areas where competitive strategies based on scale and intensification are more difficult to be adopted. Conventional farms present a rather flat trend. On the contrary, all diversified and/or differentiated farms present an upward trend, even if the results obtained in the case of differentiated farms have to be read with some caution since the distribution is particularly dispersed for this group of farms. The sustained income growth in these farm types signals diversification and differentiation strategies are successful for farms specialized in permanent crops. This finding supports the evidence that these strategies are not survival strategies; on the contrary they can be an effective and successful solution for those farms that either cannot gain from the adoption of conventional modes of production or want to regain an autonomy in decisions and free themselves from the dependence on the decisions (technology, timing, varieties, etc.) imposed by the monopolistic agents at the end of the supply chain (distribution and supermarkets).

CONCLUSIONS

In this chapter we analyzed the development strategies set up by Italian commercial farms and provided evidences about structures and economic results for five types of farms defined on the basis of the farms' effort of differentiation and/or diversification.

The first step has been to highlight that the diversification strategies, traditionally interpreted as survival strategies for small or marginal farms, can in fact be a way to turn some farm specificities into features that consumers perceive as important and that they are keen on rewarding with a premium price. In other words, differentiation and diversification strategies can be a source of competitive advantage and effective in improving farm profitability. Given this interpretation of diversification, we could read some specific features of Mediterranean and Italian agriculture with new lenses. For example, the observed permanence in Italy of a large number of farms characterized by small size, reduced use of industrial inputs, specialization in permanent crops, large use of family-based labor, rather than signals of backwardness and lagging behind in respect to the modernization process can be interpreted as a reaction of farmers to the modernist paradigm that has pushed farms into a process of production intensification and specialization. This has often resulted in technological dependence, loss of autonomy in decisions, as well as loss of economic and environmental sustainability of farms. In the attempt to react to the decline of farm income induced by the cost-price squeeze, as well as to regain autonomy in the decisions about what, when, and how to produce, farmers have started to experiment new activities, some of them connected to farming, some others far from it, all expected to lead to higher and more stable incomes.

The result of this diversification process has led to the co-existence in Italy of the dominant model of agricultural development, based on the concepts of specialization and modernization, with new different models that have been differently catalogued as post-modernist, neo-modernist, and multi-functional.

In this chapter we used the information collected by the Italian FADN to provide evidence about the structural and economic characteristics, as well as about the economic performance over time, of farms that have adopted alternative strategies to the so-called conventional one. For this purpose we divided the population of Italian commercial farms into a fivefold farm typology based on the extent of diversification and differentiation strategies adopted by farms.

Our first finding is about the diffusion of the different diversification profiles. Conventional farms, which are farms that make little or no use of diversification and differentiation strategies, are still by far the largest category within the population of Italian commercial farms, accounting for 73% of the total. A large number of farms (15%) are micro farms, which are farms characterized by very small size and little connections to the market. Finally, 13% of the total commercial farms can be classified as differentiated and/or diversified. This was somehow expected since these strategies can be interpreted as innovation niches, hence they can be associated with a small fraction of the whole population, that is the most innovative farms.

A second finding is related to structure and economic results. Micro farms are on average the lowest income farms in the population, though this cannot be straightly interpreted as a symptom of poverty of the households, since they often use farm income only to complement other sources of off-farm income. According to the productivistic approach, this category of farms can only grow or die. Our findings show a progressive structural and economic marginalization of this group of farms. As a matter of fact, the financial aid they receive through the CAP might artificially "freeze" their position in the sector, slowing down their economic marginalization and preventing their eventual exit from the sector. A revision of public intervention, with the introduction of policy tools targeted to the specific need of micro farms, would correct this problem, at the same time helping the dynamic ones to definitely abandon a survival and subsistence dimension.

As for conventional farms, we found they are the best off in terms of economic results. This suggests they respond well to the prescriptions of the productivistic mode of production and to the public support mainly designed to answer their needs, namely to accompany and sustain their modernization.

As for the differentiated and/or diversified farms, we observed large negative changes in structures of these groups of farms. This suggests they are still looking for an optimal resource allocation that would allow them to increase their efficiency and profitability. This is partly due to the fact that products of farms adopting diversification strategies (multifunctional services, quality products, etc.) are often unique, since they rely on highly specific attributes of farms and on the combination of reciprocal effects between farms and the local context. As a consequence, each diversified farm has to find its own optimum. Under this circumstance the search of the optimal structure requires time, since the solution is farm specific and, contrary to the dominant conventional model of development, there are not well-defined rules to refer to. In consideration of this, more flexibility at the farm level for access to Pillar II of the CAP support, rather than relying on the regional menus as in current programs, might facilitate them to address their specific needs.

The increase in profitability observed in differentiated and diversified farms highlights that these farms have been able to reduce their dependence on off-farm inputs, capturing in this way a quota of value added that otherwise would have been passed to other agents in the food chain.

Another interesting finding of our analysis is that differentiated farms, and also farms adopting both differentiation and diversification strategies, have resulted to follow a more sustained income growth path in comparison to conventional ones. This suggests that the on-going structural change and the experience gained over time in producing and marketing products with new attributes are leading these farms toward increasing levels of profitability. This, in turn, can be taken as a signal of the success of diversification and differentiation strategies that definitively prove not to be solely survival strategies. On the contrary, they can be an effective and successful solution for those farms that cannot gain from the adoption of conventional modes of production and for those farmers who want to regain some extent of autonomy in decisions and free themselves from the dependence on the decisions (technology, timing, varieties, etc.) imposed by the monopolistic agents at the end of the supply chain (distribution and supermarkets).

In conclusion, diversification does not appear to be a necessarily optimal solution to the low-income problem faced by many farms. Future research is necessary to investigate further the issues analyzed here. One relevant topic, here just sketched, is the relationship between diversification strategies and policies. Is diversification actually supported and enhanced by the EU policy? If so, is the current (and future) policy really effective in re-addressing farm strategies toward diversification, so that it becomes a relevant part of them?

Another crucial theme to work on is the future path of agriculture in developed countries: we are witnessing a deep process of change in agriculture that puts the primary activity at the center of a more general rethinking and that see farms and farmers dealing more and more with "other" issues, such as environment, social concerns, integration with territory, all in all sustainable development. A deeper analysis of the internal and external drivers of this change would help to better understand the future role of agriculture in post-modernist economies.

NOTES

1. Indeed, pluriactivity is only one way to deal with low returns on agricultural activities. Other diversification strategies involving the off-farm deployment of resources other than labor are the leasing of land and buildings, as well as renting out of machinery.

2. Stratification allows greater sample efficiency, minimizing the number of sample farms required to represent the variety in the field of observation.

3. The Farm Accounting Data Network of the European Union defines a farm as specialized in a TF if the Standard Gross Margin (SGM) for the particular production covers more than two thirds of the total SGM. SGMs are calculated as the balance between a standard value of production and a standard value of certain specific costs, determined for the various crop and livestock characteristics within each region, and are stratified in European Size Units (1 ESU = 1.200 Euros of SGM).

4. This typology discriminates farms in 10 categories: (i) micro (GP less than 15,000 euros), (ii) multifunctional micro, (iii, iv) small and large conventional, (v, vi) small and large conventional with quality, (vii, viii) small and large differentiated and (ix, x) small and large diversified.

5. This will not be possible when we analyse the balanced panel, as already mentioned in the previous section.

6. Though in Table 3 we present both mean and median values, then, when commenting the results, we often refer to the median only since this statistic is less sensitive to the presence of outliers in the distribution and helps giving a clearer view of the characteristics shared by almost all farms in different groups.

7. The percentage changes shown in Table 4c are the mean of percentage changes calculated at the farm level. Since the data are not normally distributed, these means are different from the percentage changes calculated using the averages presented in previous tables.

REFERENCES

Aguglia, L., Henke, R., & Salvioni, C., (Eds.). (2009). *Multifunctional agriculture. Entrepreneurial behaviour and strategies in the search for diversification*. Napoli, Italy: INEA Studi & Ricerche, Edizioni Scientifiche Italiane.

Arkleton Trust. (1983). *Institutional approaches to rural development in Europe*. Report of a seminar held in Scotland, 17–22 October 1982, Langholm, Dumfriesshire.

Ascione, E., Carillo, F., & Vagnozzi, A. (2011). *Verso la consulenza alla gestione attraverso la RICA. Creazione di gruppi omogenei di imprese e verifica dei risultati economici*. Working Paper. Rete Rurale Nazionale, Giugno.

Belletti, G., Brunori, G., Marescotti, A., & Rossi, A. (2003). Multifunctionality and rural development: A multilevel approach. In G. van Huylenbroeck & G. Durand (Eds.), *Multifunctional agriculture. A new paradigm for European agriculture and rural development*. Aldershot, UK: Ashgate.

De Benedictis, M. (Ed.). (1990). *Trasformazioni agrarie e pluriattività in Italia, INEA Studi & Ricerche*. Bologna: Il Mulino.

De Benedictis, M. (Ed.). (1995). *Agricoltura familiare in transizione*. Rome, Italy: INEA Studi & Ricerche.

De Benedictis, M. (2002). L'agricoltura del Mezzogiono: "La polpa e l'osso" cinquant'anni dopo. *QA-La Questione Agraria*, *2*, 199–236.

De Benedictis, M. (2005). L'agricoltura del Mezzogiorno ieri e oggi: Fattori di crisi e di rilancio. *Rassegna economica*, *2*, 13–42.

Esposti, R., & Finocchio, R. (2008, August 26–29). Determinants of farm diversification and interaction with the CAP; An application to FADN of Marche region (Italy). *XII EAAE conference on People, Food and Environments: Global Trends and European Strategies*, Ghent, Belgium.

European Commission. (1997). *Agenda 2000 – For a stronger and wider union*. Bulletin of the European Union, Suppl. 5.

Evans, N. J., & Ilbery, B. W. (1993). The pluriactivity, part-time farming, and farm diversification debate. *Environment and Planning A*, *25*(7), 945–959.

Fabiani, G. (1986). *L'agricoltura italiana tra sviluppo e crisi*. Bologna: Il Mulino.

Fabiani, G. (Ed.). (1991). *Letture territoriali dello sviluppo agricolo*. Milan: Franco Angeli.

Fuller, A. M. (1990). From part time farming to pluriactivity: A decade of change in rural Europe. *Journal of Rural Studies*, *6*(4), 361–373.

Gasson, R. (1988). *The economics of part-time farming*. New York, NY: Longman.

Gasson, R., & Errington, A. (1993). *The farm family business*. Wallingford: Cab International.

Henke, R., Macrì, M. C., & Storti, D. (2005). Multifunzionalità e sviluppo rurale nell'UE: Un'analisi comparata. *QA–La Questione Agraria*, *2*, 105–133.

Henke, R., & Povellato, A. (2012). L'attività di diversificazione nelle aziende italiane. *Agriregionieuropa*, *31*, 24–29.

Henke, R., & Salvioni, C. (2008). Multifunzionalità in agricoltura: Sviluppi teorici ed evidenze empiriche. *Rivista di Economia Agraria*, *1*, 5–34.

Henke, R., & Salvioni, C. (2011). La diversificazione dei redditi nelle aziende agricole italiane. *QA–Rivista dell'Associazione Rossi-Doria*, *3*, 25–56.

Hill, B. (1999). Farm households incomes: Perceptions and statistics. *Journal of Rural Studies*, *15*(3), 345–358.

Hill, B. (2000). *Incomes, wealth and agricultural policy* (3rd ed.). Ashgrove: Ashgate.

Istituto Nazionale di economia Agraria (INEA). (2010). *Le politiche comunitarie per lo sviluppo rurale. Il quadro degli interventi in Italia.* Rapporto 2008/2009. Osservatorio Politiche Strutturali, INEA, Rome.

Jongeneel, R. A., Polman, N. B. P., & Slangen, L. H. G. (2008, January). Why are Dutch farmers going multifunctional? *Land Use Policy*, *25*(1), 81–94.

Marsden, T. K. (1990). Towards the political economy of pluriactivity. *Journal of Rural Studies*, *6*(4), 375–382.

Mardsen, T., & Sonnino, R. (2008). Rural development and the regional state: Denying multifunctional agriculture in the UK. *Journal of Rural Studies*, *24*, 422–431.

Matthiews, A. (2012). Greening the CAP: The way forward. *QA–Rivista dell'Associazione Rossi-Doria*, *4*, 37–59.

McNally, S. (2001). Farm diversification in England and Wales – What can we learn from the farm business survey? *Journal of Rural Studies*, *17*(2), 247–257.

Meert, H., Van Huylenbroeck, G., Vernimmen, T., Bourgeois, M., & van Hecke, E. (2005). Farm household survival strategies and diversification on marginal farms. *Journal of Rural Studies*, *21*, 81–97.

Organisation for the Economic Cooperation and Development (OECD). (2005). *Multi-functionality in agriculture: What role for private initiatives?* OECD, Paris.

Organisation for Economic Co-operation and Development (OECD). (2009). *The role of agriculture and farm household diversification in the rural economy: Evidence and initial policy implications.* TAD/CA/APM/WP(2009)1/FINAL. OECD, Paris.

Perali, F., Salvioni, C., & Tommasi, N. (2005). *La tipologia delle imprese agricole basata sui dati dell'indagine ISMEA. Agricoltura e ruralità.* Milan: Franco Angeli.

Ravenscroft, N., & Taylor, B. (2009). Public engagement in new productivism. In M. Winter & M. Lobley (Eds.), *What in land for? The food, fuel and climate change debate.* London: Earthscan.

Saraceno, E. (1985). Il part-time nell'agricoltura dei paesi occidentali: Line evolutive e strumenti di intervento. *La Questione Agraria*, *18*, 47–62.

Sotte, F. (2009). *La politica di sviluppo rurale 2007–2013. Un primo bilancio per l'Italia, Quaderni del Gruppo 2013.* Roma: Edizioni Tellus.

Van der Ploeg, J. D. (2002). Revitalizing agriculture: Farming economically as starting ground for rural development. *Sociologia Ruralis*, *40*(4), 497–511.

Van Der Ploeg, J. D., & Roep, D. (2003). Multifunctionality and rural development: The actual situation in Europe. In G. Van Huylenbroeck & G. Durand (Eds.), *Multifunctional agriculture. A new paradigm for European agriculture and Rural Development* (pp. 37–53). Burlington, VT: Ashgate.

Wilson, G. A. (2007). *Multifunctional agriculture. A transition theory perspective.* Cambridge, MA: CABI.

Wilson, G. A. (2008). From "weak" to "strong" multifunctionality: Conceptualising farm-level multifunctional transitional pathways. *Journal of Rural Studies*, *24*, 367–383.

Zahrnt, V. (2009). *Public money for public goods: Winners and losers from CAP reform.* ECIPE Working Paper No. 08/2009. ECIPE, Brussels.

CHAPTER 10

TRANSFORMATIONS IN CROATIAN AGRICULTURE AND AGRICULTURAL POLICY: CHALLENGES AND OPPORTUNITIES WITHIN THE EUROPEAN CONTEXT

Ramona Franić and Ornella Mikuš

ABSTRACT

During the past two decades Croatia has faced numerous challenges: gaining independence, war conflicts, political and economic transition and the process of European Union (EU) accession. Despite rich and diversified landscapes and cultural heritage, it is still faced with problems limiting the economic development. So the purpose of this chapter is to point out the pragmatic reasons of Croatian delay in the process of adjustment to European business and agricultural policy standards.

Based on statistic and literature analysis, the study determines specific characteristics of Croatian regions, rural areas, rural population and agriculture. Agriculture after independence shows increase in utilized

Agriculture in Mediterranean Europe: Between Old and New Paradigms
Research in Rural Sociology and Development, Volume 19, 233–261

ISSN: 1057-1922/doi:10.1108/S1057-1922(2013)0000019012

area, but the production is still below pre-war level and results with unsteady and modest value. Harmonization with Common Agricultural Policy (CAP) standards is slow; big steps have been made in establishing new institutions in agriculture and preparing adequate legislative framework, so there are no significant formal differences between Croatian and European agricultural policy. However, European agricultural policy models cause problems. There is a daily debate about a low degree of self-sufficiency of the domestic production, low competitiveness and uncontrolled import of farm products. Farmers still often expect for the government to organize the production and guarantee the purchase prices as were in former, socialistic system.

Due to these reasons, a fear was expressed by farmers that they could not be able to meet the strict criteria for the European financing. Despite this, a large part of farmers see the possibilities for their existence in rural areas, mostly through development of non-agricultural activities.

Keywords: Croatia; agriculture; agricultural policy; transformations; Common Agricultural Policy

INTRODUCTION

Croatia is not a large country; however, it covers a diversity of landscapes: Mediterranean, Central European, mountainous and flat, coastal and continental. Its diversities and wealth of nature can be compared with much larger European countries. This is why Croatia ranks as one of the top five European countries with regard to biodiversity, with some parts being among the richest areas of their kind in the world. Richly indented wooded shore line with numerous islands and preserved old Mediterranean towns, green coastal meadows with dry stone walls, mountainous areas abound with wide expanses of woods and forests, karstic landscapes with deep gorges and canyons, rich water resources and wide plains on the north contribute to Croatia's natural diversity. The most treasured parts of Croatia's natural heritage comprise 447 different protected areas covering about 10% of its land area.

Besides the diversified natural resources, its geographical position was a fertile ground for wide and intensive historical, political and cultural influences from different parts of the world, shaping diversified regional, ethnic and folklore patterns. These are still obvious, especially in Croatian rural areas, where lifestyles are considerably different among regions. Croatia has managed to preserve the rural and indigenous architecture, which clearly testifies to its cultural diversity. Today we can still see old wooden houses and churches in the flatlands of Pannonia and the unique elongated Slavonian houses with large yards in the Danube region, picturesque huts on hills in the north and shingle-covered old houses in the snowy mountains of Croatia. On the other side, there are the Mediterranean villages with houses made of stone on the mountain peaks in Istria, or by the sea on the Croatian coast and islands.

The diversities are evident that large European civilizations had converged here, and that it is right here that the influence of Europe's largest ethnic groups of Slavs, Romans and Germans had come together. Moreover, the country's Eastern borders were the former western and eastern borders of Christianity, and Western and Eastern European culture. This is where the last line of defence used to be for the overcoming Islamic civilization in the late Middle Ages, as evidenced by many mighty works, and monuments of the rich military history (Jačan Obratov, 2010).

During the past two decades Croatia has faced numerous challenges: gaining independence and war conflicts, a process of political and economic transition from a centrally planned, socialistic to capitalistic, market-oriented system focused on the accession to the EU. These elements together with natural and cultural differences reflect also on current production and business practices, socio-cultural capital and people's expectations from the future. Trying to respond to the expectations of recently gained independence and requirements of basic strategic commitments (European integration), Croatian strategies to adjust to Western European modernization processes merge with the attempts to keep its Central European and Mediterranean identity and traditional values. However, in contrast with the integration processes of the rest of the Mediterranean EU countries, the modernization paradigm (Karanikolas & Martinos, 2007; Moreno-Perez, Arnalte-Alegre, & Ortiz-Miranda, 2011) is no longer the dominant model; instead, the diversity of strategies characterizes the actual dynamics of European agriculture. From this point, this chapter discusses the recent transformations of Croatian agriculture and agricultural policy, particularly considering the potential opportunities arising from the valorization of specific natural and cultural attributes and focusing on how changing EU policies are affecting these transformations.

During the last two decades researchers have emphasized on numerous problems limiting the development of the overall economy and especially rural areas in Croatia: the inherited problems of agricultural land in the context of privatization, land registers, inheritance system, agricultural markets, taxation, etc. In spite of difficult political and economic conditions imposed on rural areas after the World War II and dramatic changes which were often not progressive, some factors facilitated survival of villages: natural capital in rural areas of Croatia, possibility of diversification of activities, entrepreneurial skills and social and economic heterogeneity, including the knowledge and experience of the local population (Franić & Žimbrek, 2003; Grgić, Žimbrek, Tratnik, Markovina, & Juračak, 2010; Mikuš, Mesić, & Cerjak, 2010).

From the beginning of the process of Croatian accession to the EU towards the end of the last century, the underdevelopment of Croatian rural areas was even more strongly emphasized. There is a great difference between the standards in relation to development, organization and protection of rural areas in the EU and those in Croatia. The differences are primarily a consequence of specificity in historical development; rather long period of socialistic regime and centrally planned economy neglected rural areas and private farmers. It caused disorientation within the market economy rules and private initiative after independence, so the stereotypes from the past reflected in legislative framework.

Since the accession process requires adjustment of agricultural policy to the Common Agricultural Policy (CAP) standards, this intensified particular issues that have been investigated in EU member states as well. Researchers in Croatia stress out the need of agricultural structural adjustments based on production competitiveness for the part of vital, economically successful farms and regional approach in using natural resources in the most productive way (Franić, Tratnik, Bašić, & Svržnjak, 2007b; Radinović & Žutinić, 2007; Tratnik, Franić, Svržnjak, & Bašić, 2008). With the purpose of maintaining and developing agricultural production within European competition, they put an emphasis on principles of sustainable development, food safety and environmental standards. Suggestions are in favour of production practices based on these principles and for agricultural policies to set priorities within the available resources, selection of models of support and adjustments of policy measures according to regional requirements (Franić, Mikuš, & Andabak, 2011; Mikuš et al., 2010).

However, considerable part of actual number of farms probably wouldn't survive international competition and there is reasonable 'fear' of loosing small farms, traditional rural products and values and even disappearance of entire villages (Grgić et al., 2010). Therefore, researchers suggest European models based on searching for additional sources of income in rural areas, multifunctionality and its contribution to public goods (Franić et al., 2011; Grgić et al., 2010) having examples in European economies (Moreno-Perez & Ortiz-Miranda, 2008; van der Ploeg, Laurent, Blondeau, & Bonnafous, 2009; Pinto-Correia and Breman, 2009). Special attention is put to rural development issues and rural policy as an alternative opportunity for sustaining rural livelihoods in Croatia (Franić, 2006; Mikuš, Franić, & Grgić, 2012), pointing out the most important constraints similar to those in other European countries – misinterpretations of the concept and the practice, prevalence of agricultural interests at the expense of wider rural concerns (Papadopoulos & Liarikos, 2007) and low participation of farmers in rural development programmes (Borsotto, Henke, Macrì, & Salvioni, 2008; Franić, Bokan, & Kumrić, 2007a).

Having in mind the above-mentioned characteristics and influences, the intention of this analysis is to point out the pragmatic reasons of Croatian delay in the process of adjustment to European (business, agricultural and agricultural policy) standards and rather disorientation in creating (agricultural) policy goals and measures for their realization.

CHARACTERIZING REGIONAL DISPARITIES

As mentioned earlier, besides difficulties associated with the period of transition to a market economy, Croatian agriculture suffered considerable destruction caused by war activities in the first half of 1990s. Problems of agricultural land administration (unsolved land property rights and cadastre) are additionally burdened by still large surface of farm land under mines (according to official data, more than 95,000 hectares are contaminated, mostly forests and arable land; MAFRD, 2009).

Concrete differences are still visible in Croatian agrarian and rural policy with regard to the developed European countries, because of which rapid administrative adjustment encounters numerous difficulties in application. Systems of inheritance have left deep differences in the form of parcelled land in Croatia, and the search for better life and work conditions, as well as forced industrialization and de-agrarization during the socialist period, in the form of excessively large acreage of abandoned and disused land. This is why even today there are unsolved ownership relationships, denationalization of State-owned land is not finished, and land purchase mechanisms are still too complex because of the unbalanced situation in the land registry. Land policy problems are the biggest obstacle in the adjustment of both policy and the whole sector to the desired European standards.

Croatia is characterized by different climates, natural resources and economic heterogeneity. Its territory is divided into three large natural, geographical and agro-ecological regions: Pannonian, Mountainous and Mediterranean (Bašić & Franić, 2003). Various natural pre-conditions and also recent political events have resulted in this region being significantly different with regards to both demographic and economic characteristics, and these differences are additionally accentuated in the rural regions. The aim of the following sections is to describe regional, natural and socio-economic characteristics of Croatian rural areas through three different regions (Pannonian, Mountainous and Mediterranean) as a basis for analysis of transformations in Croatian agriculture during the last two decades.

The Pannonian Region

The northern, continental part of Croatia, the Pannonian region, is for the most part bordered by two great rivers, Drava and Sava. This region encompasses about 47% of the total area of the state and about 64% of the Croatian population, making it the most densely inhabited region in Croatia. An abundance of water, woods and fertile soils, navigable rivers and significant landward European highways are natural predispositions which enable a developed economy, transport, trade and culture. The quality of the soil, climate and favourable annual rainfall schedule ensures that this region has good natural potential for agricultural production, which is also the basis for a food industry. Surfaces covered with quality woods, especially high-quality trees, are a favourable basis for the development of forestry or a strong wood-processing industry.

Thermal springs, preserved surroundings and unique ecological regions, numerous game reserves, excursion points and historical monuments offer the basis for a successful combination of agricultural, forestry and tourist activities. Production capacities in the primary part of agriculture make it Croatia's granary, while outstanding results are also being achieved in the winegrowing and winemaking industry. The northern and north-western parts of this region have recently developed successful business relations with the largest urban centre in the country, the city of Zagreb, which as the largest consumer centre offers great employment opportunities.

Today, the Pannonian region is characterized by two opposing realities: on the one hand, rich natural resources and traditional success in agriculture, and on the other, the destructive consequences of recent wartime events, especially in eastern and western Slavonia with destroyed production capacities and great rural areas still not cleared of mines. Additional challenges to the development of this region are the search for solutions to transitional problems, like the privatization of former socialist agricultural conglomerates and the question of ownership and use of State agricultural land and other capital goods.

Mountainous Region

The central part of the Croatian territory is the mountainous region, a heterogeneous area with karst areas in its north to mountain massifs to the south. The economic development of this region is based on rich natural resources – oil and gas fields, clay and stone quarries, fertile soils, forests, rivers and navigable water potentials, thermal springs, and diverse plant and animal world. The main economic activities of this region are forestry and agriculture, wood-processing industry and tourism due to its national park, the Plitvice Lakes. Agricultural production in this region is adapted to the typical mountain climate with abundant rainfall and heterogeneous soils: on small family farms livestock production dominates, while the crop farming is limited due to the sharp climate. It is because of this that early types of corn, potatoes, rye and vegetables dominate. Agricultural production is traditionally extensive with proportionally small use of mineral fertilizers and protective agents, which offers the possibility of the development of ecological (organic) agriculture in combination with eco-tourism.

Exploitation of the natural resources is limited due to the sparse population density and still insufficiently developed road infrastructure, and additional efforts are being invested in the reconstruction of production capacities in rural areas which were destroyed by the war. The greater part of the mountainous region and the surrounding region may therefore utilize additional budgetary resources intended for stimulating the development of the areas of special state concern.

Mediterranean Region

The long coastal and island belt from Istria in the northwest to Konavle in the southeast is the Mediterranean region of Croatia, a region of attractive beaches and diverse plant and animal world, both on land and in the sea. The Mediterranean climate, warmer and drier towards the south, are good pre-conditions for the cultivation of high-quality Mediterranean crops like olives, figs, vineyards and Mediterranean fruit. However, the backbone of the economy is, of course, tourism, while agriculture and fisheries are traditional activities in the flatter regions. Agricultural areas are mostly located in plains and river valleys, tablelands and karst fields, while in terms of structure small family agricultural farms dominate. The diverse tourist and catering offer is supplemented by numerous traditional products and services, so a large percentage of the regional GDP is achieved by the tourism sector. The Mediterranean region is the 'most urban' region in Croatia, with a small share of agricultural population over the total, significantly lower than the national average.

The northern part of the Mediterranean region (Istria) leads in an economic sense: it offers better employment opportunities, developed infrastructure, as well as entrepreneurial spirit in the population. Contrary to this, the central and southern part of Dalmatia are still feeling the effects of war: numerous production capacities are damaged or destroyed; however, in recent times additional incentives for the development of the economy have improved traffic infrastructure and renewed tourist activities. Sparsely inhabited islands, as the areas of weaker production possibilities, receive additional aid from budgetary resources so as to renew agricultural production and life on them.

THE RECENT EVOLUTION OF CROATIAN AGRICULTURE

Croatian agriculture is characterized by a large number of small, fragmented family farms, owing 77% of total agricultural area. According to the last agricultural census (2003), there was more than 450,000 family farms, with a considerable share of so-called mixed farms (with diversified production), making agricultural activities just one of the income sources. However, only about 140,000 farms are officially registered as market producers, utilizing about 850,000 hectares of agricultural area, which is only 27% of total agricultural area in Croatia. Out of the total number of registered farms, 98% are family farms, namely, unlike other Central and Eastern European countries, even in the period of socialistic regime in Croatian agriculture, large, collective farms were not a predominant form.

Official statistical data show that in 2010 Croatia utilized 1.334,500 hectares of total agricultural area[1]; while the majority corresponds to arable crops (about 66%), considerably smaller share is under permanent crops (less than 7%) and permanent pastures (slightly more than 26%). However, during the past decade, although with fluctuations, total utilized agricultural area was increasing (Fig. 1).

When agriculture is exposed to international market influences, and especially in countries where average farm size is small, land tenure is perceived as the most adequate mean for expanding the agricultural production. After gaining independence, former 'socially owned' land is declared as 'state-owned'. The procedure of giving the tenure of the state land is under the competence of local government and the preference in this process is given to family farms. About 34% of the state agricultural land was given for disposal till 2009 (out of 60.8% planned) and the total utilized agricultural area was increased for about 30% during the past decade (2000–2010). This resulted in increased number of farms registered for agricultural production to 18% as well as increased average farm to 9.5% (Svržnjak, 2012).

According to the estimations prepared for the strategic programmes (MAFRD, 2009), the number of employed persons in agriculture, fishery, forestry and food-processing industry is continuously decreasing. By the end of 2000, it was about 120,000 (8% of the total number of employed persons); in 2008, the number of employed in this sector was about 70,000 (Fig. 2).

Although regional diversity and natural conditions provide conditions for production of rather wide range of agricultural products, destructive political and economic phenomena since the beginning of the 1990s reflected on rather modest agricultural results. Even plant production was increasing since 1995, it is currently at about 80% of pre-war level due to variety of

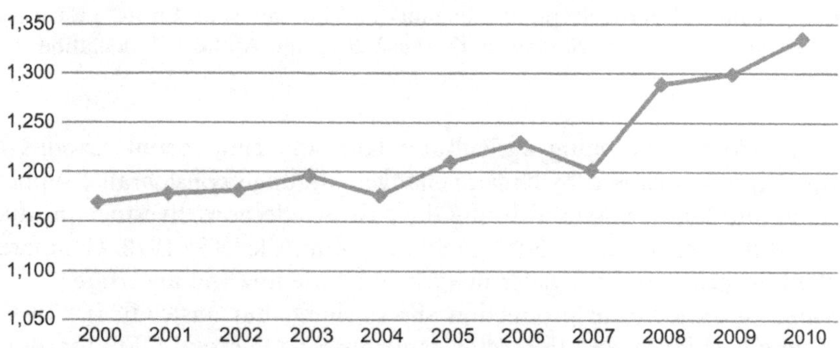

Fig. 1. Utilized Agricultural Area (in Thousands of Hectares). *Source*: CBS and *Statistical Yearbook* 2011.

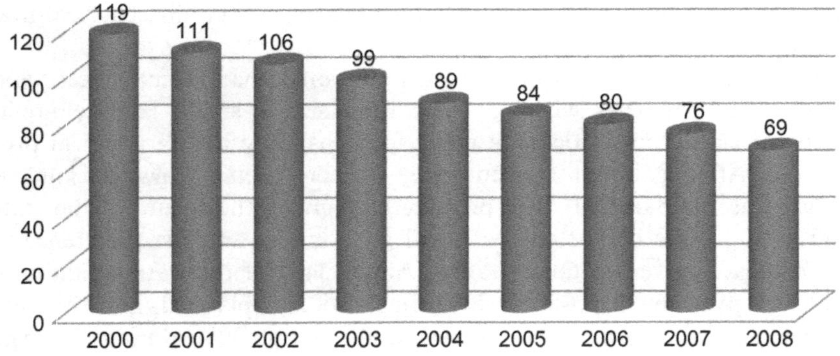

Fig. 2. Number of Employed in Agriculture (in Thousands). *Source*: MAFRD (2009).

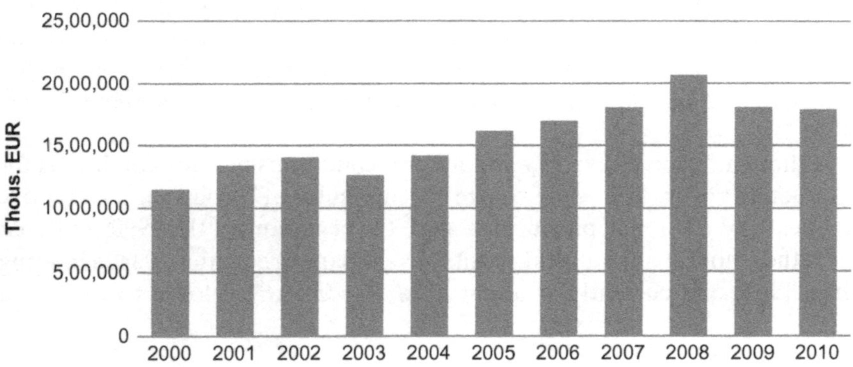

Fig. 3. Value of Agricultural Production (in Thousands of Euros, at Current Prices). *Source*: CBS, *Statistical Yearbook* 2011 and Authors' Calculation.

reasons, from abandoning agricultural land and rural people exodus to natural disasters caused by climate change. Although considerable support was put into livestock sector during the past two decades, livestock production is still lower for about 30% compared with its level in 1990. Compared to the European average, yields in agriculture are low and uncertain.

Value of agricultural production shows slight, but unsteady rise in the past decade (Fig. 3), and the sector contributes to the total GDP for about 6% in 2011. The greatest part in total agricultural value has crop production, mostly cereals, while livestock products contribute with about 40% in total value of agricultural production. As a positive consequence of agricultural policy measures carried out through several action plans (setting up of permanent crop plantations, vegetable growing, cattle production, pig production), areas under vineyards and olive plantations were increased in the Mediterranean region.

The market situation (supply–demand) is not balanced and foreign trade balance is in long-term deficit. Data for the last two decades show that Croatia was a net importer of agricultural goods and food products (Fig. 4) and the greater share of imports comes from the EU markets (Italy, Germany, Austria and Hungary).

Market liberalization was intensified after 2002, so, together with numerous bilaterateral arrangements, it changed foreign trade relations and the structure of export markets considerably. The most important export destinations are countries of Former Yugoslavia, especially Bosnia and Herzegovina, followed by the EU countries.

International agreements clearly define the foreign trade regime of agricultural and food products: reducing support and tariff protection in agriculture, possibilities for free export of agricultural-alimentary products to countries of the EU, and on the import side further gradual liberalization of Croatian agricultural market. The application of market concessions for the products from EU countries in some cases had also negative influences on our market, and through exhaustive negotiations and consultations with the European Commission representatives, an adjustment of conditions was arranged, which fixed those negativities to some extent.

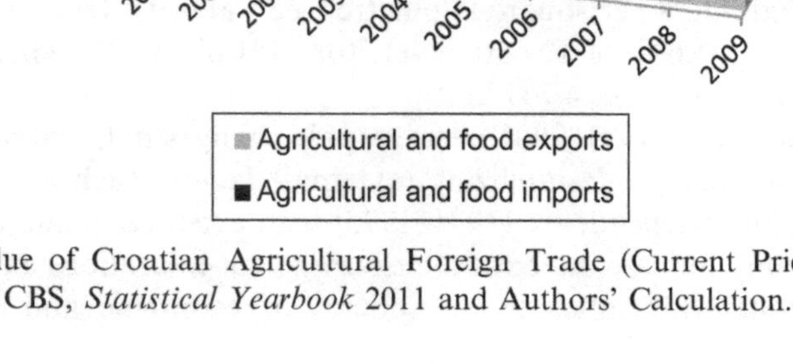

Fig. 4. Value of Croatian Agricultural Foreign Trade (Current Prices). *Source*: CBS, *Statistical Yearbook* 2011 and Authors' Calculation.

Presuming that Croatia will become a member of the EU in 2013, we cannot forget the fact that it is a way of losing the sovereignty in bringing the choices about domestic agriculture. Furthermore, one should take into consideration that the CAP will continue to develop as an answer to the new circumstances – expansion of the EU, economic crisis, global climatic and ecological changes etc. The perspectives of the current budget end in 2013 and before that further reform suggestions can be expected – possibly within a few years after the Croatian accession in the EU.

SOCIO-ECONOMIC SITUATION OF RURAL POPULATION

Since the beginning of the 1990s, with the declaration of independence, Croatia has shifted its political system from the real-socialist one party to a capitalist liberal representative democracy, i.e. from communist to liberal/capitalist ideology. Accordingly, this led to the abandonment of the centrally planned economy and the introduction of market-based economy rules.

Thereby, as the concept of *rurality* was not precisely defined, one could understand rural areas as areas extending across territory with natural countryside, including villages, small settlements etc., with agriculture and forestry as its main economic characteristics. In the frame of methodology of Croatian Bureau of Statistics (CBS) implemented during the Census in 1991, rural/urban areas and population were approximated by considering all areas under 'towns' as urban areas, and those under 'municipalities' as rural areas.[2] That was the so-called administrative criteria and data showed that Croatia had 45.7% of population living in rural areas and 54.3% of population living in urban areas (Table 1).

In 2001, the methodology was complemented with OECD criteria by defining rural areas as local administrative unit level 2 (LAU 2) with less than 150 inhabitants per square kilometre (see Table 1). Due to that, rural areas occupied 91.6% of Croatian territory (51,872 km^2), whereas urban areas occupied 8.4%, i.e. 4.731 km^2.

During the period 1991–2001 demographic changes in Croatia have been strongly influenced by destabilizing (external) factors such as aggressions during War for Independence (1991–1995) with extensive material damage, the loss of many human and natural resources, out-migrations and post-war and transitional difficulties in all spheres of economic and social life.

Table 1. Rural Population in Croatia in 1991 and in 2001.

Population	1991		2001			
	Administrative Criteria		Administrative Criteria		OECD Criteria (LAU 2)	
	Number of Inhabitants	%	Number of Inhabitants	%	Number of Inhabitants	%
Rural	2,187,060	45.7	1,966,005	44.3	2,112,085	47.6
Urban	2,597,205	54.3	2,466,455	55.7	2,325,375	52.4
Total	4,784,265	100.0	4,437,460	100.0	4,437,460	100.0

Source: Censuses 1991 and 2001.

Although in both rural and urban areas population growth was negative in recent years, depopulation was much stronger in rural areas compared to urban regions. Natural population growth rates differed strongly among the regions. In number of extreme rural areas the vital index (live births per 100 deaths) during the years 1998–2002 amounted to less than 50%. Also, there is a much higher share of population of the 65 and older age group in total population (an increasing trend) of rural and especially 'extreme' rural counties,[3] i.e. regions with rural population between 90% and 100% compared with predominantly urban areas (between 14% and 16%). A much lower share of the work contingent (percentage of females 15–59 years old and males 15–65 years old) in total population lives in rural areas (approximately 60%) compared with urban areas (approximately 66%). Furthermore, a much lower percentage of young women (ages 20–29) in total population live in rural regions as compared to urban areas. A generally much lower proportion of women in reproductive age (between 18 and 49 years of age) lives in rural areas compared with urban areas, which causes decline in birth rate. In 'extreme' rural counties the respective coefficient equals to only 35%, i.e. about 10–14 percentage points lower than in the majority of urban regions (48–49%). All these factors have caused the increase in the age index[4] of the rural population from 0.79 in 1991 to 0.96 in 2001.

The methodology used by OECD allows comparison of evolution during 2001–2011. OECD considers three types of regions: predominantly rural, significantly rural and predominantly urban (see Table 2).

Results of OECD classification show that 81.5% of inhabitants live in rural areas (predominantly rural and significantly rural), which cover 98.9% of Croatian territory. Inclusive, 18.5% of inhabitants live in predominantly

Table 2. Rural Population in Croatia in 2001 and 2011.

OECD Criteria	2001		2011	
	Number	%	Number	%
Predominantly rural area	2,110,988	47.6	1,678,194	39.1
Significantly rural area	1,547,327	34.9	1,819,543	42.4
Predominantly urban area	779,145	17.5	792,875	18.5
Total	4,437,460	100.0	4,290,612	100.0

Source: MA (2012); censuses in 2001 and 2011.

urban regions which cover only 0.01% of Croatian territory, and only one county (the *city of Zagreb*) could be described as predominantly urban region in 2001 and in 2011 calculations as well.

Employed or unemployed population in total working age population (1,952,619 persons), due to the census from 2001, does not differ significantly with regard to whether they live in rural or urban area. There are 79.3% of employed persons in total rural and 79.8% employed persons in total urban working age population.

Rural areas in Croatia are generally characterized by a poor access of local population to a basic infrastructure. All major indicators which describe a level of endowment of rural population with infrastructural facilities or access to the basic infrastructure, e.g. number of telephones per 1,000 population, number of post offices per 1,000 of population, and roads and railways density (per 100 km^2), are much lower in rural than in urban regions. Also, living conditions of the rural people are in the majority of the regions much worse compared with those in urban areas. Living space of rural residents is smaller than that of urban ones, and access to housing facilities of rural households is poor compared to urban households.

AGRICULTURAL POLICY MEASURES DURING TRANSITIONAL SHOCK

In the beginning of the 1990s, most of the Croatian agricultural policy measures were taken from the ex-Yugoslavian legislation with very general objectives: economic growth and progress provided by the supply of high-quality food offered at acceptable prices and the export of products with comparative advantage; concentration of family farms to a level which provides an adequate standard of living; providing necessary funds through credit, benefits in input procurement, implementation of new technologies and information through advisory services; increasing investments in agriculture; protection from foreign competition; converting the ownership and management structure; and development of market infrastructure and adaptation of institutional framework in accordance with market trends (Mikuš et al., 2010).

Although experiencing great socio-economic problems and changes in its structure due to the war, Croatian political aspirations were 'pro-Western' and 'pro-EU' oriented. It has considered harmonization of legislation with *acquis communautaire* and a kind of imitation of theory and practice in economic policy of EU members.

Common Agricultural Policy (CAP), which sets out the rules for farming, environmental and rural development activities and EU agricultural markets, has always played an important role in the EU budget. As it absorbs a significant amount of European budget (European Commission, 2011) and deals with sensitive topics, CAP has experienced several turning points since its establishment. Namely, CAP caused a number of negative effects such as high level of price protection followed by agricultural overproduction, shift of incomes towards large producers and a high budgetary burden. In the 1990s, society became increasingly concerned about the environmental sustainability of agriculture and food safety especially due to the emergence of mad cow disease and dioxin affair (Lovec & Erjavec, 2012). Through two significant reforms in the 1990s, intervention price levels were significantly reduced towards international market prices, direct payments to farmers were established in 1992 and a serious debate about the role of agriculture in the European society was initiated. Later on, especially with Agenda 2000 and Mid-Term Review in 2003, the strategy was to gradually decouple supports away from production and to turn subsidy towards rural development.

Following the new direction of the CAP, Croatia also started to adjust its agricultural policy to the changed circumstances. In the mid-1990s an adequate legal framework for governing the protection and allocation of resources and economic processes of privatization was created with the document 'Development Strategy of Croatian Agriculture' (1995). The first important acts (Law on Agricultural Land – Official Gazette 66/01, 87/02, 48/05, and the Law on Agriculture – Official Gazette 66/01, 83/02) came into force in 2001 and were the beginning of an initiative to regulate and harmonize domestic legislation in the sphere of agriculture and rural development according to the European standards and principles (Franić, 2006).

Due to the growing awareness that the rural economy is more than just agriculture, the goals of the future development were formally established through the strategy of agriculture and fishery of the Republic of Croatia in 2002 (Official Gazette 89/02). The agricultural measures were no longer focused only on improving the agricultural activity (along with hunting, forestry and fishing), but extended to the entire rural area. In addition, a new emphasis was given to food security and organic farming as a special system of sustainable management and the possibilities that this system provides (Franić & Žimbrek, 2003).

THE SALIENT FEATURES OF THE NEW AGRICULTURAL POLICY

The Law on State Support in Agriculture, Fishery and Forestry (Official Gazette 87/02, 117/03, 82/04, 12/05) introduced new models of support. The precondition for its implementation was introducing the Register of Agricultural Holdings as the administrative basis for managing agricultural finances. The intention was to make the support system in agriculture more simple and transparent, and to gradually adjust agricultural policies with the EU policy. Besides the changed production support model (which consisted of price aid, coupled area payments in crop production and per head in livestock production), three new models of support were introduced: income support, capital investment and rural development. The aim was to re-allocate funds in favour of these new models, with partial decrease in production support. For the first time this law distinguishes commercial and non-commercial farms. Commercial farms had to fulfil the condition of having 'three units of output'[5] in order to obtain the support in the frame of models as direct payments, capital investment and rural development. Non-commercial farms fulfilled the conditions for income support schemes and rural development schemes (Franić, 2006; Mikuš et al., 2010).

The objective of the rural development scheme was the maintenance and development of rural areas, and particularly traditional features of Croatian agriculture. This scheme comprised three programmes: (1) development of rural areas; (2) preservation of autochthonous and protected breeds; and (3) marketing of agricultural products and foodstuffs. The fundamental objective of the programme for the development of rural areas was the sustainable development of rural areas by ensuring adequate working and living conditions and the preservation of natural and cultural heritage.

Measures for the rural development encompassed aid for investments into commercial farms; processing of agricultural products; establishment of machinery rings (machinery co-operatives); young farmers; education; retraining and training; environmental protection measures in agricultural and forestry areas; renewal of fire affected forests and agricultural areas; land improvement measures; reconstruction and development of the countryside; preservation of cultural property; rural customs and their manifestations; various agricultural and other activities aimed at realizing additional or alternative sources of income; improvement in rural infrastructure connected with the development of agriculture; promotion of wine roads and other tourist routes; and promotion of autochthonous products and development of public services in rural areas.

During the same period, some region-based acts were adopted, trying to respect regional, local and territorial specificities (Franić, 2006). The Law on Mountain Areas (Official Gazette 12/02 …, 80/08[6]) determines conditions for demographic renewal, economic growth and sustainable development and puts the priorities on measures for keeping the population in the zones with negative demographic trends and provides the framework for more sensitive resource use. The other specific act was the *Law on Areas of Special State Concern* (Official Gazette 26/03 …, 57/11) directed towards the areas which were particularly damaged during the recent war, with the purpose of removing the war consequences, faster return of refugees and displaced persons and overall reflation. Also regionally sensitive is the *Law on Islands* (Official Gazette 34/99 …, 33/06), defining Croatian islands as natural 'treasure' that requires particular state protection. Due to this law, the government is obliged to support programmes and activities for sustainable development of islands. These acts can be understood as forerunners of actual regulations, since these issues are now under the scope of state support in agriculture and rural development legislation regarding less favoured areas.

The analysis of territorial differences in the early 2000s, across geographical regions and counties in Croatia, performed on several occasions, showed significant discrepancies in economic performance and standard of living between urban and rural areas.[7] There are several significant negative indicators of structure of rural population: depopulation trends and out-migrations, low level of educated rural population, a high dependence of rural population on social support, poverty and high rate of unemployment in majority of rural areas, a poor access of local rural population to a basic infrastructure etc. (Grgić et al., 2010).

One of the reasons for that situation could be found in the fact that policies and programmes of existing institutional structures dealing with rural areas were not coordinated. Also no individual, area-based or bottom-up approaches were implemented considering agro-ecological diversity and features. Many institutions were insufficiently endowed with professional and financial resources to fully understand the development needs of family farms themselves and in the context of rural economies.

The first definition of rural areas according to OECD (<150 persons/km^2) and detailed screening of the situation in rural areas considering geographical differences were submitted in 2004 through the draft of the document *A Strategy for Rural Development in Croatia (2004–2010)*. The socio-economic analysis of the rural areas in Croatia was carried out at the level of three main agro-ecological and geographical regions: (1) Pannonian region, (2) mountain region and (3) Mediterranean region. Strategic objectives of rural development in Croatia were defined as follows (UN FAO and MAF, 2003):

1. Rural residents' standard of living and their quality of life should improve continuously and the difference between rural and urban must not be increased.
2. In future, rural communities have to tend to become vibrant, sustainable and attractive places to live and work, with equitable economic opportunities for all residents of rural communities regardless of their income, status or gender.
3. Rural areas should contribute to the development of the overall economy by maintaining dynamic links to villages, small towns and urban areas.
4. Rural areas have to become able to adapt to on-going economic, social, cultural, environmental and technological change and to be fully integrated into the market economy.

For unknown reasons *A Strategy for Rural Development in Croatia (2004–2010)* was never officially adopted, although it was prepared upon the base of the EU principles. Furthermore, the *National Plan of Agriculture and Rural Areas 2006–2008* – the continuation of *The National Plan of Agriculture and Rural Areas 2003–2005* – was another crucial document created by a national team of experts, and not adopted by national government. One might suspect it was due to the inconsistency between the two opposite political parties that gained access to the government in 2000 and 2004 and would not adopt the suggestions or the strategy of development of the previous one.

In the meantime, Croatia also started the EU accession negotiation process in 2005; two years after the CAP once again experienced changes through *Mid-Term Review* reform (2003). A full move to farm-based area payments without links to current production but with greater environmental commitments and strengthening of rural development measures was the main things that Croatia was supposed to adopt gradually, but steadily.

LEARNING CAP LESSONS THROUGH PRE-ACCESSION PROGRAMMES

Regarding rural development and agricultural competitiveness issues, in the beginning of 2006, the European Commission (EC) adopted the *SAPARD Programme 2005–2006 – Agriculture and Rural Development Plan* prepared by Croatian Ministry of Agriculture, Forestry and Water Management (MAFWM, 2006) as a framework for financing rural development measures under the SAPARD (Special Accession Program for Agriculture and Rural Development). Two measures for improving the competitiveness of the sector were stipulated (MAFWM, 2006): (1) investment in agricultural holdings, and (2) investment in the processing and marketing of agricultural and fishery products. There were 4 public tenders and 37 applications were approved and paid from SAPARD funds in the amount of EUR16.1 million.

An assessment of the degree of alignment driven by European Commission in 2007 (EC, 2007) noted the lack of harmonization and serious activities in the field of rural development. Among all, the unsatisfactory dynamics of disbursement of SAPARD funds (26% until 2007) was interpreted as lack of initiatives to follow the EU practice of implementation of agricultural policy measures.

In 2007 the *IPARD Programme 2007–2013 – Agriculture and Rural Development Plan* was prepared within IPARD (Instrument for Pre-accession Assistance, component V – Rural Development) by the Ministry of Agriculture, Fisheries and Rural Development (MAFRD, 2009), followed after SAPARD as the basis for utilizing the pre-accession funds. The programme was adopted by the EC in February 2008 and a total of 41.8 million Euros has been approved. IPARD consists of seven measures which are to be implemented, within three priority axes: (1) improvement of market efficiency and implementation of community standards, (2) preparatory actions for implementation of the agri-environmental measures and local rural development strategies and (3) development of rural economy. The SAPARD instrument, and later on the IPARD (utilized 38% until October 2012), was supposed to enable Croatia to accelerate the alignment with the *acquis* in the agricultural and rural development sector.

Later, an analysis showed (Mikuš et al., 2010) that the reasons for such a small number of applications could be found in the fact that Croatian farmers are mostly not creditworthy, and there are often complications with gathering ownership documents, financing consultants and investment projects, as well as poor information dissemination (e.g. lack of workshops and seminars on the subject of pre-accession programmes). An additional reason for the low absorption of SAPARD was an overlapping with national measures that had lower requirements and a milder control system compared to the EU programmes.

Other criticisms familiar with the overall situation in Western Balkan agricultural policy (Volk, Rednak, & Erjavec, 2010) remarked the lack of a stable agricultural policy and a true strategy of reforms and adjustment to EU requirements with the pragmatic ad hoc approach for defining measures. With no defined priorities in agricultural policy in accordance with national interests and CAP framework, Croatia was not able to allocate SAPARD funds specifically and due for the given period (2006–2009). Unfortunately, that could also be expected with the anticipated EU funds in the future.

Another important issue was connected to the analytical and consulting support. Reliable dataset of the situation in agriculture and rural areas was not fully prepared or publically available until the last few years (from 2009). Moreover, experts from the academic community had hardly been involved in monitoring, evaluating and programming of agricultural policy measures during the negotiation process; moreover, it should include far more than only the agricultural ministry as it was in practice (Volk et al., 2010).

ADJUSTMENT TO DECENTRALIZED WAY OF THINKING AND RAPID DECISION-MAKING PROCESS

The new bottom-up, LEADER or area-based approach in the frame of CAP reforms, starting in the 1990s and accentuated in the following reforms, was also a kind of innovation for Croatia. Bearing in mind that until the 1990s Croatia as a part of Yugoslavia had centrally planned economy, top-down approach of decision making, it took some time for the local communities to realize it should initiate its own development.

The first initiatives of civil society to enhance rural development came into force in 2006 by the establishment of *The Croatian Rural Development Network (HMRR),* after more than two years of informal work. The network was founded by seven civil society organizations, coming from different parts of Croatia, which linked in the wish to 'jointly contribute to improving conditions for integrated development and good quality of life in rural areas'.[8]

Activities during the project 'Sustainable Future for Rural Areas in Croatia' (2008–2010), developed in cooperation with partners from Amsterdam, resulted in the first Local Action Group (LAG) in Croatia in 2008. The establishment of LAGs has continued in the following years. As made up of public and private partners, LAGs suppose to apply a bottom-up approach regarding the formulation of objectives and priorities, expressing the views of specific area and, if necessary, to modify proposals made by the central administration.

The role of the local administration is also emphasized in the Leader approach. It should facilitate the application of the local development programme, which brings the need to reach certain level of fiscal and administrative autonomy. Unfortunately, in the Croatian case, the differing political interests of governments lead to discord in the public administration hierarchy. Also, chairmen (head, president) of the village committees frequently volunteer and do not pay full attention to current problems (Mikuš et al., 2012).

Nonetheless, in 2008, the *National Strategy of Rural Development 2008–2013*[9] was finally officially adopted due to pre-accession obligations, although in other form (less regionalized analysis) than one of the UN FAO and MAF (2003). General goals of the strategy were set up as follows: the improvement in the competitiveness of the agricultural and forestry sectors, the preservation, protection and sustainable use of the landscape, the cultural and natural heritage, the improvement in the quality of life in rural areas and diversification of rural activities (tourism, processing of agricultural products and wood, production of renewable energy) and the improvement in the institutional efficiency.

Although engaged in the negotiation process and accelerating procedures of harmonization, but mostly on paper, central administration did not make enough effort to explain and implement necessary policy reforms. A lot of energy, time and budget were spent on maintaining social peace, often disrupted by demonstrations by farmers. Therefore, Croatian agricultural policy measures are still focused on direct payments, which are outdated in CAP, instead of using pre-accession funds for rural development, environmental protection, animal welfare and less favoured areas. Due to the unpreparedness of beneficiaries for CAP-like mechanisms, undesirable consequences could be expected after the accession in 2013 (Volk et al., 2010; Mikuš et al., 2010).

Crucial political decisions were often taken ad hoc, and the example of such behaviour is the procedure of adoption of the *Act on State Support in Agriculture and Rural Development* in 2009 and 2010 (OG, 83/2009, 92/2010) and its three amendments (OG, 153/2009, 60/2010, 124/2011) coming quickly into force in the years 2009–2011, after the long hiatus of seven years. The intention behind those acts was to change the previous structure of incentives and to introduce first basic flat rate payments and later a Single Payment Scheme in 2012 – one year before the entrance.

In the acts mentioned in this chapter, the measures of rural development are defined by commitments contained in IPARD, which include support for farm insurance against damages, support for keeping indigenous and protected species, capital investment support, income support, and support for organic and integrated production and for less favoured areas. As Croatia is going through an economic recession and crisis of the national budget, it still lags behind in direct payment for beneficiaries. This phenomenon is obstructing the timely implementation of rural development measures.

The comparison of the agricultural budget structure in 2004 and 2010 shows an overall increase in allocation, especially in favour of structural policy and public services. A decrease in budget share is evident in the market support category (Fig. 5).

In addition, initiatives in rural areas of Croatian LAG associations evolved over time (since 2008), as familiarity with the Leader approach grew. Local actors have already been involved in various projects for education, dissemination of information, foundation of business incubators, involvement of women or young people in community actions, etc. Since April 2012, LAGs have been connected through the *Leader Network*. The goals of the Leader Network are linkage, capacity building, exchanges of experiences and public advocacy of interests of LAGs, research and development institutions and organizations dealing with rural development in Croatia. The intention of the Leader Network is to facilitate local goals in accordance with the government support – The most important goal is the connection of the Leader Network with available pre-accession funds, post-accession structural funds and with other similar rural stakeholders in European rural networks.

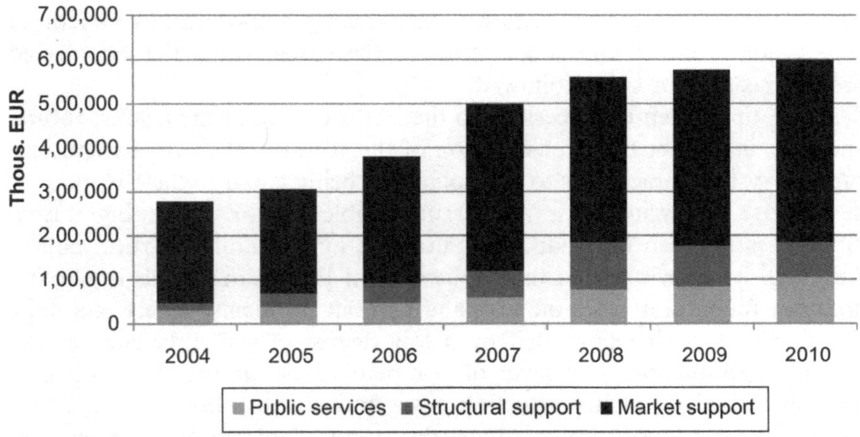

Fig. 5. Agricultural Budget Structure in Croatia (Current Prices). *Source*: MA (2011).

CONCLUSION: TRANSFORMATION OF CROATIAN AGRICULTURE

There are two basic elements which define the domestic agricultural policy at present: (1) the obligations taken over by international contracts and (2) the assumption that Croatia will become a member of the EU in 2013.

Up till now, big steps have been made in Croatia in establishing new institutions in agriculture (Paying Agency for Agriculture, Fisheries and Rural Development; Croatian Agricultural Agency as a body for implementing measures brought by the Ministry of Agriculture, Croatian Chamber of Agriculture responsible for the interests of small private farmers, Croatian Centre for Agriculture as a research and advisory body), whose task is the implementation and supervision of the application of new production, business and administration standards and procedures (phytosanitary standards, standards in protection of animal health and generally in the food production, Register of Agricultural Holdings, ARKOD, implementation of incentive system, tax obligations and so on). Considering the fact that most of these institutions have been founded rather recently, time is necessary for the formally adopted procedures to be applied with full intensity in practice, and for farmers themselves to accept new models of business behaviour. This is especially related to the adoption of business responsibility standards (as well by the farmers and the administration which implements the procedures), introducing a more strict system of supervision, control and sanctioning of the cases where the established procedures are not being followed.

At the time when the accession to the EU is certain in the nearest future, and especially after the intensification of the adjustment of the agricultural support system (practically to the point of it being equal to the EU models) in the last two years, it is almost impossible to talk about big formal differences between Croatian and European agricultural and rural policy. Although it is obvious that the application of European models in Croatia does not function in a simple way and without problems – there is a daily debate about problems caused by a low degree of self-sufficiency of the domestic production, low level of competitiveness of the domestic production caused by high production costs and uncontrolled import with a simultaneous impossibility of domestic products' placement; farmers are repeatedly protesting, expecting greater involvement of the central authority in the regulation of prices, purchase, import and export etc. often to the extent that they cannot be pleased because of the limitations of the international agreements. The unclear stand of the government in the recent past (concerning consequences of the Croatian War of Independence, undefined working criteria in the production and performance of State support, insufficiently clear and rigid financial obligations and farmers' rights etc.) did not ensure the restructuring of the sector on sound economic principles; the farmers do not have management skills, knowledge of economy and will to cooperate, associate and launch initiatives on a local level, which are necessary for efficient usage of the models of rural development support. Different needs are being met by equal or similar measures and instruments: on the saturated European market, policy reforms have logically resulted in the separation of support from production, but it is not efficient if the aim is to increase production and self-sufficiency – which are still the declared aims expected from domestic agriculture. Furthermore, the adoption of the liberal market principles is also logical in the countries with saturated markets; in Croatia, on the contrary, increase in production and self-sufficiency and consequently government protection of domestic producers is expected, which is contrary to the principles of liberal economy.

In Croatia, farmers still too often expect the state and its 'strong hand' to organize the production, to define and guarantee the purchase prices. That means that we will probably witness for a long time protests and political pressure exerted by farmers, before they accept the market economy principles. Furthermore, although in Croatia there are various more or less efficient models of farmers' association into cooperatives, associations and unions, their influence in the production and market is not very significant; considering the possibilities which the European models ensure for the manufacturing organizations, it is necessary to make extra effort in raising the awareness regarding the economy (Maietta & Sena, 2008; Tratnik, Radinović, & Žutinić, 2007), and even more social usefulness of these associations (Ortiz-Miranda, Moreno-Perez, & Moragues-Faus, 2010). The same evaluation is valid for the application of European models of financing the rural development, for which the local farmers are not prepared, due to a lack of their own capital and the lack of ability and will for the cooperation and initiative needed to launch local (rural) community projects.

In 2011, negotiations for the accession to the EU membership were finished for the chapter Agriculture and Rural Development. The negotiators point out that they have accomplished significant possibilities of financing the Croatian agriculture: 373 million Euros of financial envelope for direct payments to the farmers which should support and maintain the agricultural production and stabilize the farmers' income, the possibility of co-financing the support from the national budget (if the domestic budget is strong enough), and 333 million Euros in the financial envelope for rural development programmes, with special measures of support for self-supplying farms (which could save a vast number of Croatian farmers and villagers).

On the other hand, a strong fear was expressed by the local farmers that they could not be able to meet the strict criteria of the European financing. Therefore, there is a lot of work left to inform and train them, not only about the new criteria, demands, limits, possibilities and risks in the production, but also about different forms of behaviour to access new sources of agricultural income. The aims of the agricultural policy remain: accomplishing the highest level of possibility of self-sufficiency in agricultural production (but also within the common European market), competitiveness in production, acceptable food prices with respect to health regulation, fulfilment of environment protection standards, good agricultural practice codes and contribution to the public goods. These aims are planned to be accomplished by means of a regulated land policy and proper market infrastructure, and with intensive farmers' association, which are measures that go beyond the possibilities of agricultural policy and include numerous structural measures of educational and social policy.

The most recent research point out that a large part of Croatian farmers see the possibilities of their survival and growth in non-agricultural activities. They are aware of comparative advantages of Croatian rural areas (quality natural resources, traditional and cultural values, attractive regional diversities) and expect positive consequences of their entry into the EU: better economic conditions for the rising productivity, better market organization and functioning and more effective use of state support for agriculture and rural development. They see opportunities of farm income increase in achieving added value through organic production, traditional practices or new products development. However, the chance of growth of Croatian agriculture is above all through 'ennobling' agricultural activities by agritourism or supplying tourists' demand during the season of conventional massive tourism (Franić et al., 2011). From the aspect of government policy and economic orientation, linking the regional particularities of Croatian 'agricultural north' and 'touristic south' is the most certain option of survival for agricultural tradition and euro-Mediterranean identity and existence, but also an opportunity for modernization and economic development within the European context.

NOTES

1. The real problem of Croatian agriculture (and agricultural statistics) is a great gap between total (available) agricultural area and areas that are utilized for agricultural production.
2. *Town* is a unit of local self-government which represents a natural, urban, economic and social unity. Municipalities that are seats of counties and are with more than 10,000 inhabitants also have the town status. In certain cases, a municipality that does not meet the above stated requirements can also have the town status when there are specific historical, economic, geographic and traffic oriented and other reasons. Borders of a town are determined by borders of marginal settlements. *Municipality* is a unit of local self-government, established, basically, for a territory comprising a number of settlements that are considered a natural, economic and social entity, and are connected by common interests of inhabitants. Borders of a municipality are determined by borders of marginal settlements.
3. Counties where 90–100% people lives in rural areas.
4. Age index is the analytical index of the population age structure expressed as the relation between old (>60 years old) and young population (<19 years old). The population is considered old if the index exceeds 0.40.
5. According to formula, the unit of each production (ha, head, number) is multiplied by a certain coefficient which varies from product to product. Result for each product has to be at least 0.4 and total output has to be at least 3.0 in order to be considered as commercial.
6. Due to many changes and amendments, we indicate only numbers of the first and final act.
7. Data obtained from Department of Agricultural Economics and Rural Development, Faculty of Agriculture, University of Zagreb.

8. Retrieved from http://www.elard.eu/elard/national-leader-networks/en_GB/croatian-rural-development-network-hmrr. Accessed on 15 October 2012.
9. Retrieved from http://www.mps.hr/default.aspx?id=13. Accessed on 15 October 2012.

REFERENCES

Bašić, F., & Franić, R. (2003). Nature and man in Croatian agriculture. In M. Cvetković (Ed.), *Croatian agriculture, food and food processing industry* (pp. 14–49). Zagreb: ProTim.

Borsotto, P., Henke, R., Macrì, MC., & Salvioni, C. (2008). Participation in rural landscape conservation schemes in Italy. *Landscape Research, 33*(3), 347–363.

Croatian Bureau of Statistics (CBS). Censuses 1991, 2001 and 2011. Retrieved from http://www.dzs.hr/default_e.htm. Accessed on 15 October 2012.

European Commission – EC. (2007). *Screening report Croatia, Chapter 12 – Food safety, veterinary and phytosanitary policy*. Retrieved from http:/ec.europa.eu/enlargement/pdf/croatia/screening_report_12_hr_internet_en.pdf. Accessed on 15 October 2012.

European Commission – EC. (2011). *Common Agricultural Policy towards 2020: Assessment of alternative policy options*. Retrieved from http:/ec.europa.eu/agriculture/analysis/perspec/cap-2020/impact-assessment/full-text_en.pdf. Accesseed on 15 October 2012.

Franić, R. (2006). Politika ruralnog razvitka – nova prilika za Hrvatsku/Rural development policy – new opportunity for Croatia. *Agronomski Glasnik, 3*, 221–235.

Franić, R., Bokan, N., & Kumrić, O. (2007a). U očekivanju SAPARD-a: rješenja EU-a nasuprot domaćim očekivanjima/Expecting SAPARD in Croatia: EU Solutions Versus National Aspirations. *Društvena istraživanja, 16*(4–5), 701–724.

Franić, R., Mikuš, O., & Andabak, A. (2011). Optimistički pogled na ulazak u EU: poljoprivrednicima može biti bolje/Optimistic view on the entrance to the EU: things could be better to farmers. In: Pospišil, Milan (Ed.), *Proceedings of the 46th Croatian & 6th international symposium on agriculture*. Opatija, February 14–18 (pp. 3–10).

Franić, R., Tratnik, M., Bašić, F., & Svržnjak, K. (2007b). Sustainable allocation of land and capital resources: The case of wheat production in Croatia. *Cereal Research Communication, 35*(2), 397–400.

Franić, R., & Žimbrek, T. (2003). Premises for the inclusion of agriculture in the process of Croatian accession to the European Union. In K. Ott (Ed.), *Croatian accession to the European Union: Economic and legal challenges* (pp. 159–179). Zagreb: Institute of Public Finance and Fredrich Ebert Stiftung.

Grgić, I., Žimbrek, T., Tratnik, M., Markovina, J, & Juračak, J. (2010). Quality of life in rural areas of Croatia: To stay or to leave. *African Journal of Agricultural Research, 5*(8), 653–660.

Jačan Obratov, S. (2010). *Croatian cultural heritage*. Zagreb, Croatia: Croatian National Tourist Board.

Karanikolas, P., & Martinos, N. (2007). The modernization process in Greek agriculture: The case of investment aid. *Agricultural Economic Review, 8*(1), 37–49.

Lovec, M., & Erjavec, E. (2012). Big bang' enlargement and Common Agricultural Policy reform. *Društvena Istraživanja, 1*(115), 219–238.

Maietta, O., & Sena, V. (2008). Is competition really bad news for cooperatives? Some empirical evidence for Italian producers' cooperatives. *Journal of Productivity Analysis*, *29*(3), 221–233.

Mikuš, O., Franić, R., & Grgić, I. (2012). The evaluation of rural competitiveness in creating a policy of rural development in Croatia. *Journal of Food, Agriculture & Environment*, *10*(2), 962–969. Retrieved from http://www.isfae.org/scientficjournal/2012/issue2/abstracts/abstract176.php. Accessed on 15 October 2012.

Mikuš, O., Mesić, Ž., & Cerjak, M. (2010). Review of agriculture and agricultural policy in Croatia. In: T. Volk (Ed.), *Agriculture in the western Balkan countries* (Vol. 57, pp. 93–121). Halle (Saale): Leibniz-Institut fur Agrarentwicklung in Mittel-und Osteuropa. Retrieved from http://www.iamo.de/dok/sr_vol57.pdf. Accessed on 15 October 2012.

Ministry of Agriculture (MA). (2011). *Internal database of socio-economic indicators*. Retrieved from http://www.mps.hr/default.aspx?id=5401

Ministry of Agriculture (MA). (2012). *Internal database of budget support*. Retrieved from http://www.mps.hr/default.aspx?id=5401

Ministry of Agriculture, Forestry and Rural Development (MAFRD). (2009). *IPARD Programme – Agriculture and Rural Development Plan 2007–2013*. Retrieved from http://ec.europa.eu/enlargement/pdf/croatia/ipa/ipard_en.pdf. Accessed on 15 October 2012.

Ministry of Agriculture, Forestry and Water Management (MAFWM). (2006). *SAPARD Programme – Agriculture and Rural Development Plan 2005–2006*. Retrieved from http://ec.europa.eu/agriculture/enlargement/countries/croatia/sapard_en.pdf. Accessed on 15 October 2012.

Moreno-Perez, O. M., Arnalte-Alegre, E., & Ortiz-Miranda, D. (2011). Breaking down the growth of family farms: A case study of an intensive Mediterranean agriculture. *Agricultural Systems*, *114*, 500–511.

Moreno-Perez, O. M., & Ortiz-Miranda, D. (2008). Understanding structural adjustment in Spanish arable crop farms: Policies, technologies and multifunctionality. *Spanish Journal of Agricultural Research*, *6*(2), 153–165.

Ortiz-Miranda, D., Moreno-Perez, O. M., & Moragues-Faus, A. M. (2010). Innovative strategies of agricultural cooperatives in the framework of the new rural development paradigms. The case of the Region of Valencia (Spain). *Environment and Planning A*, *42*, 661–677.

Papadopoulos, A. G., & Liarikos, C. (2007). Dissecting changing rural development policy networks: The case of Greece. *Environment and Planning C – Government and Policy*, *25*(2), 291–313.

Pinto-Correia, P., & Breman, B. (2009). New roles for farming in a differentiated countryside: The Portuguese example. *Regional Environmental Change*, *9*, 143–152.

Radinović, S., & Žutinić. (2007). Može li Hrvatska imati konkurentnu obiteljsku poljoprivredu? Prilog istraživanju agrarne strukture. [Can croatia have competitive family farming? A contribution to the research of agrarian structure]. *Društvena Istraživanja*, *16*(1–2), 175–197.

Svržnjak, K. (2012). Mjere okrupnjavanja poljoprivrednog zemljišta kao činitelj konkurentnosti hrvatske poljoprivrede. [Land consolidation measures as a factor of Croatian agricultural competitiveness]. Unpublished PhD thesis, Faculty of Agriculture, University of Zagreb.

Tratnik, M., Franić, R., Svržnjak, K., & Bašić, F. (2008). Land rents as a criterion for regionalization – The case of wheat growing in Croatia. *Land Use Policy*, *26*(1), 104–111.

Tratnik, M., Radinović, S., & Žutinić, Đ. (2007). Zadrugarstvo Hrvatske: izazovi stabilnosti poljoprivrednih gospodarstava. [Cooperatives in Croatia: Challanges to stability of farms]. *Agronomski Glasnik*, *1*, 63–76.

UN FAO, Ministry of Agriculture and Forestry. (2003). *A strategy for rural development in Croatia (2004–2010) – Draft*. Retrieved from http://www.google.hr/url?sa=t&rct=j&q=un%20fao%20ministarstvo%20poljoprivrede%20strategija%20ruralnog%20razvoja&source=web&cd=3&cad=rja&ved=0CD8QFjAC&url=http%3A%2F%2Fwww.mps.hr%2FUserDocsImages%2Fstrategije%2FSTRATEGIJA_RR_19-05-081.doc&ei=Ur9qUdXTDMLStQb1zoHQDg&usg=AFQjCNEKsNbcQvo_y8iJ3Ix7OhQ5wRZLxA&bvm=bv.45175338,d.Yms. Accessed on April 14, 2013.

van der Ploeg, J. D., Laurent, C., Blondeau, F., & Bonnafous, P. (2009). Farm diversity, classification schemes and multifunctionality. *Journal of Environmental Management*, *90*(S2), S124–S131.

Volk, T., Rednak, M., & Erjavec, E. (2010). Western Balkan agriculture and agricultural policy – Cross-country overview and comparison. In T. Volk (Ed.), *Agriculture in the western Balkan Countries* (Vol. 57, pp. 7–35). Halle (Saale): Leibniz-Institut fur Agrarentwicklung in Mittel- und Osteuropa. Retrieved from http://www.iamo.de/dok/sr_vol57.pdf. Accessed on 15 October 2012.

CHAPTER 11

RURAL TRANSFORMATIONS AND FAMILY FARMING IN CONTEMPORARY GREECE

Charalambos Kasimis and Apostolos G. Papadopoulos

ABSTRACT

A series of changes have taken place over the past 20 years that have transformed the face of rural Greece. At the heart of these changes have been the rural farm household and the European agricultural and rural development policies.

The processes of de-agriculturalization and rural restructuring in the early 1990s have been accompanied by 'rurbanization' and socio-economic integration of rural populations. These interrelated processes have internally transformed the rural areas, thus forming a 'new rurality' characterized by contraction of agriculture, expansion of tourism and construction, increased pluriactivity, increased employment of international migrant labour and the reorganization of farm family labour and operation. However, in the environment of economic crisis, the conditions of the 'new rurality' have been affected by falling incomes, contraction of public services and by a 'back to the land' movement. This 'reverse mobility' has the elements of both modernity and tradition: engagement

Agriculture in Mediterranean Europe: Between Old and New Paradigms
Research in Rural Sociology and Development, Volume 19, 263–293

ISSN: 1057-1922/doi:10.1108/S1057-1922(2013)0000019013

with new methods of organization and work and rediscovery of traditional crops, products and cultures.

The chapter will discuss the characteristics and dynamics of the changing physiognomy of rural Greece in the past 20 years focusing upon three paths: the de-agriculturalization of the countryside, the perplexity of rural mobilities and rural resilience during the economic crisis. The chapter moves from a theoretical analysis of these paths to a detailed account of secondary sources on the transformation of agriculture and the countryside in Greece before it discusses the implications of the crisis upon the population movements and the 'rediscovery' of the economic, social and cultural values of rurality.

Keywords: De-agriculturalization; rural mobilities; new rurality; resilience; economic crisis; family farming

SOLIDIFYING A POST-AGRICULTURAL COUNTRYSIDE

Population mobility has been a critical factor for the 'urbanization' of rural areas. Rural space acquires new content as a result of the changes that have transformed rural locales into something different from the agricultural sites they used to be. New forms of populated areas emerge, which in no way correspond to the binary separation of 'urban' and 'rural'. But in the past the separation between the rural and the urban offered the basis for distinguishing between the 'old' and the 'new' social conditions of production[1]; in the current era there are increasing efforts to overcome the dichotomous division between urbanity and rurality.

Lately more effort is made on acknowledging and analysing the mixed fragmentary patterns and conflicting types of land use on (rural) space (Kūle, 2008).

In this context, rural locales contain hybrid and/or (re)negotiated socio-economic, cultural and environmental elements. A number of interpretative labels are conjoined with the term 'rural'. One might speak today of peri-urban rural areas, rural-urban fringe, dynamic rural areas, predominantly rural areas, intermediate rural areas, marginal rural areas, less favoured rural areas and so forth. For many writers on population movement, not all rural areas are likely to be receiving population; a great many of them are, on the contrary, losing population (Halfacree, 2001, 2009; Halfacree & Boyle, 1998).

The notion of mobility, in particular, transcends the urban/rural dichotomy because the rural is at least as mobile as the urban. Moreover, many rural locales have experienced significant transformations due to changes in the markets for land and housing, in local economies and cultures, in rural policy, in retailing and provision of services. The rural thus becomes a receiver and arena of different forms of mobility with many rural locales gradually being stripped off their demarcating characteristics (Bell & Osti, 2010).

In this way, rural space is as much material as it is symbolic and ideological and extends into three interconnected domains which cannot be isolated when studying specific rural locales: (a) the rural localities, which are the actual arenas for both production and/or consumption, (b) the everyday lives of the rural, incorporating both individual and social characteristics personified in the way rural life is negotiated and interpreted, and (c) the different representations of the rural, as expressed by different social actors (Halfacree, 2006).

Three themes can be integrated into the discussion. First, the gradual process of disengagement from agriculture and the ensuing *de-agriculturalization of the countryside*. This process may be seen as an expansion of the notion of the rural to include off-farm and/or multi-sectoral penetration as well as non-agricultural land uses (Kasimis & Papadopoulos, 2001; Papadopoulos, 1999; van der Ploeg, 1997) pointing towards an 'urbanization of the rural' (Lacour & Puissant, 2007).

As a counter response to productivist agricultural policies and practices in the rural regions of the economically developed nations, a movement towards post-productivism grows and places significant emphasis on the diversification of rural economies. The term 'multifunctionality of rural space', which better conveys the idea that rural space is being used in hybrid ways involving both agricultural and non-agricultural economic activities, incorporating the amenity attributes of the countryside, is adopted by an increasing number of writers (see, e.g., McCarthy, 2005; Wilson, 2009; Woods, 2009).

Second, the *perplexity of rural mobilities* as an aggregation of diverse outgoing, incoming and intergoing population movements within rural areas. Currently we witness an 'era of mobilities' in which rurality has been at the centre stage (Halfacree, 2012).

Three sets of such mobilities are relevant in the transformation of rural locales. On the one hand, there is the *out-migration/rural exodus of people* from certain rural locales, leading to the decline of agricultural activity and the depopulation of the marginal and disadvantaged rural areas considered in the past as a 'normal' process interlinked to the urban development pattern of Western capitalist countries. On the other hand, there is the *in-migration of people* to rural locales entailing two patterns: one connected with counterurbanization and the 'return to the countryside' as a new demographic trend in developed countries (Halfacree, 2009; Halfacree & Boyle, 1998; Mitchel, 2004) and the other connected to the inflow of international migrants to various rural places – a recent phenomenon in rural Europe for both agriculture and other economic activities and services (Jentch, 2007; Papadopoulos, 2011; Kasimis & Papadopoulos, 2005; Kasimis, Papadopoulos, & Pappas, 2010). Both types of movement represent novel factors that reflect the reshaping of family farming and farm structures in the countryside, thus having an important impact upon social cohesion and sustainability of rural locales. Finally, the *increasing mobility of rural residents* meaning the inclination to move for work, leisure and/or tourism, sometimes apparent in-between inhabitants of these locales. In this respect, subsistence and/or hobby farming and time-sharing farming activities reflect new hybrid forms of involvement with agriculture. All types of mobilities reflect divergent and/or convergent mobilities connected to the production and consumption aspect of the rural.

In this context, the relocalization of food production is well connected to the different population movements and cannot be dissociated from the adaptability and survival of rural communities (Fonte & Papadopoulos, 2010). On the consumption side, local food production is considered to be an alternative production pattern related to the contemporary needs of the urban population. The latter consider rural locales as symbolic goods which may be part of the lived experiences of second home owners, countryside visitors, etc. On the production side, rural locales become territories of specialized food production. Therefore, these local production spaces are identified as unique places and may be distinguished from other ordinary rural locales. Those places attract migrant labour, rural entrepreneurs, urban dwellers wishing to farm and so on.

Third, the adaptive capacity shown by rural households, rural social actors, communities and local/regional economies in the context of current economic crisis is a reflection of *rural resilience* which represents an encompassing narrative for rural development policy and practice. The notion of resilience refers to the capacity of rural areas to absorb disturbances and reorganize, while undergoing changes, so as to retain the functions, structure, identity and reflexivity of rural activities (Folke et al., 2010;

Heijman, Hagelaar, & van der Hiede, 2007; Schouten, van der Hiede, & Hiejman, 2009; Walker, Holling, Carpenter, & Kinzig, 2004).

The family farming capacity of rural households has been changing in response to the wider socio-economic conditions and global changes. Rural households use their resources flexibly in view of their survival against the worsening economic conditions caused by competition and recession. Consequently, family farming remains a defensive practice for numerous rural households in their struggle for coping with in a constantly changing global setting (Kasimis & Papadopoulos, 1997).

Despite the assertion that the global reshapes the local, the opposite is also true (Massey, 2005). The global is constructed through the local as a contested space, in which processes of negotiation, manipulation and hybridization are conducted among local and non-local actors and produce outcomes that are not pre-determined (Woods & McDonagh, 2011).

Thus, the projection of the discussion on post-productivist and multi-functional rural space leads us to the discussion on the 'commodification of the countryside', based on the idea that rural areas are locales where a wide range of non-traditional rural commodities, services, lifestyle products and experiences are offered to consumers (MacKay, Perkins, & Espiner, 2009). On this basis, both farming and non-farming activities are part and parcel of rural commodities as well as of 'rural gaze' as a symbolic landscape. The meaning of the 'rural' is continually reconstituted by local and non-local actors as well as by residents, visitors and distant consumers (Perkins, 2006).

RURAL TRANSFORMATIONS IN GREECE

Greek rurality has been transformed, modernized and reshaped over the past few years. Rural areas have been affected by farm size polarization and land concentration as well as by the arrival and increased activization of newcomers as either hobby farmers or farmers with an origin from other non-agricultural high-status professions (e.g. public employees, doctors, lawyers etc.). Changes in the perception of what farming is and who the farmer is are challenging the evolutionary thinking of agricultural transition towards a less agricultural countryside. Rural Greece can be considered as an illustrative case of a constantly changing countryside with discontinuities and regressions. In other words, drawbacks and delays are as equally important as are progressions and/or innovations (Kasimis & Papadopoulos, 2001, p. 200).

There are clear signs of the solidification of a post-agricultural countryside in Greece. On all three themes mentioned before, the de-agriculturalization of the countryside, rural resilience and perplexity of rural mobilities, Greece has been a good example of how the rural remains vivid while constantly changing, reshaping its agricultural and non-agricultural mix in rural locales.

Today, more than ever before, a number of questions hold true and call for our attention. How, for example, agriculture will be reconfigured in view of the changing paradigm of agricultural production and the functions of farming in favour of the defensive strategies of rural households within an increasingly hostile socio-economic environment? How will it be possible to re-connect agriculture to the social needs of people in rural areas? How can social cohesion be achieved in rural areas, given the increasing heterogeneity of their population and the diverging needs of residents in rural areas? How can agriculture play a bridging role between the productionist and post-productionist phase?

Despite the recent changes that have taken place in the countryside, this chapter argues that agriculture still remains an important component of present-day rural Greece. Opposing and contradicting processes on the one hand restrict and on the other reinstate farming as a functional element of rural households, rural communities and regions contributing to the reshaping of rurality. In order to understand all these conflicting trends, an overall picture of Greek agriculture and social transformations is depicted in what follows along the broad axes of the three paths of de-agriculturalization, rural mobilities and rural resilience mentioned before.

The Long and Winding Road of De-Agriculturalization

Farm Structures: 'Nothing Has Changed But Nothing Remains the Same'

In the whole of the post-war period agriculture experienced a continuous decline in its size. More particularly, agricultural, rural and regional development policies implemented in the last two decades, as well as external factors, such as the international markets and competition, have all dramatically contributed to that decline. However, given the expansion of employment in the non-agricultural sectors, agriculture has maintained its supporting role for the livelihoods of rural population.

Today the sector still accounts for 11 per cent of all employment (against 4.5 per cent in EU-25), contributes 3.5 per cent to the country's GDP and is responsible for 25 per cent of exports.

According to Eurostat (Farm Structure Survey), the number of holdings between 2003 and 2010 fell by 14.3 per cent from 824,460 to 706,400. Nearly 80 per cent of these holdings had a size less than one European Economic Size Unit (ESU).

These holdings occupy 4 million hectares of utilized agricultural area (UAA) while the majority of them are small-sized (average size 5.0 hectares/holding). More than two thirds of them (70 per cent) cultivate less than 5.0 hectares and the majority (71 per cent) cannot ensure full employment for more than one employed person.

Small size and high fragmentation (exceeding 6 parcels per farm) of farm holdings epitomize the historic, cultural, geographic and geomorphologic construction of modern Greek agriculture. More specifically, the late land reform of the 1920s, the socio-cultural traditions of inheritance and dowry, as well as the unfavourable geomorphologic environment of narrow, often hilly, land zones have contributed to the survival of a late farm structure resulting in increased costs of production and low competitiveness.

Contracting competitiveness in Greece could be further explained by the decline of private investments, the irrational management of EU funds (directed towards consumption and/or investments outside agriculture and the countryside) and the ineffective structural policy implemented by Greek governments since the accession of the country in the EU.

The fragmented small-sized farms, the high cost of production, disproportionally high compared to countries such as France, Italy and Spain (in cost of credit, cost of irrigation and cost of agricultural insurance above the average of the Eurozone), that lead to lower productivity and competitiveness have been indicators of accumulating structural problems and an undergoing crisis already evident in the mid-1990s (Alpha Bank, 2009).

Professional family farming of large scale in Greece (i.e. of farms over 10 hectares) or very intensive production (i.e. horticulture, vineyards, glasshouses etc.) cover in 2007 Eurostat Farm Structure data just over half of the agricultural land (54.6 per cent) representing nearly one tenth of the total number of farms (10.7 per cent). This professional agriculture seems to be close to the average of the EU in terms of labour intensity; it has a better age structure involving exclusive or mainly full-time farmers. At the other end, *small-scale family farming* (i.e. of less than 5 hectares) covers nearly 26.5 per cent of the land and concerns 76.2 per cent of the farms. In the middle of these two types of farming, however, lies 13.1 per cent of Greek farms encompassing medium-sized farm holdings which occupy 18.9 per cent of the agricultural land and lean either towards the 'professional' or the 'small-scale' family farming.

The above discussion states some of the reasons for the delayed and contained agricultural modernization in rural Greece. We argue, however, that family farming should not be considered as the outcome of an incomplete agricultural modernization process, but rather as a systemic characteristic of households, communities and local economies. The structure and characteristics of family farming are considered to be part and parcel of the adaptive capacity and survival propensity of numerous rural households (Kasimis & Papadopoulos, 2001; Papadopoulos, 1998), while multifunctionality reflects the dynamic relationship of family farms with the local/rural labour markets (Efstratoglou, Papadopoulos, Efstratoglou, & Kouroussi, 2004; Goussios, 1999; Vaiou & Hadjimichalis, 1997).

Demographic ageing, underemployment, extensive presence of holdings in the LFAs have further contributed to structural deadlocks in agriculture. Many of these socio-economic and geographical constraints have played a major role in the containment of agricultural modernization in Greece, while on the other hand they have increased resilience pathways of family farms discussed later in the chapter (Kasimis & Papadopoulos, 2001).

The Mosaic of Employment Structures

In the past 30 years employment in the agricultural sector was reduced almost by half. This decline has been largely the outcome of the rural exodus of the 1950s and 1960s that slowed down and stabilized in the early 1980s as a result of the reduction of employment opportunities both inside and outside the rural areas.

In 1981, 1,083,000 people were employed in the agricultural sector. The numbers were reduced to nearly 500,000 in 2012, 31 years later, despite positive fluctuations in the first two years of crisis (Fig. 1).

Between 2008 and 2010, and despite the continuous contraction of total employment, the number of those employed in the sector increased by 37,400 (7.3 per cent). This increase was explained by some analysts as an indication of a recovery of employment in agriculture (PASEGES, 2011, pp. 26–29; Tsiforos & Stavropoulou, 2011, p. 7).

After 2010, the decline of employment in agriculture started again and is expected to continue falling, despite the emergence and intensification of different mobilities, in the period of economic recession that have somehow contained and redirected outflows.

Another factor related to the decline of agricultural employment and de-agriculturalization is the extended underemployment which concerns nearly two thirds of the farm family population with average annual employment close to 60 labour days.

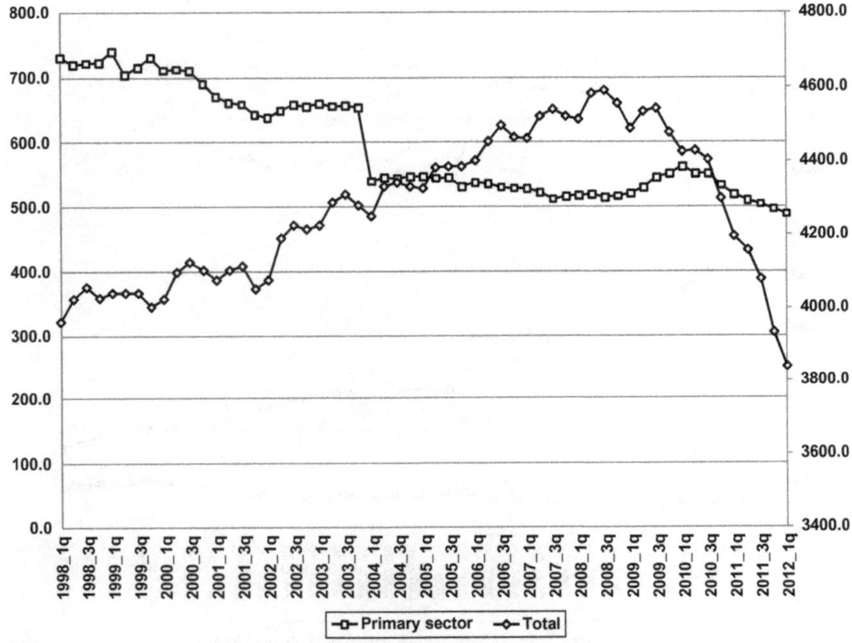

Fig. 1. Employment in the Primary Sector and the Country, 1998–2012 (in Thousands). *Source*: ELSTAT, Labour Force Surveys, 1998–2012. Calculations made by the Authors. *Note*: The left-hand axis records the number of jobs in the agricultural sector, while the right-hand axis jobs in all sectors of the economy. Employment is registered quarterly for the whole 1998–2012 period, reflecting annually the seasonal fluctuation of employment. Each quarter of the year is recorded as follows: 1998_1q, 1998_2q, 1998_3q, 1998_4q. For 2012 the available data is only for the first quarter.

Underemployment is particularly evident in the plains and coastal areas, where modernization (i.e. mechanization and agricultural investments) has played a significant role in reducing labour on the family farm.

According to Eurostat, 71 per cent of agricultural holdings in Greece employ less than one Annual Work Unit (AWU) and 7 per cent two or more AWU.

The contraction of agricultural employment in the period 1998–2012 concerns mainly unpaid family workers, almost exclusively women (Fig. 2).

Unpaid family workers were reduced from 37 per cent to 17.3 per cent, substituted by wage labour. Reduced were also self-employed/own account

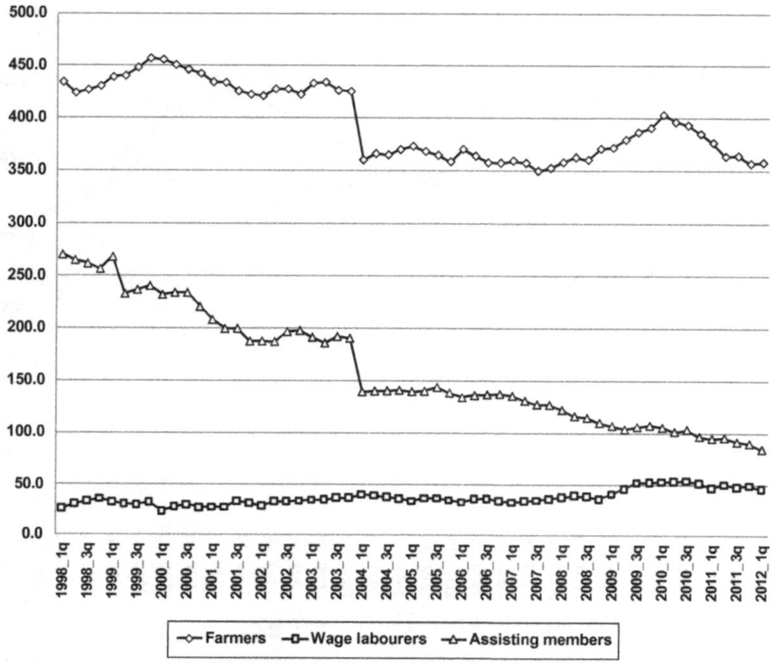

Fig. 2. Position in Employment in the Primary Sector, 1998–2012 (in Thousands). *Source*: ELSTAT, Labour Force Survey, 1998–2012. *Note*: The category farmers include both employers and own account workers/self-employed. Employment is registered quarterly for the whole 1998–2012 period, reflecting annually the seasonal fluctuation of employment. Each quarter of the year is recorded as follows: 1998_1q, 1998_2q, 1998_3q, 1998_4q. For 2012 the available data is only for the first quarter.

workers contracted by 19 per cent while employers remained more or less stable in the period under examination.

Salaried workers were the only category with a steady increase. Their numbers increased by 78 per cent in the period 1998–2012. Since 1998 they have increased from 3.6 per cent to 9.5 per cent of the total labour force in 2012. Permanent wage workers increased nearly five times. Equally important was the increase of seasonal labour as well as the number of farms employing seasonal labour (Table 1). Both permanent and seasonal wage labour increased from 7 per cent of total labour time in agriculture in 1991 to 18 per cent in 2007, while the average number of seasonal labour days per farm increased from 49 to 57.

Table 1. Employment of Family Farm Members and Wage Labour.

Year	Number of Farms	Employed (Family Members)	Farms with Permanent Wage Labour	Numbers of Permanent Wage Workers	Farms with Seasonal Wage Labour	Numbers of Labour Days of Seasonal Hired Labour in Agriculture
1991	861,623	1,570,533	3,881	6,189	248,611	12,077,955
2000	817,059	1,434,504	7,247	10,465	293,650	17,295,225
2003	824,458	1,483,450	16,086	24,230	355,252	25,790,849
2005	833,590	1,501,840	16,752	24,861	363,804	23,477,884
2007	860,154	1,478,590	19,295	29,582	367,497	20,910,841

Source: ELSTAT, Census of Agriculture Livestock 1991, 1999/2000. ELSTAT, Farm Structures Surveys 2003, 2005, 2007.

These developments are confirmed by a number of studies carried out in the past years in rural Greece (Kasimis & Papadopoulos, 2012; Papadopoulos, 2013) and by the data provided by the Census of Agriculture and Livestock and the Farm Structure Surveys. Migrant labour in Greek agriculture has a special weight despite the fact that, when compared to other European countries, Greece is among the four countries with the lowest percentage of non-family labour (10 per cent) in the 'professional' farms according to the FADN/RICA data.[2]

The Changing Patterns of Pluriactivity

The expansion of other, non-agricultural activities like tourism and housing construction in the rural areas, the return of urban dwellers to their place of origin and the growth of new consumption patterns connected to leisure and recreation have created a multifunctional environment in rural Greece that has increased labour demand and has created the preconditions of off-farm employment of household members and of migrant labour (Kasimis & Papadopoulos, 2001).

Pluriactivity has been growing over the past three decades being mainly confined today to farms under 5 hectares. The degree of pluriactivity among family farm heads is nearly 30 per cent, while 18 per cent of farm heads hold a main job outside their farm and agriculture. Off-farm employment has been 'diffused' to all household members whose percentage has risen to one fourth of their total.

In the mid-1990s, pluriactivity of the family members exceeded, for the first time, that of the farm head, providing some proof of the strategic adaptation of family farms to the external setting. This is done either by considering the internal labour capacity of the farm or by taking advantage of the externally available opportunities. These trends describing changes in family labour should be set against an increase in non-family farm employment and an even greater increase in seasonal migrant labour.

In other words, these tendencies may be connected to the changing family roles, the shifts in the gender division of labour, the increasing flexibility in the use of family and of cheap, non-family labour as well as the economic restructuring of the farm holding.

Pluriactivity, however, has another dimension and direction that reflect the communication and the ties developed between the agricultural and the non-agricultural sectors.

Along with the diffusion of pluriactivity in other farm family members, the increasing employment opportunities in other economic sectors in rural areas and the changes in the employment patterns and gender roles, one may note a corresponding expansion of agricultural activity in urban areas. Agriculture ceases to be an activity restricted to rural population but has rather started to be relevant for urban and/or peri-urban populations. Over the past years the non-rural population engaged in agriculture has increased to nearly one third of the total number of those involved in agriculture. The importance of this in the discussion for the shaping of 'new rurality' in Greece is revealing of the growing engagement of urban residents in farm and non-farm activities like rural tourism and small industry developed in rural areas.

Farm Household Incomes

The Eurostat Farm Structure data for 2007 show that only 1 in 10 of those employed in agriculture are employed exclusively in it and gain income exclusively from it (against 16 per cent in EU-27). Moreover, 70.5 per cent are employed for less than 50 per cent of their labour time (against 66.5 per cent in EU-27).

The combination of on-farm and off-farm work has differentiated the sources of farm households income. OECD (2004) estimates off-farm income of farm households as an average of the three available years of 1996–1998 at just under half of the total farm households income.

Despite a long period of Common Agricultural Policy (CAP)-instigated growth of farm incomes, farm households still have lower incomes (at the level of nearly 75 per cent according to the Greek Household Budget Surveys) and a lower standard of living compared to the overall households in the country.

According to Eurostat, between 2000 and 2009, agricultural income per working person was reduced by 16.9 per cent, while in the crisis years (2009–2012), the decline totalled 10.8 per cent when in EU-27 the increase was 7.5 per cent (Eurostat, 2012, p. 24).

Karanikolas and Zografakis (2008) note that there is a large differentiation of households in terms of their income inequalities and poverty. 'Predominantly farm households'[3] face higher income pressures. The authors argue that the higher the dependence of a household upon agriculture, the larger the percentage of the farm population below the poverty level. According to their estimates, one in three rural households is below the poverty line. Worse poverty problems are faced by pensioner households, many of which are located in the rural areas. Pluriactive farm households are in a much better position; the combination of farm and off-farm incomes creates a more stable environment moderating the wider income inequalities among the households.

Regional and other geographically dispersed local activities appear to complement the engagement in agriculture of the farm family members. On the other hand, the lack of crystallized social collectivities leaves significant room for 'resistance' to externally induced changes or integration movements. Even the apparent convergence to modernization patterns may be linked to the resurgence of 'informal' phenomena which express some form of 'social resistance' to the formal rules of the division of labour within a modern society. This 'resistance potential' or 'norm blindness' of family farms constitutes and is reconstituted by a set of practices which sometimes formulate strategies outside their operation within the agricultural sector. This strategic adaptability is both inter-sectoral and cross-sectoral, the result of a significant 'room for manoeuvre' on the part of small family farms reflecting their flexibility for a differentiated utilization of family and farm resources (labour, land, machines, gender division of labour, etc.) (Papadopoulos, 1998).

As a result, the adaptability of Greek family farms, which may be considered as a form of 'reflexive responsiveness' against the calls of agricultural policy and European integration process, is both the cause and the effect of de-agriculturalization of the Greek countryside. De-agriculturalization, thus, can be seen as the result rather than the cause of the transformation process of Greek countryside.

The Perplexity of Rural Mobilities

The Rural Exodus

Between 1951 and 1971 the proportions of urban and rural populations were reversed, from 38 per cent and 48 per cent, to 53 per cent and 35 per cent respectively. This can be explained because of the mass exodus from the countryside in the 1950–1973 period: 560,000 in the 1950s, 680,000 in the 1960s, 620,000 in the 1970s (Close, 2002, p. 61).

The incentives of internal emigrants were first and foremost economic. They moved from the rural areas with a view to find some stable and decent income from wage employment. Yet, this massive exodus of labour force from rural areas took place without any central state planning and at a much faster pace than the capability rate of the, at the time, developing manufacturing sector to absorb such labour force. In this respect, external emigration helped to avoid a situation of disorder, by absorbing part of the potential unemployed workforce, although, overall, this rural exodus led the urban labour markets to a confusing and unbalanced situation. At the same time, emigration directly or indirectly contributed to increasing the income of those who had remained in rural areas and agriculture either because they could expand the size of their farm by renting the land of those who had left or because of the remittances they received.

Three fourths of the case studies on migration in the period refer to declining villages (Wagstaff, 1982). This rural exodus was connected to the process of social transformation of the countryside and considered as part and parcel of the structural disintegration of the traditional rural society. This disintegration meant either the marginalization of whole rural communities or the increasing polarization within the local rural social structures. Such analyses led to an underestimation of the internal processes of rural social transformation related to the survival strategies of family farms. A number of empirical studies carried out in the 1970s attempted to shed light on the internal processes focusing on three interrelated issues: (a) the inequalities of agricultural land property; (b) the influence of the city on the rural economy through the non-agricultural engagements of farmers; and (c) the relationship between the rural exodus and agricultural structures (Burgel, 1977, p. 192). The analysis of the studies of concrete, highly diverse, Greek rural localities has resulted in a number of observations, two of which are particularly useful. The first refers to the recognition of the fact that the supplementary resources originating from off-farm activities play a significant role for the economy of the farm itself. The second refers to the consequences of the exodus for the rural localities. The fact that migration did not lead to an expected redistribution of the land in the rural localities, while it affected invariably plains and mountainous areas is quite interesting (Burgel, 1978).

Rural areas are handled by certain writers as a 'closed' social and economic system which operated by reproducing itself in isolation from the rest of the society. According to this approach, post–World War II mobility and urbanization gave the rural population an opportunity for an exodus from this isolation. In Greece, however, a large number of rural communities have had a long history of external economic relations. As a result, a rural exodus often constituted a response strategy of the family to these external economic pressures (Sutton, 1983). What differentiates the farm family from other (more traditional) forms of organization, like *tseliggato* or *zadruga* in the pastoral communities, has been historically its open, extrovert character, in other words its external relations with the surrounding economy and more specifically the market. The other forms of organization remained 'closed' relying on self-sufficiency and an internal division of labour under a patriarchal structure. Sutton (1988) considered modern urbanization as an initial phase of a long historical period which does not necessarily imply a break with the past. She writes that any type of migration is a strategic engagement of the rural household to the wider economic system.

For Vergopoulos (1975) the rural exodus was the consequence of the commercialization of agricultural production and of the 'unequal exchange' which lead to a rural household budget deficit, while for Kavouriaris (1974), increased productivity in agriculture, as a result of mechanization and worsening terms of trade, intensified the proletarianization of the peasantry and increased 'latent' labour. Kavouriaris's 'push' approach did not, nevertheless, ignore the 'pull' factors; imitation, social values, the attraction of city life and the absorbing capacity of German labour market were also taken into consideration but were given secondary importance, complementary to 'push' factors.

In the immediate post-war period, land structure was not seriously altered as a result of the conditions which defined the demand and supply of agricultural production. The position of the family farm as a unit of production was strengthened but some interesting differentiations had appeared in terms of mechanization, labour productivity and types of production. Labour power surpluses were very high even at the time when rural exodus had already started.

According to the National Statistical Service of Greece, the population of rural areas dropped considerably between 1961 and 1991, increased by 3 per cent in the 1990s and stabilized as a percentage of the total population to just over 25 per cent. The support of CAP and the arrival of international migrants in the early 1990s have been crucial in this direction.

An 'ex deus machina': In-Migration of Wage Labourers and Implications for the Farm Family Division of Labour

The arrival of large numbers of migrants in the past two decades coincided with regional and global developments with important implications for South European rural areas in general, and its agricultural sector in particular. These developments have had a particular effect on Greece's family farming, which still holds an important position in both the economy and society.

More specifically, after a period of substantial rise in agricultural incomes due to the support of the CAP, the first consequences of the fall in the competitiveness of Greek agriculture appeared as a result of increased costs of production, cutbacks in private investments, irrational management of EU funds, and ineffective structural policy (Demoussis, 2003;

Louloudis & Maraveyas, 1997). Furthermore, the unequal distribution of income and subsidies in agriculture were considered a deficiency of CAP leading to reduced effectiveness with regard to its welfare role of ensuring 'a fair standard of living for the agricultural community' (Karanikolas & Zografakis, 2008). However, the level of the 'fair standard of living' is differentiated significantly between the various member states and/or regions within the member states.

The pressures for technological modernization and the restructuring of agriculture towards intensive crops resulted simultaneously in the growth of off-farm employment of family members and in greater seasonal demands for labour (Kasimis & Stathakis, 2003; Zacopoulou, 1999).

This development was reinforced by the demographic deficiency of rural labour created by the massive rural exodus of the 1950s and 1960s and by the restructuring of rural areas that has exerted pressures for types of labour not available from the indigenous population (Kasimis & Papadopoulos, 2001).

Such labour deficiencies are explained not only by demographic and structural factors but also by social factors connected to the rejection of low-status, unskilled and badly paid jobs in rural areas by the younger generation. The improved level of education and standard of living as well as the spreading out of urban consumption patterns led to the creation of high expectations by the younger generation, who looked for jobs outside agriculture and outside rural areas. Moreover, the integration of women into the labour market, the accompanying changes in family structures and the lack of adequate social infrastructures have resulted in increased demands for domestic support work (King, 2000; Lazaridis & Psimmenos, 2000; Papadopoulos, 2006).

Non-family labour becomes more and more crucial for the operation of the farm, an indication of the fact that Greek agriculture needs more and more flexible labour hands that migrants only can provide.[4]

The presence of migrant labour is differentiated in accordance with the local production systems in various areas of the country. It is higher in the plains and areas of intensive agriculture as well as in the island regions of multifunctional economies where they seem to play a multifunctional role themselves adjusting their individual and family division of labour to fulfil the particular seasonal needs of the local economies (Kasimis & Papadopoulos, 2005).

Migrants are estimated to have provided more than one fifth (22.2 per cent) of the total labour expended in agriculture in AWU in 2007 (PASEGES, 2011) and despite an increase in their cost of labour in the pre-crisis years (the result of their regularization and integration), today they are back to low wages and long working hours restraining the cost of production. The main working nationalities are those of Albanians, Bulgarians, Indians, Pakistanis and Bangladeshis.

A number of empirical studies have already revealed both the structural role of migrant labour in Greek agriculture and the implications for the labour relations and family division of labour (Kasimis & Papadopoulos, 2005; Kasimis et al., 2010; Lawrence, 2007; Labrianidis & Sykas, 2009; Papadopoulos, 2009, 2011).

The extensive participation of non-family labour in the farm work has contributed not only to the disengagement of the farmer from heavy manual work, but also of his spouse and, to a significant degree, of the rest of the family members. More than 50 per cent of male farm heads declared they reduced their workload or withdrew from heavy farm work and devoted more time to the organization and management of their farms – particularly those with large size farms – while their spouses either reduced their farm work or deserted it altogether to return exclusively to housework or to seek off-farm employment when conditions allowed it (Kasimis & Papadopoulos, 2005).

In sum, during the last years we observe a number of trends which reveal the de-agriculturalization of Greek countryside: the decline of agricultural employment, the growth and diffusion of pluriactivity to other than the head family members, the increase in the flexibility of labour and especially of female labour within the family farm, the increasing inter-relationship between urban and rural labour markets and the changing gender roles within the family farm are only an indication of the tendencies which were reinforced by the extensive employment of migrant labour in the past 20 years. The arrival of migrants, seen in retrospect, has offered solutions to these pressing problems and generated, at the same time, new demands for labour and new job positions in agriculture and the countryside in general. Additionally, it gave the family the opportunity to 'redesign' its allocation of family labour on and off the farm. This sharpened the survival strategies of the farm families and contributed to a (post-)modern transformation of Greek countryside.

The Implications of the Economic Crisis and the Resilient Mobilities

Greece Through the Economic Crisis

This section of the chapter concentrates on the discussion of the implications of the crisis upon a number of indicators of the agricultural sector like employment, GDP and the balance of agricultural trade in order to better understand the incoming changes and the opening of possibilities in view of a 'recovery' in rural areas.

The crisis in Greece appeared initially as a crisis of the private financial sector, turned first into a fiscal crisis and then into a debt crisis. Soon it affected different types of institutions and the political structure itself. Although it had as starting date 2008, no termination date was foreseeable. In Table 2 we get a vivid impression of the main indicators of the crisis affecting the Greek economy and society.

In Greek agriculture, for example, a simmering crisis had been already evident since the 1990s. Structural problems like the small and fragmented size of holding, the ageing of farm and rural population, the low levels of employment, the high cost and low competitiveness of agricultural production, the stagnation and decline of productivity as well as wide regional and sectoral inequalities had accumulated various deadlocks which were, however, until recently, blurred by the CAP support and the continuous arrival of cheap mass migrant labour in the past two decades.

Implications for Agriculture and Rural Regions

Since the beginning of the crisis in 2008, more than 600,000 jobs were lost in the economy and unemployment reached 22 per cent by 2012 while recession was officially expected to reach 4.5 per cent of the GDP in 2013.

Both quantitative and qualitative evidence, however, suggest that agriculture and rural areas 'resist' better than other sectors of the economy and are increasingly turned into a 'refuge and laboratory' of ideas and initiatives for a large part of urban dwellers directed towards the countryside and agriculture by either necessity or choice. This is happening

Table 2. Main Economic Indicators in the Years of Crisis.

Indicators	2009	2010	2011	2012
GDP at market prices (% changes)	−3.2	−3.5	−6.9	−5.3
Gross fixed capital formation (% changes)	−15.2	−15.0	−20.7	−13.4
Unemployment rate (%)	9.5	12.5	17.6	21.2
General government financial balance (% of GDP)	−15.6	−10.5	−9.2	−7.4
General government gross debt (% of GDP)	134.0	149.6	170.0	168.0
Current account balance (% of GDP)	−11.1	−10.1	−9.8	−7.8

Source: Compiled from various OECD reports in May 2012.

against a background of heavy cuts in public spending in rural areas with serious implications in the provision of health, school, transport and social services.

Drawing from the data of the Labour Force Survey, for the period 2008–2011 the most recent evidence and the conditions created in the Greek primary sector and rural areas in the environment of crisis are analysed.

The return to agriculture and the countryside is connected to the overall developments in the labour market considering that an increase in the size of employment in either agriculture or the rural areas, in a period of high unemployment in urban areas, will mark a change in the tendency of a primary sector contraction.

Employment Changes

Evidence has indicated that rural areas show lower unemployment rates both before and after the crisis (Fig. 3). Unemployment in rural areas before the crisis was 5.5 per cent against 7.8 per cent in the urban, while during the crisis years it soared to 18.4 per cent, but remained still below that of urban areas (23.8 per cent in 2012).

Within the rural areas the most vulnerable group to become unemployed are the migrants in comparison to the Greeks. The situation was reverse before the outbreak of the crisis in 2008. Additionally, older, regular and more integrated migrants (Albanians mostly) are more vulnerable than irregular, low-paid and uninsured Asians who are prepared to work for even lower wages.

Women, compared to men, and university graduates compared to those having only primary education, are more likely to have a higher probability to become unemployed.

In the first three years of the crisis (2008–2010) both the primary sector and the rural areas performed better in employment, reversing the declining trend of the pre-crisis period. Overall, rural areas had the lowest job losses (7 per cent or 62,000 jobs against a loss of 16 per cent or 495,000 jobs in urban areas). Again the primary sector lost only 4 per cent of its employment (mostly family withdrawal) when the secondary sector lost 26 per cent, construction 41 per cent and the tertiary 9 per cent.

Even in the worst recession year 2011 (the year of the sharpest fall of employment), the decline in employment was smaller for the primary sector than for the other two sectors of the economy. The decline in employment was equally small in rural areas, in 2011, compared to that of the urban areas.

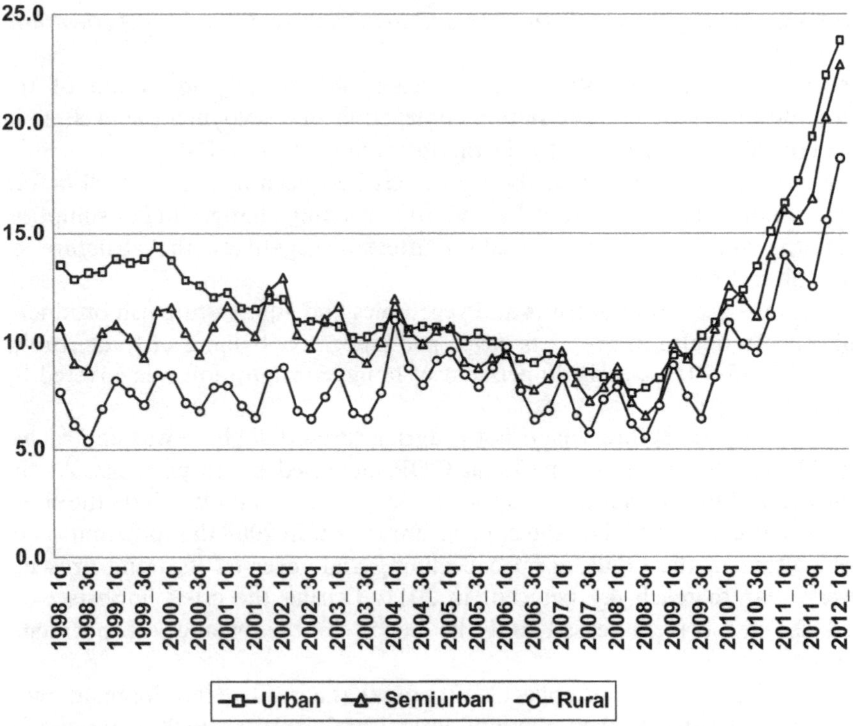

Fig. 3. Evolution of Unemployment Rates by Urbanity Index, 1998–2012. *Source*: ELSTAT, Labour Force Survey, 1998–2012. *Note*: The Greek Statistical Authority defines as 'rural' settlements with a population size of less than 2,000 inhabitants, as 'semi-urban' those with a size between 2,000 and 10,000 inhabitants and as 'urban' those with a population of over 10,000 inhabitants.

Within the rural areas the tertiary sector suffered by far the lowest employment losses. Lost jobs were limited (only 5,000 compared to 28,000 for the secondary and 30,000 for the primary sector). As a consequence, the tertiary sector became the main job provider in rural areas pushing the primary sector to the second position. In other words, agriculture is not the main employer in rural areas anymore.

All these developments are very much affected by diverse intra-regional and inter-regional mobilities. The analysis of the Labour Force Survey data has indicated that unpaid family labour and wage labour show high mobility within rural areas and the primary sector.

Gross Domestic Product Formation and the Balance of Trade of Agricultural Products

In the whole period before the crisis, a continuous contraction of the agricultural sector was registered, contraction in employment as well as in the contribution of the sector in the formation of the GDP.

Dependency on imported food products had been increasing well before the economic crisis. Improved standards of living, changes in consumption patterns and CAP implementation affected negatively the structure of agricultural trade.

With the exception of fruits and vegetables and aquaculture fish products, all other agricultural products developed a negative balance of trade. In the period 2005–2011, nearly 40 per cent of home consumption was covered by imports.

In the decade before the crisis the agriculture GDP share was contracted by 22 per cent, while the national GDP increased by 16 per cent. At the outbreak of the crisis it had dropped to 3.5 per cent, but after 2008 the share of agriculture in the GDP and exports improved. In 2008 the contribution of agriculture in the GDP stopped declining and reversed its trend growing every year to reach 4.1 per cent in 2011. During the crisis imports were reduced, exports increased and the deficit in the balance of payments improved.

Zografakis and Karanikolas (2012) note that a positive development amid the crisis is related to the growth of agricultural exports which they estimate contributed to the creation of 40,000 new employment positions between 2009 and 2011.

Reverse Mobilities, Old Ties

What is of interest during the past years of the economic crisis is the mobility and characteristics of new entrants in the agricultural sector. Working on the Labour Force Survey data, we discover that new entrants originate mostly from urban areas. One in two of them come from the worst hit, in terms of unemployment, region of Attica/Athens (45 per cent) and the rest from Central Macedonia and Thessaloniki (9 per cent), Thessaly and Central Greece (7 per cent), Eastern Macedonia (6 per cent), Crete (5 per cent), etc. When we examine their employment conditions one year before their entry in agriculture, we discover that nearly one third of new entrants held no jobs (i.e. 21 per cent of them were unemployed, 24 per cent students, 22 per cent serving their military service and 32 per cent were housekeepers). Two thirds moved into the sector from other sectors of employment (nearly all originating from commerce, leisure and public services).

The new entrants are of a relatively younger age and a better educational background. Their age classifies them among the youngest group in the agricultural sector (i.e. they are on average 41.7 years against 47.3 for wider population of the employed in the agricultural sector), while their largest part belonged to the 35–39 year age group (19 per cent). Their educational level was higher than the rest of the farming population, since one third of them received tertiary education (24 per cent hold a higher education degree, 9 per cent postgraduate degree) and nearly half secondary.

In rural areas, changes in employment positions reflect an internal restructuring of employment in the period 2008–2011. With the exception of wage employment, all other categories of employment are declining in numbers.

In short, in rural areas 'self employed/own account workers' and 'unpaid family workers' suffered most the implications of the crisis. Two thirds of the jobs lost were jobs of family unpaid labour, i.e. of women assisting family labour.

The picture is more balanced when we look at new entrants in the primary sector. New entrants originating from urban areas – it is estimated that 17,000 people moved to rural areas – are more or less evenly distributed in all employment positions with 'self employed/own account workers' absorbing most of them.

More than half of the new entrants in agriculture become independent farmers; the majority of them are working in crop production. Important is their engagement in agriculture as wage labourers, too. Wage labourers seem to be the only category with a positive denomination in rural employment in the past years (Papadopoulos, 2013).

The tendencies for return to the land and the countryside studied through the Labour Force Survey statistics have been well supported by a recent research (KAPA Research, 2012) carried out in a representative sample in Athens and Thessaloniki in which more than two thirds of those questioned (a sample of 1,286 people) claimed that they had thought of leaving the cities to settle in the 'Greek countryside' (i.e. villages or small towns).

For 77 per cent of those who said that they would like to settle in the countryside, this decision can be realized to the extent that they get permanent employment. Almost 60 per cent of them are in the age group below 40 years and 9 out of 10 seek employment and a better quality of life in the countryside. What seems to facilitate their movement to rural areas is the fact that many of them declare they have agricultural property which they would like to exploit. For 39 per cent of those who have thought of settling in the countryside, their first option is to start their own business while 31 per cent opt for a wage job in the private sector.

Agriculture attracts the attention of only one third of the respondents while the remaining two thirds are interested in other sectors of the rural economy like tourism, culture and communications. Despite a mass media 'bombardment' for very sophisticated crops (hippophaes, spirulina, asparagus, aromatic plants etc.), most respondents choose more conventional crops with a better business organization. Their options in agriculture are hierarchically: olive trees (47.6 per cent), organic crops (42.7 per cent), vegetables (37 per cent) and aromatic plants (32 per cent).

CONCLUSIONS

In the contemporary 'era of mobilities', in which the countryside experiences a perplexity of movements, both rural areas and agriculture are simultaneously becoming recipients and/or providers of population movements. These mobilities may be interpreted in a modernity frame focusing on the de-agriculturalization process of the countryside as a phase of the increasing integration of rurality into the wider economic system and the inevitable domination of the urban characteristics in space, or else, in view of the post-modernity approach, which places emphasis on the fragmentation of rural space and the acknowledgement of a multifunctional countryside that allows for the adaptability, transformability and resilience of the socio-economic and institutional characteristics of rural areas.

In this context, rural Greece is considered both modernized and resilient due to the fact that modernization tendencies have nurtured and corroborated the de-agriculturalization process, whereas on the other hand, the latter has been a sign of the maintenance of rural structures, functions, activities and identities through the reorganization and reshaping of rurality and more particularly of the farm and off-farm mixture in rural areas.

Contemporary 'reverse mobility' towards the countryside has elements of both modernity and tradition, of necessity and choice: return to the land of origin, new methods of organization and employment, rediscovery of traditional and development of new crops (legumes, aromatic herbs, pomegranates, mushrooms, spirulina, hippophaes, snails etc.) and methods of cultivation (organic farming, greenhouse production, producer groups etc.), elements of a new space and time division of living and working between rural and urban areas are some of these elements.

Moreover, intra-regional and inter-regional mobilities and transformations reflect the interweaving of rural and urban characteristics on particular rural locales. The intensification of mobilities between rural and urban areas reinforce both the disengagement of rural residents from (conventional) agriculture and the engagement of former urban residents with local food production, alternative farming, subsistence-hobby-lifestyle farming etc. (local food product processing, organic farming, aromatic herb cultivations etc.).

The idea of 'rurality' itself, either as a novel synthesis of 'rur-urban' traits or as a nostalgic 'back to the land/roots' movement, has become more and more familiarized and fashionable. On the other hand, the economic crisis has led to the reinstatement of a 'new rurality', in which agriculture seems to attract new attention re-enhancing its role.

Given the size of non-farm income (40 per cent of the total rural household income) as a pillar for rural development, it is expected that the decline in the non-agricultural sectors – due to the economic crisis – would eventually imply a decrease in the total household income. Consequently, the importance of farm income and of CAP support is expected to rise as a result of the quest for new opportunities.

'Return to the land and the countryside', however, is not an easy task. The first enthusiasm is followed by the difficulties of implementing farming ventures and living in the countryside. Recently, the new austerity programme passed through the Greek Parliament in November 2012 imposes upon the farming population strict measures connected to diesel taxation, abolition of tax exemption, reduction of pensions etc. which, in combination with the cuts in public expenditure in rural areas, are expected to raise the costs of production and create a negative environment in the immediate future. For some analysts, however, the problem will be overcome by the worse conditions created in urban areas where unemployment and poverty have hit the roof and push people to seek refuge in rural areas despite the new measures.

The return to the countryside seems to be easier for the younger and the better educated. For the rest it is either a rural household strategy plan involving a new family division of labour on and off the farm or a return of urban households to the countryside. This 'return' movement does not necessarily lead to an involvement of the ex-urban family with agriculture. These movements are not accompanied by a policy framework and a strategy for rural development.

The Ministry of Rural Development and Food trapped for years in clientelistic politics has failed to restructure the agricultural economy and re-orientate rural development (Papadopoulos & Liarikos, 2007). The Ministry announced in 2012 the implementation of a programme of public land distribution of 100,000 hectares at a very low rent for young people interested in becoming farmers in an attempt to boost the revival of agriculture and rural areas. The programme, however, has been implemented very slowly and no tangible results have so far been achieved.

CAP reform discussed these days brings forward again the issue of the strategy of rural development. In the years of crisis, however, challenges are bigger and time is shorter.

Nevertheless, the prospects of Greek agriculture and rural areas seem more favourable than those of the other sectors of the economy. In a period of intense fiscal changes, agriculture and rural areas have ensured a more stable fiscal framework of more than 2 billion euros annually until 2017 through the new CAP reform. Possible benefits are pending while the most negative implications of the crisis have already been experienced in rural areas.

However, this is not enough. Today, more than ever, the reform of the institutional environment of agricultural and rural development is imperative, while clientelism and corporatism have led to the collapse and bankruptcy of farmers' cooperatives and organizations.

In many parts of the country the agricultural paradigm is identified with intensive monocultures like cotton and tobacco. Producer price reductions, on the one hand, and increased costs of production, on the other, have pushed many producers to the exit gate of agriculture and the countryside. Inputs to agriculture are controlled by suppliers and farm holdings are dependent upon food processing industries and food retailers. In other words, the food sector has become an arena of oligopolistic competition incomprehensible by the producer or the consumer but with important implications for both of them. That is why new entrants remain still vulnerable despite their early enthusiasm.

Changes in education, research, farmers' collective forms of organization, support services and policy implementation are urgently required if the country is to avoid a deeper food and humanitarian crisis.

Rural areas have been the main focus of our analysis despite the fact they are constantly reshaped by the movements of indigenous population and incomers. The rights of the 'local' people are challenged by the arrival of incomers, who are often excluded from 'the local' milieu. Thus, rural space becomes 'politicized' by being reconstructed through the interaction between the 'locals' and 'non-locals'. This is because the 'social' and the 'spatial' are inextricably intertwined (Massey, 1995). Places and communities are being increasingly opened up to external influences from the remotest parts of the globe and a number of tensions are in fact observed between the various (sub)facets of rural space. These tensions drive the dynamism of rural places, providing the impetus for rural restructuring creating the conditions for 'politicizing the rural' (Andersson, 2010; Massey, 2005; Woods, 2011).

ACKNOWLEDGEMENTS

The authors wish to acknowledge that the part of the chapter analysing the implications of the crisis draws to a certain extent from recent joint work of C. Kasimis with S. Zografakis.

NOTES

1. In the past this transition from the rural 'past' to the urban 'present' was connected to an attempt to exempt the 'rural' as a meaningful category. The search for new concepts – which makes sense in the context of the post-agricultural and multifunctional era – followed the path of reconceptualizing the 'rural' against the notion of 'urban'. The different representations of rurality actually substituted for the lack of a conceptual clarity of the notion of rurality, which has been pinpointed by a number of writers (Cloke & Goodwin, 1992; Halfacree, 2001; Hoggart, 1990).
2. One needs to be very careful with the use for comparison of data from different sources. For this reason we are limiting the use of data from different sources only for wage labour and some other qualitative characteristics of Greek countryside.
3. The farm head defines himself as a farmer and declares that the largest part of his income comes from agriculture.
4. The economic crisis is expected to lead parts of the indigenous population in search of such jobs, but so far we have not seen a worth mentioning switch of Greeks to wage labour in agriculture.

REFERENCES

Alpha Bank. (2009, June). The agricultural sector in Greece: Subsidies and provisions instead of exploitation of our comparative advantages. *Economic Bulletin 109*, 3–23.

Andersson, M. (2010). Provincial globalization: The local struggle of place-making. *Culture Unbound*, *2*, 193–215.

Bell, M. M., & Osti, G. (2010). Mobilities and ruralities: An introduction. *Sociologia Ruralis*, *50*(3), 199–204.

Burgel, G. (1977). Rural research in Greece: Part I. *The Greek Review of Social Research*, *30/31*, 190–213. (in Greek).

Burgel, G. (1978). Rural research in Greece: Part II. *The Greek Review of Social Research, 33/34*, 190–213. (in Greek).

Cloke, P., & Goodwin, M. (1992). Conceptualizing countryside change: From post-fordism to rural structured coherence. *Transactions of the Institute of British Geographers, 17*, 321–336.

Close, D. H. (2002). *Greece since 1945. Politics, economy and society*. London: Longman.

Demoussis, M. (2003). Transformations of the CAP and the need for reorganizing agricultural policy in Greece. In C. Kasimis & G. Stathakis (Eds.), *The reform of the CAP and rural development in Southern Europe* (pp. 173–185). Aldershot: Ashgate.

Efstratoglou, S., Papadopoulos, A. G., Efstratoglou, A., & Kouroussi, E. (2004). Greece. In J. Bryden & K. Hart (Eds.), *A new approach to rural development in Europe: Germany, Greece, Scotland and Sweden*. Mellen Studies in Geography (Vol. 9, pp. 151–222). Lewiston: Edwin Mellen Press.

Eurostat. (2012). Agricultural income per AWU in the EU-27 up by 7.1% in 2011. *Statistics in Focus*, (24).

Folke, C., Carpenter, S. R., Walker, B., Scheffer, M., Chapin, T., & Rockström, J. (2010). Resilience thinking: Integrating resilience, adaptability and transformability. *Ecology and Society, 15*(4), 20. [online]. Retrieved from http://www.ecologyandsociety.org/vol15/iss4/art20/

Fonte, M., & Papadopoulos, A. G. (Eds.). (2010). *Naming food after places: Patterns of food relocalization and knowledge dynamics in rural development*. Aldershot: Ashgate.

Goussios, D. (1999). Countryside, rural space and small town: From ruralisation to local development. In D. Economou & G. Petrakos (Eds.), *The development of Greek cities* (pp. 157–207). Athens: University Press of Thessaly/Gutenberg. (in Greek).

Halfacree, K. (2001). Constructing the object: Taxonomic practices, 'counterurbanisation' and positioning marginal rural settlement. *International Journal of Population Geography, 7*, 395–411.

Halfracree, K. (2006). Rural space: Constructing a three-fold architecture. In P. Cloke, T. Marsden & P. Mooney (Eds.), *Handbook of rural studies* (pp. 44–62). London: Sage.

Halfracree, K. (2009). Rurality and post-rurality. *International Encyclopedia of Human Geography, 9*, 449–456.

Halfacree, K. (2012). Heterolocal identities? Counter-urbanization, second homes, and rural consumption in the era of mobilities. *Population, Space and Place, 18*, 209–224.

Halfacree, K., & Boyle, P. (1998). Migration, rurality and the post-productivist countryside. In P. Boyle & K. Halfacree (Eds.), *Migration into rural areas: Theories and issues* (pp. 1–20). Chichester: Wiley.

Heijman, W. J. M., Hagelaar, J. L. F., & van der Hiede, M. (2007). Rural resilience as a new development concept. In M. M. Ševarlić & D. Tomić (Eds.), *Development of agriculture and rural areas in Central and Eastern Europe*. 100th Seminar of the EAAE. Novi Sad, Serbia.

Hoggart, K. (1990). Let's do away with the rural. *Journal of Rural Studies, 6*, 245–257.

Jentch, B. (2007). Migrant integration in rural and urban areas of new settlement countries: Thematic introduction in migrant integration in rural areas. Evidence from new countries of immigration. *International Journal of Multicultural Societies, 9*(1), 1–12.

KAPA Research. (2012). *Return to the countryside: Employment and quality of life*. Public Opinion Research [online]. Retrieved from http://www.agronews.gr (in Greek).

Karanikolas, P., & Zografakis, S. (2008). Greek farm households: Income inequality, poverty and distributional impact of farm income. *Social Cohesion and Development*, *3*(1), 27–42.

Kasimis, C., & Papadopoulos, A. G. (1997). Family farm and capitalist development in Greek agriculture: A critical review of the literature. *Sociologia Ruralis*, *37*(2), 209–227.

Kasimis, C., & Papadopoulos, A. G. (2001). The de-agriculturalisation of the Greek countryside: The changing characteristics of an ongoing socio-economic transformation. In L. Granberg, I. Kovacs & H. Tovey (Eds.), *Europe's green ring* (pp. 197–218). Aldershot: Ashgate.

Kasimis, C., & Papadopoulos, A. G. (2005). The multifunctional role of migrants in Greek countryside: Implications for rural economy and society. *Journal of Ethnic and Migration Studies*, *31*(1), 99–127.

Kasimis, C., & Papadopoulos, A. G. (Eds.). (2012). *Immigrants in Greece: Employment and incorporation in local societies*. Athens: Alexandreia. (in Greek).

Kasimis, C., Papadopoulos, A. G., & Pappas, C. (2010). Gaining from rural migrants: Migrant employment strategies and socioeconomic implications for rural labour markets. *Sociologia Ruralis*, *50*(3), 258–276.

Kasimis, C., & Stathakis, G. (Eds.). (2003). *The reform of the CAP and rural development in Southern Europe*. Aldershot: Ashgate.

Kavouriaris, M. (1974). Some thoughts on the causes and implications of migration. In M. Nikolinakos (Ed.), *Economic development and migration* (pp. 24–62). Athens: Kalvos. (in Greek).

King, R. (2000). Southern Europe in the changing global map of migration. In R. King, G. Lazaridis & C. Tsardanidis (Eds.), *Eldorado or fortress? Migration in Southern Europe* (pp. 1–26). London: Macmillan.

Kūle, L. (2008). Concepts of rurality and urbanity as analytical categories in multidimensional research. *Proceedings of the Latvian Academy of Sciences, Section B*, *62*, 9–17.

Labrianidis, L., & Sykas, T. (2009). Geographical proximity and immigrant labour in agriculture: Albanian immigrants in the Greek countryside. *Sociologia Ruralis*, *49*(4), 394–415.

Lacour, C., & Puissant, S. (2007). Re-urbanity: Urbanising the rural and ruralising the urban. *Environment and Planning A*, *39*, 728–747.

Lawrence, C. M. (2007). *Blood and oranges: Immigrant labour and European markets in rural Greece*. New York, NY: Berghahn Books.

Lazaridis, G., & Psimmenos, I. (2000). Migrant flows from Albania to Greece: Economic, social and spatial exclusion. In R. King, G. Lazaridis & C. Tsardanidis (Eds.), *Eldorado or fortress? Migration in Southern Europe* (pp. 170–185). London: Macmillan.

Louloudis, L., & Maraveyas, N. (1997). Farmers and agricultural policy in Greece since the accession to the European Union. *Sociologia Ruralis*, *37*(3), 270–286.

MacKay, M., Perkins, H. C., & Espiner, S. (2009). *The study of rural change from a social science perspective: A literature review and annotated bibliography*. Lincoln: Lincoln University.

Massey, D. (1995). Making spaces or, geography is political too. *Soundings*, *1*(1), 193–208.

Massey, D. (2005). *For space*. London: Sage.

McCarthy, J. (2005). Rural geography: Multifunctional rural geographies – reactionary or radical? *Progress in Human Geography*, *29*(6), 773–782.

Mitchell, C. J. A. (2004). Making sense of counterurbanization. *Journal of Rural Studies*, *20*, 15–34.

OECD. (2004, October). Farm household income: Towards better informed policies. *OECD Observer*, 1–8.

Papadopoulos, A. G. (1998). Flexible agriculture: Survival strategy and resistance to marginalisation or form of integration into the agro-food system. In C. Zioganas (Ed.), *Competitiveness and integrated development of the rural sector: New challenges for Greece* (pp. 255–268). Thessaloniki: ETAGRO. (in Greek).

Papadopoulos, A. G. (1999). Revisiting the rural: A southern response to European integration and globalization. In C. Kasimis & A. G. Papadopoulos (Eds.), *Local responses to global integration* (pp. 245–271). Aldershot: Ashgate.

Papadopoulos, A. G. (2006). Rural immigrations and female employment. In B. B. Bock & S. Shortall (Eds.), *Rural gender relations: Issues and case studies* (pp. 196–213). London: CABI.

Papadopoulos, A. G. (2009). Begin from the bottom to move on: Social mobility of immigrant labour in rural Greece. *Méditerranée, revue géographique des pays méditerranéens*, *113*, 25–39.

Papadopoulos, A. G. (2011). Transnational immigration in rural Greece: Analysing the different mobilities of albanian immigrants. In C. Hedberg & R. M. do Carmo (Eds.), *Translocal ruralism: Mobility and connectivity in European rural spaces* (pp. 163–183). Berlin: Springer.

Papadopoulos, A. G. (2013). Which is the agricultural proletariat? Between the theoretical approaches and the empirical reality. In A. Moissidis, D. Papadopoulou & G. Petraki (Eds.), *Sociology and social transformation in contemprorary Greece: Honorary Volume for the Sociologist D. G.Tsaoussis* (pp. 391–422). Athens: Gutenberg. (in Greek).

Papadopoulos, A. G., & Liarikos, C. (2007). Dissecting changing rural development policy networks: The case of Greece. *Environment and Planning C: Government and Policy*, *25*(2), 291–313.

PASEGES. (2011). *The situation in the agricultural sector: Developments, projections, priorities.* Athens: PASEGES. Retrieved from http://www.paseges.gr/. (in Greek).

Perkins, H. C. (2006). Commodification: Re-resourcing rural areas. In P. Cloke, T. Marsden & P. Mooney (Eds.), *Handbook of rural studies* (pp. 243–257). London: Sage.

Schouten, M., van der Hiede, M., & Heijman, W. (2009). Resilience in social-ecological systems in European rural areas: Theory and prospects. Paper presented in the 113th EAAE Seminar, Belgrade, Serbia. Retrieved from http://agenconsearch.umn.edu/bitstream/57343/2/Schouten%25Marleen%20cover.pdf

Sutton, S. B. (1983). Rural-urban migration in Greece. In M. Kenney & D. Kertzer (Eds.), *Urban life in mediterranean Europe*. Urbana: University of Illinois Press.

Sutton, S. B. (1988). What is a village in a nation of migrants? *Journal of Modern Greek Studies*, *6*(2), 187–215.

Tsiforou, I., & Stavropoulou, C. (2011). *Recent developments in agricultural employment.* Athens: PASEGES. Retrieved from http://www.paseges.gr/. (in Greek).

Vaiou, D., & Hadjimichalis, C. (1997). *With the sewing machine in the kitchen and the poles in the fields: Cities, regions and informal labour.* Athens: Exantas.

van der Ploeg, J. D. (1997). On rurality, rural development and rural sociology. In H. de Haan & N. Long (Eds.), *Images and realities of rural life: Wageningen perspectives on rural transformations* (pp. 39–73). Amsterdam: Van Gorcum.

Vergopoulos, C. (1975). *The agrarian question in Greece: The social integration of agriculture.* Athens: Exantas. (in Greek).

Wagstaff, M. (1982). *The development of rural settlements*. London: Avebury.

Walker, B., Holling, C. S., Carpenter, S. R., & Kinzig, A. (2004). Resilience, adaptability and transformability in social–ecological systems. *Ecology and Society*, *9*(2), 5. [online]. Retrieved from http://www.ecologyandsociety.org/vol9/iss2/art5

Wilson, G. (2009). The spatiality of multifunctional agriculture: A human geography perspective. *Geoforum*, *40*, 269–280.

Woods, M. (2009). Rural geography: Blurring boundaries and making connections. *Progress in Human Geography*, *33*(6), 849–858.

Woods, M. (2011). *Rural*. London: Routledge.

Woods, M., & McDonagh, J. (2011). Rural Europe and the world: Globalization and rural development. *European Countryside*, *3*, 153–163.

Zacopoulou, E. (1999). Pluriactive population and agriculture: First unravellings of a complex phenomenon. In C. Kasimis & L. Louloudis (Eds.), *Countryside in Greece: Late 20th century greek rural society* (pp. 115–147). Athens: EKKE/Plethron. (in Greek).

Zografakis, S., & Karanikolas, P. (2012). Tracing the consequences of economic crisis in rural areas: Evidence from Greece. In R. Adisa (Ed.), *Rural development* (pp. 311–335). Rijeka: InTech Publications.

CHAPTER 12

AGRICULTURE IN MEDITERRANEAN EUROPE: CHALLENGING THEORY AND POLICY

Dionisio Ortiz-Miranda, Ana Moragues-Faus and Eladio Arnalte-Alegre

ABSTRACT

This concluding chapter presents a transversal reading of some key elements of Mediterranean agriculture and its pathways of transformation, in part under the lens of main theoretical paradigms, where, as the first section discuss, it has had a peripheral character. In this regard, the chapter tackles the impure nature (where allegedly productivist and non-productivist dynamics coalesce) of some agricultural pathways, at both the farm and the territorial levels. As it is discussed later, this relates to the role played by policies and the way Mediterranean decision-makers mediate EU regulations. Subsequently, three interwoven dimensions of the recent evolution of Mediterranean agriculture are analysed: the dynamics of family farming and the role played by family networks, the position of migrants in the evolution of agriculture and some hypothesis and clues about the implications of the current economic crisis for Southern farming

Agriculture in Mediterranean Europe: Between Old and New Paradigms
Research in Rural Sociology and Development, Volume 19, 295–310

ISSN: 1057-1922/doi:10.1108/S1057-1922(2013)0000019014

systems. In the two last sections, the chapter ends discussing the theoretical and policy challenges posed by Mediterranean small-scale farming systems.

Keywords: Mediterranean agriculture; agricultural policies; small-scale family farms; foreign migration; economic crisis

A CONSTELLATION OF AGRARIAN TRANSITIONS ON PARADIGMS' LANDSCAPE

The fields of rural studies and agrarian change in Europe have been a very fertile academic framework in the last 30 years. Indeed, theoretical and empirical contributions have followed one another in order not only to explain the rapid transformations at play, but also aiming to outline the negative implications of certain dimensions of those changes and to indicate which models of agrarian change would result more desirable. This mixed theoretical-normative character is intimately related to the aspiration of paradigms to impregnate and guide policy. Moreover, the dynamism of these debates has surpassed national borders to become an international arena of debate where academic and political spheres coalesce. Furthermore, the topic and the debates around agriculture and potential development pathways have gained prominence and acquired a new scope after the world food crisis of 2008. This price hike and its subsequent effects jointly with the climate change stress the doubts and the debates about the sustainability of the actual agro-food system.

In such a process, theoretical paradigms have shown to be useful and inspiring devices in several ways. Firstly, more recent theoretical approaches have contributed to challenge many initial assumptions and previous conceptualisations on agrarian change, opening the focus to include realities that were marginalised and misunderstood under more restrictive lens, but that showed a striking capacity of resistance. Secondly, in doing so, the theoretical debate has also enriched with new research and empirical evidences, which have contributed to a better knowledge and understanding of the deep transformations taking place in European rural areas. Thirdly, the academic debate has shown to be able to permeate public policies. Indeed, the objectives of rural and agricultural development policies, its modes of intervention or evaluation criteria have integrated, at least in certain cases, some aspects stemming from the academic and research sphere. Furthermore, different actors participating in policy arenas have also reinforced and shaped paradigms' propositions in different debates and negotiations, ranging from civil society organisations such as environmental NGOs, agrarian unions, farmers' associations to private interests such as seed and fertilisers companies or the retail sector.

In this regard, theoretical paradigms and policy discourses on agrarian and rural change have become intertwined to constitute a mixed academic/policy oriented domain, taking place in a European supranational arena. And this combination sometimes has played, despite the efforts of many authors to feed and reshape that arena with Mediterranean research, as a constraint for contributions from the 'periphery' (both academic and political). In other words, Mediterranean agriculture has had a complicated and contested adaptation to theories of agrarian change as well as to agricultural and rural policies designed mostly 'in the North'.

In this context, the challenge of theoretical paradigms formulated in Europe is outstanding, since they aim to accommodate and explain a diversity of agrarian systems and a growing diversity of transformation pathways. This volume aspires to contribute to this quest by providing different types of depth analysis of Mediterranean countries. Needless to say that the case studies included in this volume do not represent a comprehensive sample of all Mediterranean farming systems; however, they do reflect thoroughly their diversity, as well as the underlying complexity behind their transformations.

Consequently, the aim of this chapter (and of the volume as a whole) is to extract from the case studies key transversal elements for a comprehensive reading of the dynamics that Mediterranean agricultures are undergoing. This integrative analysis allows discussing old and new transformation pathways not only under the focus of main theoretical paradigms, but also highlighting crucial aspects that by and large have been by-passed in other explanatory frameworks.

IMPURE FORMS OF AGRICULTURAL PATHWAYS: FROM FARM-LEVEL DECISIONS TO TERRITORIAL DYNAMICS

Agrarian change is the aggregated outcome of myriads of decisions carried out by farmers, collective organisations and agricultural enterprises. They are individual decisions, but at the same time, they are strongly interconnected among them. Indeed, the farm decision level cannot be fully understood if dissociated from the territorial context where it takes place, since this is the level where interaction between farmers becomes evident. The Mediterranean literature highlights the importance of territory to understand agrarian dynamics. It is precisely the adoption of this territorial approach a necessary analytical tool to unveil and understand the diversity of actors, pathways and interrelations that finally constitute agriculture in a place. In this regard, this type of analysis overcomes partial case studies on one type of farmer or outcome leading to a more complex but also better informed picture of agricultural dynamics.

In this sense, the chapters of this volume have shown how, at these two levels (farm and territorial) practices and activities belonging to confronted paradigms integrate in an interdependent manner, as Moragues-Faus et al. discussed in Chapter 2.

At the farm level, a number of examples have illustrated how individual decision makers put the available farming and management practices at the service of their own logic. In this way, they integrate in a 'coherent' manner activities that, according to dualistic models, would seem to correspond to irreconcilable combinations. This has been illustrated in two distinct Mediterranean farming systems. On the one hand, Pinto-Correia and Godinho have shown how an important part of land managers' decisions in the extensive and environmentally valuable *montado* system (which would include all the conditions to be considered a magnificent example of multifunctional agriculture) keep anchored in and oriented to conventional production. On the other hand, Moreno-Pérez illustrated the capacity of Spanish intensive horticulture to respond to new market signals and to integrate new environmentally labelled production methods in an undeniable productivist path. In doing so, most farmers might contest the normative charge of these practices (what correspond to good or bad agricultural development models) and accommodate the available business options to the main objective, in most cases the (financial, family and ecologic) viability of their holdings.

It is not to say that farmers have not normative assumptions of what a 'good farmer' should be. For instance, Pinto-Correia and Godinho illustrate how most holders have internalised productivist logics, so that the 'good farmer' would be the one that is capable to follow modernisation premises, either if they do or not. A smaller percentage of farmers maintain a discourse and practice which does respond to multifunctionality paradigm. The combinations of discourses and practices are manifold and sometimes contest the pathways designed under current theoretical paradigms.

At the territorial level, this interdependence between supposedly productivist and non-productivist practices has become evident in one of the major components of agricultural change in the Mediterranean: the expansion of quality food production. For instance, Arnalte-Alegre and Ortiz-Miranda and Moreno-Pérez show how the development of PDO has not been only compatible with, but it is a main explanatory driver of agricultural intensification. So that quality orientation has led sometimes to monocropping expansion and the transformation of traditional agricultural landscapes, as well as to a growing pressure over water resources. Also, this quality differentiation has been increasingly driven by international markets. The expansion of organic agriculture in the South of Europe is, to great extent, a response to the market signals received from Northern European consumers. De Devitiis and Maietta put forward that, despite the proliferation of PDOs in Italy, there is a concentration of turnover in few and internationally recognisable cases of export-oriented products.

In short, the expansion of 'labelised' food quality production in Mediterranean Europe, rather than constitute a sign of reversal of the agricultural development path, is showing in many cases the capacity of the most dynamic agriculture to respond to new market signals, while maintaining business models and farming practices fully embedded in the productivist logic.

Similarly, Laurent argued that, sometimes, direct selling – a form of short food chain that seems to be frequent in some Southern regions (see Chapters 3 and 8) – can also hide a knowledge gap (real ecological footprint or salaried work conditions) behind the image of the farmer selling directly his/her production to consumers he/she knows personally.

Even in the most idyllic conditions, multifunctional agriculture – as well as productivism – does not seem a solution for all. In some cases neither desired by the farmers themselves (as Pinto-Correia and Godinho showed) nor possible in terms of sustaining economically the farming activity. Furthermore, in some cases, when deepening into the building blocks of initially successful multifunctional outcomes, such as quality products, we might find neo-productivist logics operating or poor labour conditions that clash with assumptions of new paradigms and furthermore, social justice issues.

PARADIGMS, POLICIES AND POLICY-MAKERS

As argued above, public policies and theoretical approaches are indissolubly associated in this field. In this regard, agricultural and rural policies implemented in Mediterranean countries are almost exclusively those of the European Union. Actually, EU policies have a 'pull' effect over these countries, which has undoubtedly contributed to the improvement of policy making in many domains (e.g. biodiversity conservation, fight against climate change, etc.). But this fact also has a double component.

On the one hand, policies are also inspired by the academic mainstream of the moment, and contributions from the periphery have shown, as stated above, limitations to permeate dominant theoretical approaches. This has meant that, very often, Southern agricultural stakeholders complaint about the supposedly lack of sensibility of European policies to the Mediterranean distinctiveness.

Sometimes this situation turns Mediterranean policy making into a continuous race to reach European standards that are also continuously going ahead. This race was evident in countries which accessed to the European Community in the 1980s (Greece, Portugal and Spain), and it is still with pre-accession countries, as clearly the Croatian chapter has illustrated. The point is that this can lead to the creation of institutions and agencies and the elaboration of plans and programmes that are difficult to be implemented in certain realities. Also, EU requirements have played against small farms, unable to adapt from one day to the other to the increasingly demanding regulations on health and quality standards. This has been illustrated by Franić and Mikuš for Croatia – a country in the middle of the process of accession, but it is also found in countries with more than 25 years in the EU, as Moreno-Pérez showed for small ruminant systems in Spain.

On the other hand, EU policies are mediated by Southern national and regional policy-makers, who have been said to 'distort' the, at least in theory, more multifunctional spirit of community regulations (see Chapter 2). It is widely acknowledged that, within the room of manoeuvre the European legislation allows, Mediterranean implementation of agricultural and rural development policies have emphasised the productive dimension of agriculture at the detriment of other more environmental approaches. For instance this has been evidenced by the reluctance to full decouple farm subsidies, the weight given to productive investments into the Rural Development Programmes or the special attention that cross-compliance pays to the resource base of agricultural production (land and water). Nevertheless, this does not only reflect policy-makers' views or priorities, but it also seems to be supported on the public attitudes towards the role of agriculture in Mediterranean countries.

In any case, the chapters have also shown how farmers distinctly respond to some policy instruments. That is to say, how policy stimuli and constraints affect individual behaviour, and how its aggregation shape different processes of agrarian change at the landscape scale, precisely where the ecological integration of farming becomes more evident and which constitutes a relevant policy decision unit.

Several chapters have shown how the aggregation of individual decisions deeply modifies landscape dynamics and alters the ecological balances in Mediterranean farming systems, by and large strongly integrated in high value nature areas. Furthermore, a number of examples have illustrated how policy signals have contributed to this. The regulation of livestock premiums has fuelled the substitution of sheep with bovine cattle in the Portuguese montados, which has increased overgrazing and the degradation of this fragile agro-ecosystem (Chapter 4). The model of application of milk quotas has contributed to farm land abandonment in the Spanish region of Galicia – reflected on both the disappearance of farms and the dissociation between farming practices and the management of other rural land, provoking in many areas serious problems of forest fires (Chapter 5). Restructuring subsidies for new vine plantations (Chapter 6) and production subsidies for olive production (Chapter 3) clearly played in favour of intensification of these farming systems in Spain and the resulting transformation of their traditional landscapes. At the same time, most of these farmers participate in agri-environmental schemes aimed to protect these agro-ecosystems or resort to afforestation aids. These processes are another facet of the way farmers accommodate the available, and sometimes contradictory, economic and managerial options in order to ensure the feasibility of the holding and comply with the embedded believes of what a farmer should be.

Besides the sometimes contradictory signals sent by policies, contributors have shown other regulatory failures: a lack of clear land planning in Galicia which obstructs land transfers and the subsequent restructuring of viable farms, non-effective rural and agricultural policies in Greece due to clientelistic decision-making, or the difficulties of rural development policies implemented in Italy to adapt the diversity of needs of different types of holdings (as argued in Chapter 9).

In short, these policy failures highlight the danger of oversimplifying the debate on rural and agricultural policies by focusing on 'how much' the budget will be for each country or region (as it appears often in the media and seems to be the flag of stakeholders). A more rich and multidimensional debate is needed to address current policy pitfalls and future challenges in a context of austerity. In this regard, these policies should be part of a more integrated planning that addresses actual and future roles of Southern

European agricultures in a framework of climate change vulnerability and desertification in the Mediterranean basin.

IS MEDITERRANEAN AGRICULTURE STILL FAMILY FARMING?

The family character of Mediterranean agriculture and the roles played by family ties have been also a transversal and constant feature in most of the chapters. Nevertheless, we should note at this point that, as it has been also the case of Northern countries, Southern family farming has gradually individualised, breaking in this way the family work group. As explained in the next section of this chapter, the availability of migrant workers has enormously facilitated this.

However, this is not to say that family ties have lost relevance in the dynamics of many Mediterranean farming systems. On the one hand, some chapters have highlighted how, in certain cases, strategic decisions (and the practices they give rise to) are taken into a complex and diverse framework of (often rural-urban) family relationships and networks. This has been illustrated for Greece in Chapters 3 and 11.

On the other hand, family networks have found to be crucial for land property relationships. Indeed, some Mediterranean farming systems are dominated by a multiplicity of absent small landowners, who maintain different degrees of relationships with their land. The dense and active rural/non-rural family networks in Greece would be facilitating farmland to be exploited under a diversity of informal cessions, so that a hidden farm restructuring would be questioning the officially rigidity of this agriculture. On the contrary, the same situation is playing in Galicia (Spain) as a driver of farmland abandonment, due to the blockage it produces in land markets.

MIGRANT WORKERS IN MEDITERRANEAN AGRICULTURE: A CONSOLIDATED TRAIT

The question of the role of foreign migrants in the development paths of Mediterranean agricultures deserves an outstanding place. The massive arrival of migrants from Africa, Latin America, Asia or Eastern Europe took place in these countries later than in the North, but it was so rapid and intense that in few years, foreign workers became easily available for agricultural firms. Therefore, the presence and the importance of foreign workers has become a characteristic trait of many these agricultures.

It is about a working population perfectly adapted to the needs of intensive farming systems (particularly those specialised in fruits and vegetable): flexible supply of labour to adapt to the important volume of work concentrated in short peak times. In addition, due to their social conditions (foreigners in irregular and precarious administrative situation), they are very vulnerable and are willing to accept bad working conditions (low salaries, hard operations and long working days).

The utilisation of migrants has allowed Southern agricultures to replace the decline of domestic labour (absorbed by the economic growth of these decades), as well as to overcome the constraints that the disengagement of family labour meant for many farms (see Chapter 11 for Greece and Moreno-Pérez, Arnalte-Alegre, & Ortiz-Miranda, 2011). But also, it became a factor of competitiveness reducing labour costs.

The 'good' adaptation of this labour force is being corroborated in the current economic crisis, when Southern countries undergone rising unemployment rates, work conditions worsen and salaries decrease. In Greece, Kasimis and Papadopoulos showed in Chapter 11 how Asians, willing to accept lower salaries, are less risky to unemployment than the more integrated and regularised Albanians. In Spain, the weight of migrants in agricultural occupied population rises even during the crisis (MAGRAMA, 2012). In short, this has become a consolidated structural trait of Mediterranean, as data show: 22% of total AWU in Greece is carried out by migrants, 24% of occupied population in Spanish agriculture is foreigner.

Besides that function of key competitive factor in intensive agriculture, the analyses of this volume have also shown other functions and the 'compatibility' of the utilisation of this labour force in other types of agriculture. Indeed, they are not only present in the 'vegetable factories' and the little attractive 'fields of plastic' of Spanish South-East; also we find this in the multifunctional agriculture of Southern France (see Chapter 7). However, in this second case, the model tries to avoid the visibility of these situations, reducing the maximum period of stay according to the employer needs (to have the work without the worker), as a way 'to preserve a positive brand image of Mediterranean agriculture' (Chapter 7).

Nevertheless, we should not forget at this point the role played by the arrival of migrant population in the revitalisation of many inland rural localities. Not only contributing to the rejuvenation and the economic dynamism, but also providing care for elder and dependent local population.

Finally, two main elements can be distilled from the studies of the volume with regards to agricultural labour. First, the disparities of regulations among European countries – and particularly the more under-regularised South, impact the geography of labour relationships, since it allows Northern farmers to count on workers contracted by agencies from other countries with less demanding regulations.

Second, the move of migrants to their countries of origin (in France under the official agreements of the OFII, in Spain through the state aids to go back to Latin-America in the period of crisis) once they have carried out their function, means a sort of transfer of the social costs associated to these workers. The other side of the coin is the emergence in the developing world of rural areas organised around diverse forms of temporary migrations to far-away precarious and instable labour markets. C. de Grammont (2009) describes this type of situation, spread in rural Mexico, although examples can be also found in Africa. In doing so, rural areas in developing countries would be assuming the function of training and reproducing work force, available to be flexibly utilised in the global economy of de-regulated labour markets.

WHAT ROLE FOR MEDITERRANEAN AGRICULTURE AND RURAL AREAS IN THE PERIOD OF CRISIS?

Southern European economies are confronting in the last years a new scenario of deep recession and very high unemployment rates. This is totally different than the situation of the previous decades, when the analyses included in this volume are framed. The question here, as put forward in Chapter 2, is which processes are taking place in Mediterranean agriculture and rural areas in this new context, as well as to assess their resilience or capacity to adapt these new conditions. This has become a fashionable topic in the media, where there is a growing interest in the allegedly role of the countryside as providing employment and lifestyle alternatives for urban population. Even some international organisations have also paid attention to this subject. The CIHEAM[1] recently published a note (Brun, 2012) where some data were enthusiastically interpreted as the 'beginning of a trend which will question on the long run the model of society and of land planning' (p. 4).

However, a sound assessment faces some difficulties. On the one hand, statistical series do not cover yet the whole period of crisis, so no conclusions can be extracted from them. On the other hand, there is still a lack of empirical and representative studies. These difficulties explain some of the chapters of this volume have not addressed the crisis. Therefore, what we can discuss at this moment is about prospects. In this sense, we could put forward two main hypotheses regarding the impact of the crisis and the responses of agriculture in the Mediterranean.

The first one – clearly set out in the chapter of Kasimis and Papadopoulos – suggests agriculture and rural areas resist better the impacts of the economic deterioration. So that they are 'increasingly turned into a "refuge and laboratory" of ideas and initiatives for a large part of urban dwellers directed towards the countryside and agriculture by either necessity or choice'.

The second one, more pessimistic, focuses on the rural development model that has been advancing in Europe, where a greater integration with urban labour markets and the urban demand of new rural good and services (quality food, leisure and amenities) have become key drivers of the rural economy. Therefore, that model can be more sensible and fragile to the employment crisis and the decrease of urban income, more than when rural areas were less 'interconnected' to urban dynamics. In addition, rural development has been also linked to the consolidation of the welfare state and the extension of social services (e.g. health and education ones) and infrastructures improving accessibility. The deep fiscal crisis Southern states are experiencing will also impact in these aspects.

It is indeed too soon to have a comprehensive picture of the processes at play to soundly corroborate or reject these hypotheses. Nevertheless, there are some clues than can point out some preliminary lines of discussion. Firstly, data show that the expansion of food quality production (both organic and territorial) continues during the crisis. In both cases, what seems to constitute the main driver of this trend is precisely the export vocation of these differentiated products (the case of Italian PDO constitutes an outstanding example[2]), what makes them less affected by the decline of domestic urban income.

Secondly, there is being a notable reduction of public expenditure in rural areas. Kasimis and Papadopoulos confirm the effects of austerity programmes on health, education, transport and other social services in Greece. In Spain, the 'restructuring' of health services in rural areas is particularly affecting certain regions,[3] as well as the cuts of school transport or elder care. And this takes place in rural areas undergoing serious problems of social sustainability (Camarero et al., 2009).

Thirdly, as stated above, an outstanding debate has arisen around the 'return to the land' topic, i.e. to what extent the crisis is pushing people to move to the countryside to look for new employment opportunities as well as for a new lifestyle. Kasimis and Papadopoulos showed this is an option considered by many Greeks, favoured by the ties they still maintain with rural land. These authors argue that rural newcomers are young and well educated, but that, after the initial enthusiasm they collide with 'the difficulties of implementing farming ventures and living in the countryside'.

In Spain, the crisis has interrupted the population entrance flows. The positive flows (between 2002 and 2008 newcomers doubled rural out-migrants), began to decrease to practically balance in 2011. The rural geographer Angel Paniagua (in Corcuera, 2012) associates this evolution to the growing difficulties to face the high investments needed to initiate new ventures in the countryside.

As a consequence, the performance of agricultural employment is not showing a clear trend. In Greece, it seems that is resisting better than that of other sectors (even growing in the first years of the crisis). On the contrary, in Spain the crisis has only slightly altered the pace of the declining trend of agricultural employment. Actually, employees remaining in the sector are the most vulnerable groups of migrant workers, willing to accept salaries and work conditions which deteriorate in a period of crisis and social regression.

THE SMALL-SCALE FARMING CHALLENGE

The transformations (and the persistence) of small-scale farming has been another constant throughout the volume. In this regard, we highlight two crucial elements, which also constitute two challenges for theoretical and conceptual developments of agrarian change, as well as for public policies.

First, most Southern European farming systems are dominated by small-scale family farms. The volume has shown how this double component plays in different contexts. However, what is the basic determinant of the evolution, the situation and the perspectives of these systems, is that holdings are small, not that they are family based. And this is so in spite of the prominence that the 'family label' has acquired in the global food political arena.[4]

Size strongly conditions farms' dynamics. In this regard, small farms, which have shown serious difficulties to enter into the classical modernisation logic and which have been massively driven out from the agricultural sector and the territories they occupied, also find serious constraints to resort to other models of multifunctional development. Productive differentiation and non-agricultural diversification has proven to be, at least by the time being, effective mechanisms of business development (see Chapter 9 on Italy). But we find many of these practices closely associated to large and capitalised holdings, managed with a strong entrepreneurial orientation and linked to specialised and professionalised market networks.

In other words, the evidence of this book shows that the multifunctional model of agriculture, at least as it is currently shaped, is neither the reason that could explain the apparent resistance of certain farming systems, nor a feasible option for the majority of Mediterranean small holdings. And this is so despite the voluntarism of certain theoretical approaches, which assume these multifunctional practices to be inherently interwoven with the nature of small farms and their 'historical' resistance.

Closely related to this, progress in the understanding of Mediterranean agriculture is further hampered by the weaknesses of sound statistical information. Several chapters have shown the limits of official agricultural statistics to capture major questions of agrarian change in the Mediterranean (farmland abandonment processes, the amount, profile and conditions of the labour really involved in agriculture or the informal land cession shaping the 'real' production units). The problem is that, although some changes have been introduced into the information that official surveys collect (diversification, direct selling, quality production, and the like), statistics do not reflect well this fluid reality. Actually, this information could pass from being blurred to be completely invisible. The last Agricultural Censuses carried out in 2009/2010 have modified (higher threshold) the criteria a holding should fulfil to be included into the statistical universe. This means that a part of very small holding simply disappears from the official picture. And this is not only a problem for researchers. It becomes also a weak (and risky) information basis for agricultural and rural policy makers.

The second challenge comprises the evolution and unfolding of multiple agricultural spaces. On the one hand, there are areas of dynamic agriculture, where farmers are able to adapt to the several and complex market and regulatory signals they receive, combining intensification, differentiation, diversification and concentration in profitable ways. This would correspond to one of the types of farming systems (that made up of consolidated family-based, modern and professionalised holdings) that Arnalte-Alegre and Ortiz-Miranda identify in Chapter 3.

On the other hand, these authors also classified two types of systems dominated by small-scale farms remaining in many regions of the South of Europe. The first would constitute the most fragile one, located in remote or mountain areas and still witnessing the disappearance of small farms and the abandonment of land. The case of Galicia illustrates this type of system and its associated processes thoroughly. For Italy, statistics show a drop of 20% UAA and 40% of holdings in mountain areas between 2000 and 2010 (Arzeni & Pecci, 2012). This type of system and its current dynamics constitute a serious policy challenge, since the decline of agriculture is in most cases giving rise to processes of degradation that are threatening the ecological value of these areas.

The second type of small-scale holding system identified is linked to the existence of a dense network of outsourcing/contracting relationships and/or the presence of strong collective action. In this case, studies might need to reconsider the individual farms as the basic production unit; instead the farming system as a whole emerges as the most appropriate level to understand and explain the dynamics of this type of agriculture.

Nevertheless, this reconsideration would be a necessary but not a sufficient condition for the feasibility of these farming systems. The very pillar supporting agriculture in most of these cases is non-agricultural economy. Indeed, some Mediterranean regions have witnessed how agriculture, paradoxically, have benefited from a de-agriculturalisation of the rural economy. Farming activity contributes (although marginally and irregularly) to family incomes (by means of markets, subsidies or even self-consumption). However, there are the incomes from non-agricultural sources[5] and even non-rural activities (thanks to the growing rural-urban mobility), including the transfers from the welfare state (particularly pensions), which have contributed to maintain these farming systems. In this sense, besides the patrimonial value of farmland, social and cultural factors by and large linked to the productive specialisation of most of these agricultures (olive oil, vineyard, fruits) come at play to also explain holders' willingness to maintain some low-profitability farming systems that would have been, otherwise, unfeasible.

The question here is to what extent the current economic crisis in Southern countries will threaten the mainstay of this supporting non-agricultural economy: high unemployment rates reducing the opportunities for pluriactivity or the decrease and tightening of pensions.

CONCLUDING: FILLING THE GAPS?

Baptista (2010) argued that the rural Portugal was no longer agricultural, but it was not yet anything else. This seems to be also a good description for the situation of many small-scale farming systems in Mediterranean Europe.

As shown, in some cases the post-agricultural transition is leading to no place in rural areas that are still undergoing depopulation and farmland abandonment. In other cases, where agriculture is highly dependent on the welfare state or urban labour markets, rural areas seem not to be yet something on their own. They are both realities that seem to be misrepresented or even excluded in mainstream theoretical approaches.

Furthermore, there are relevant policy challenges related to these realities. On the one hand, current policies have not been able to halt the decline of lagged areas. On the other hand, agricultural and rural development policies do not seem to offer appropriate responses to the characteristic of farming systems whose future depend more upon non-agricultural drivers and which agricultural/non-agricultural ties are not yet well understood.

Urgency posited by long-term projections on food security and current problems derived from the agri-food system call for more research in different domains. Undoubtedly this will lead to normative formulations on what type (or types) or model of agriculture is needed and desirable. However, it is paramount to root those desired models of development or aspirations in non-idealised analysis that tackle the lights and shadows of current agricultural dynamics. Only by doing so we can aim to orchestrate public and private action in order to walk transition pathways towards a more sustainable, inclusive and overall fair agri-food system.

NOTES

1. International Centre for Advanced Mediterranean Agronomic Studies.
2. See http://www.gustiditalia.com/1561/approfondimenti/italiano-produzioni-dop-italia-e-leader-mondiale-parmigiano-batte-grana-bene-le-mele-trentine/. Accessed on February 2013.
3. See Corcuera (2012) for Castilla y León. Also the removal of medical emergencies in rural villages of Castilla y La Mancha on January 2013 is provoking mobilisation of rural population.
4. As demonstrated by the United Nations' declaration of 2014 the International Year of Family Farming.
5. For instance Chapter 9 stressed in this regard the role of pluriactivity.

REFERENCES

Arzeni, A., & Pecci, F. (2012). L'agricoltura che cambia: Il sesto Censimento tra passato e futuro, Agriregionieuropa, anno 8, numero 31, dicembre 2012.

Baptista, F. (2010). *O Espaço Rural. Declinio da Agricultura.* Celta Editora, Lisboa.

Brun, M. (2012, June 11). Europe du sud: un retour à la terre pour affronter la crise économique? *Les notes d'alerte du CIHEAM*, 82.

C. de Grammont, H. (2009). México: Boom agrícola y persistencia d ela pobreza rural. In J. Graziano, S. Gómez & R. Castañeda (Eds.), *Boom agrícola y persistencia de la pobreza rural. Estudio de ocho casos* (pp. 225–263). Rome: FAO.

Camarero, L. A. (Coord.), Cruz, F., González, M., Del Pino, J. A., Oliva, J., & Sampedro, R. (2009). *La población rural de España: de los desequilibrios a la sostenibilidad social.* Fundación La Caixa, Barcelona.

Corcuera, A. (2012). Me voy al pueblo. El País Semanal, n° 1889, 9 diciembre 2012, pp. 28–38.

MAGRAMA. (2012). Empleo en el sector agroalimentario. *Análisis y Prospectiva*, 7. Ministerio de Agricultura Pesca y Alimentación, Madrid.

Moreno-Pérez, O., Arnalte-Alegre, E., & Ortiz-Miranda, D. (2011). Breaking down the growth of family farms: A case study of an intensive Mediterranean agriculture. *Agricultural Systems*, *114*, 500–511.

ABOUT THE AUTHORS

Eladio Arnalte-Alegre is Professor of Agricultural Economics in the Department of Economics and Social Sciences, at the Universitat Politècnica de València (Spain). He has led and participated in several research projects on agricultural change, agricultural policies and rural development, with a special focus on European Mediterranean countries. In particular, many of his publications are related to structural adjustment in agricultural holdings and the modes of organization of family farming.

Elisa Ascione is Researcher at the Italian National Institute of Agricultural Economics (INEA). Her interests cover the presence of young people in agriculture and the analysis of farm typologies, particularly using quantitative methods to provide policy guidelines. Her main research areas are EU Agricultural Policies; Rural Development policies; Income diversification in Agriculture; Extension Services.

Biagia De Devitiis is Senior Researcher in Agricultural Economics at the Università degli Studi di Foggia (Italy). She holds a Master of Science in Agricultural Economics and Politics, at the Centro per la Formazione in Economia e Politica dello Sviluppo Rurale in Portici, and a PhD at the Università degli Studi di Foggia. Her research interests revolve around the agricultural policies and the organization of food markets with a specific focus on the marketing of agro-food firms.

Ramona Franić is Associate Professor at the Faculty of Agriculture of the University of Zagreb, Croatia. She graduated in agricultural economics at the same institution and defended the PhD thesis in the field of measuring effects of agricultural policy. Her research interests are agricultural economics and policy, rural development and European integration processes. She participated in a large number of domestic and international research projects, published the results of these researches in scientific and professional publications and presentations in congresses, seminars and workshops. At the Faculty of Agriculture in Zagreb she is a teacher of agricultural policy and legislation, coordinator of BS study in Agricultural Economics and the head of the Department of Agricultural Economics and Rural Development.

Sérgio Godinho is PhD student at the Institute of Mediterranean Agrarian and Environmental Sciences, at the University of Évora (Portugal). His main research area is on the spatio-temporal dynamics of Montado landscape change and their biodiversity conservation. He intends to assess the synergistic effects of climate change and socioeconomic drivers on the Montado landscape change using geo-statistics and GIS modelling techniques.

Roberto Henke is Head of Research at the Italian National Institute of Agricultural Economics (INEA). He is currently the scientific coordinator of the INEA research field 'Agricultural Policy Analysis' and of the INEA 'Observatory on EU agricultural policies'. His main research areas are EU Agricultural Policies; CAP reform; Multifunctionality; Income diversification in Agriculture; International Trade of Italian Agri-food Products, Agri-food Made in Italy Trade.

Charalambos Kasimis is Professor of Rural Sociology at the Department of Agricultural Economics and Rural Development of the Agricultural University of Athens and Director of the Postgraduate Studies MSc. Programme 'Integrated Rural Development and Management of Rural Space' (2010–2014). His research interests focus on issues of rural transformation and development in Greece, the Balkans and the Mediterranean region and more particularly on family farming, employment and rural change. Migration and, more precisely, the implications of international immigration on rural regions have become main research topics in the past years. Recently, he has turned his interest in the study of the implications of the economic crisis upon the rural areas and the 'return to the land' movements in Greece.

Catherine Laurent is a Senior Scientist at the French National Institute for Agronomic Research (INRA). She is a political economist in the research department 'Sciences for Action and Development' (SAD). Since the early 1990s she has been publishing on structural change and multifunctionality of agriculture. Her current projects focus on the transformations of the regimes of knowledge and their consequences for farm advisory services, and for policies design and evaluation.

Edelmiro López-Iglesias is Associate Professor at the Department of Applied Economics in the University of Santiago de Compostela. His research focus is on agricultural economics and rural development, with special attention to farm structures, dairy sector and the economic diversification of rural areas. He has published several books and many articles in Spanish and international journals, and acted as consultant to the Ministry of Agriculture of Spain and international organizations such as the EU Commission and Inter-American Development Bank. From 1999 to 2005 he was Director of the Institute of Studies and Development of Galicia (IDEGA). He is currently the coordinator of the research group Agricultural and Environmental Economics, Rural Development and Social Economy at the University of Santiago de Compostela.

Roberto Lorenzana-Fernández is a Researcher at the Department of Applied Economics of the University of Santiago de Compostela. He completed his PhD thesis on the structural change in cattle farms and subsequently in the dairy farm economy.

Ornella Wanda Maietta is Assistant Professor at the Università degli Studi di Napoli 'Federico II'. She holds a Master Degree in Land Economy, at the University of Cambridge (UK) and a PhD in Agricultural Economics, at the Università degli Studi di Siena. She won the prize ERAE/EAAE 2000 for the best article published by young researchers on ERAE given by the European Association of Agricultural Economists. Her research topics include the measurement of productivity growth and of firm efficiency, the economics of cooperatives, the adoption of innovation and the analysis of consumer preferences with reference to ethical attributes.

Terry Marsden is Director of the Institute for Sustainable Places and Professor of Environmental Policy and Planning in the Cardiff School of Planning and Geography. He is also Co-Director of the UK Economic and Social Research Council's Research Centre for Business Relationships, Accountability, Sustainability and Society (BRASS). He is Dean of the University Graduate College. With more than 25 years experience working in the field of sustainability, Professor Marsden has a broad range of research and policy experience based around the themes of international rural development, sustainability and the rural environment.

Ornella Mikuš is Assistant Professor at the Faculty of Agriculture of the University of Zagreb, Croatia, Department of Agricultural Economics and Rural Development. She graduated in agricultural economics and defended PhD thesis in the field of rural policy at the same institution. Her research interests and current projects focus on agricultural and rural development policy with special emphasis on competitiveness of rural area and Croatian harmonisation processes in the context of European integration. On behalf of Faculty of Agriculture, she is a member of Management Committee of National Rural Development Network.

Ana Moragues-Faus is a Postdoctoral Researcher in the School of Planning and Geography (Cardiff University). Her research revolves mainly around sustainable agri-food systems, having developed most of her work in Mediterranean European countries and Latin America. Her publications and participation in national and international research projects and conferences tackle aspects of sustainable rural development, civic food networks, urban-rural linkages, food governance and social movements. Engagement is also central to her research work, being involved in participatory-action research and knowledge brokerage projects.

Olga M. Moreno-Pérez is an Assistant Professor at the Department of Economics and Social Sciences of the Universitat Politècnica de València (Spain). She has participated in several research projects tackling the patterns of change of agriculture in Spain and their implications for the rural territories. Her research interests are particularly focused on the analysis of the structural dynamics of agricultural holdings at the micro-level and its connection with different aspects of the organization of farm families.

Dionisio Ortiz-Miranda is Associate Professor of Agricultural and Rural Economics in the Department of Economics and Social Sciences, at the Universitat Politècnica de València (Spain). He has participated in a number of national and international research projects tackling the linkages between the dynamics of agriculture and rural change. A part of this activity has been focused on both the Northern and the Southern shores of the Mediterranean. His fields of interests revolve particularly around rural institutions and governance, as well as the analysis of rural and environmental public policies.

Apostolos G. Papadopoulos is an Associate Professor of Rural Geography and Rural Sociology at the Department of Geography of Harokopio University of Athens, Greece. He is currently Vice-Rector of Economic Affairs and Development (2011–2015) and has been Director of the Postgraduate Studies (MSc.) Programme entitled: 'Applied Geography and Spatial Planning' (2009–2011) at the same university. He has long research experience as senior researcher and project leader. He has co-edited different books, as well as published numerous articles in scientific journals, collected volumes and congress proceedings. He has done extensive research on family farming, rural development, local food and immigration to rural areas. His research interests include a number of topics: socioeconomic transformation of rural and peripheral areas, local labour markets, social stratification, social and spatial mobility, migrant integration, rural geography, local food geography, rural development policy and civil society actors.

Teresa Pinto-Correia is Associate Professor in Geography, at the Department of Landscape Environment and Planning, and Head of the Institute of Mediterranean Agrarian and Environmental Sciences, both at the University of Évora (Portugal). Her research is on the dynamics of agricultural landscapes in the Mediterranean, their provisioning of multiple functions, and the decision making process at different scales, for the management of these landscapes, in particular the landscapes resulting from High Nature Value extensive farming and agro-forestry systems.

Cristina Salvioni is Associate Professor of Agricultural Economics at the Department of Economics of the University of Chieti-Pescara in Italy. She has recently participated in several research projects held by the Italian National Institute of Agricultural Economics on income diversification in Italy using the FADN data. Her main research areas are Common Agricultural Policies; diversification and pluriactivity in agriculture; organic farming.

Francisco Sineiro-García is Associate Professor of Agricultural Economics at the Department of Applied Economics in the University of Santiago de Compostela (Spain), and was formerly researcher in animal production at the Spanish Institute of Agricultural Research. He firstly worked on beef and sheep production systems in hill land and later on the economics of dairy production and the food industry.